Excelsior: Memoir of a Forester

Laurence C. Walker

with appreciation for J. Herbert Stone, a regional forester who never forgot his men.
Laurence C. Walker

Forestry History Series
College of Forestry
Stephen F. Austin State University
Nacogdoches, Texas

Printed in the United States of America

First edition, 1995

Library of Congress Cataloging-in-Publication Data

Walker, Laurence C., 1924-
Excelsior: Memoir of a Forester/Laurence C. Walker—
1st ed. p. cm.
Includes index.
ISBN 0-938361-10-4

95-068181

I. Forests and forestry—Biography. 2. Foresters—History—Ecology. 3. Forest research—silviculture. 4. Boy Scouts of America. 5. Church—Presbyterian ministry.

Cover design by S. Laurence Walker

Cover photograph:
Switzerland's der Matterhorn. The Mother of Mountains gave direction to the author as he trod its trails. Ferdinand Holder, the French artist, would not paint the scene: "It is too awesome."

In gratitude to
The Faithful Friend
who guided me all
along The Way, though
seldom was I aware
of His presence

Contents

			Page
Preface			**viii**
Chapter 1	-	Mont Alto: Matriculation to Missed Mattress	2
2	-	The Scroll Unrolls	20
3	-	"On My Honor"	30
4	-	Miss Magruder's Disappointment	46
5	-	Through the Schwartzwald to Berchtesgaden	56
6	-	A Nittany Lion	78
7	-	Junior Forester on the Sabine	94
8	-	The Forests of Deep East Texas	122
9	-	Year at Yale	134
10	-	Ulcers at Syracuse	146
11	-	Research on the Escambia	166
12	-	Georgia Challenge	172
13	-	To "The Oldest Town in Texas"	204
14	-	Litigation and Termination	266
15	-	A Few Collegians	274
16	-	"The Needed Truth to Speak"	306
17	-	Manuring Europe's Forests	318
18	-	Pacific and Asian Escapades	332
19	-	Journeys to the Near East	354
20	-	Gringos and Ticos	368
21	-	"Woods and Templed Hills"	386
22	-	The Organization Man	402
23	-	Emeritus	426
24	-	"The Higher Good"	442
APPENDIXES			
A	-	Eulogy for Dr. M. Victor Bilan	465
B	-	Bibliography of the Author	468
Index			481

PREFACE

Who but a celebrity in politics, sports, exploration, war, or theater would have the audacity to write about oneself? Why do I do this? I doubt that my children or grandchildren will read it, assuming their geneological interest is no greater than mine.

I write this story of my life because, perhaps in a fit of flattery, my former colleagues—yet always my friends—asked me to. They'd heard the tales over coffee conversations through the years. Prevailing upon Dean Scott Beasley, a student in an undergraduate class I taught three decades earlier, he graciously provided the secretarial services. Otherwise, arthritic fingers hunting and pecking on a 1942-vintage Royal portable typewriter (on which many drafts of two theses were typed) would have discouraged the effort. But when Robin Miller, the computer whiz, had completed typing from my scribbles the first draft of two chapters and I suggested she may need help, her quick response was, "No one else is going to get to type this story!" That was the encouragement I needed to keep going. Thank you, Robin.

Rehearing and re-telling stories over nearly seven decades makes them at times suspect. Are they true tales or tall tales? I hope the former, but if some exaggeration has wormed its way into the narrative that follows, dear reader, please forgive. Essentially all of the materials following the boyhood years are documented in logs, reports, published articles, office files, oral histories, or reviews by colleagues. The latter include Cynthia Parker, J. Reid Parker, Marjory Blubaugh, Nancy King Pase, Joe Pase, Dean Joe Gerber, Mrs. Gertrude Gerber, Thomas Croker, Richard Clark, David Kulhavy, Leonardo Testa, Ralph Read, John T. Lewis, William Bos, Lowell Halls, Jerry Holbert, Montague Whiting, and Lorraine Seegrist Donohue.

Ms. Louise A. Arnold-Friend, reference historian for

the U.S. Military History Institute at Carlisle Barracks, has been especially helpful in locating records of Army units with which I served. Amy S. Doherty, Syracuse University archivist, retrieved some names. Mont Alto's Marjory Blubaugh sent photographs from the old school's archives; Yale's David Smith commented on Chapter 9; and David Kulhavy, although a renowned forest entomologist, used his computer skills in formatting the manuscript for desktop publishing. Gordon Holley aided in this effort as did William Ross, Graeme Seibel, Katy Bugg, and Joy Attebury. Kenneth Kennamer applied his editorial expertise to a final draft. My everlasting appreciation goes to Robin Miller who patiently "patched" the typescript during editing and subject-matter critiques. I'm grateful, too, to Richard White, vice-president for environmental services of TUSERVICES, and to the electrical utility company for financially supporting the printing of this story.

LCW

* * * * * * * * * * * * * * * * * * *

Proceeds from the sale of this
book benefit solely the
Scholarship Fund
of the
College of Forestry
Stephen F. Austin State University
Nacogdoches, TX 75962

1

MONT ALTO—MATRICULATION TO MISSED MATTRESS

A second-growth mixed-hardwood stand in which we carried out "practicums" in mensuration and silviculture (Mont Alto archives).

A lone hiker packing Boy Scout gear, I paused at the stone-pillared entrance to the Penn State Forest School to view the rock chapel surrounded by a half-century's plantings of arboretum trees. Passing through the portal would be the single event that charted the direction of my life.

Emmanuel Chapel, built in 1854 as an Episcopal mission for the iron ore miners and others in the South Mountain area (Mont Alto archives).

I trekked with blind anticipation the two-mile stretch on the narrow blacktop road from the little farming community of Mont Alto. Excitement about the venture, especially as to whether I would succeed, had intensified since leaving home in the inner city of Washington, D.C. My younger brother, Mark, rode with me in the streetcar to the end of the Wisconsin Avenue line and waved goodbye as I raised my thumb to hitch a ride to a special kind of college.

Weistling Hall, the original part built in 1807 when a Virginia family took advantage of the surface iron ore at Pondbank and the vast forest available for charcoal production. They opened an iron furnace. Wiestling, a federal colonel at Gettysburg and an inventor, took over the foundry as ironmaster and the large home in 1864. The state bought the property in 1893, Mont Alto becoming the first Pennsylvania State Park (Mont Alto Archives).

Apparently someone observed my arrival from what appeared to be the School's main building, a big old house. Why would a lad be coming this early, two days before scheduled matriculation? That's what the portly and friendly H. Norton Cope, school director, wanted to know as he greeted the solitary student. To make sure I would have a job was my answer. And I assuredly did, as an every-other-day dishwasher in Mrs. Horsfall's spotless kitchen. The school had primitive automatic machinery.

Professor Cope introduced me to others in Wiestling Hall. (The spacious home served as a residence from 1807 to

1891; a few years later it became a dormitory for the new Pennsylvania Forest Academy, operated by the state's Department of Forests and Waters.)

Someone then showed me to my room in Conklin Hall, built of rich red sandstone quarried from an adjacent hill by students between 1909 and 1911. From the upper-floor room, I could see South Mountain, a section of the Appalachian Highlands chain that surrounded the campus. All about were second-growth forests, reestablished by man and nature following the harvest of the virgin stands. Iron ore beneath the ground and timber above the ground had supplied the demands of steel mills, the ore smelted for its metal and the timber cut for charcoal to provide heat for the task.

Soon I'd roam the Mont Alto and Michaux state forests and the Caledonia State Park nestled in these hills. To the north, the Blue Mountains cast a haze, the terpenes exuded by the conifers giving the sky a bluish tint. From my room, too, I could hear the water cascading over a falls formed by a dam just below the window. Sleep came easily with the cadence of the water bouncing on the rocks. The falls powered a grinding wheel where earlier students daily sharpened their double-bit axes before saddling horses for the day's field assignments. (They brought their own horses, required for fighting forest fires.) In the period of which I write, the School bulletin noted only that "Each student should have a woods hatchet and a box compass." Mine were Boy Scout issue.

From another window in the old stone dorm I looked southward to the state tree nursery where later we'd learn planting techniques, the use of transplant boards for moving one- and two-year-old seedlings at a density of 40 per square foot to less competitive pressures of 8 per square foot for improved vigor when out-planted. And there we would learn how time spent in seedbeds and transplant beds depends upon species and perhaps the provenance from which the seeds were collected and to which the seedlings would be returned.

How naïve I was. All I knew about forestry when I applied as an out-of-state student in the Fall of 1942, fresh from high school graduation and not of draftable age, was what I had gleaned from *American Forests* magazine. That source, with photos of western woodsworkers, suggested the need for woolen riding breeches that buttoned from the knee to the ankle, knee-high boots that lace, and a flannel shirt. Seriously depleting this kid's account at the Perpetual Building Association nearby the White House in Washington, these items were bought and packed, with a few necessities, in a footlocker shipped via Railway Express from house door to dorm door. That was the method of the times for shipping everything the Post Office would not accept. The footlocker was a graduation gift from a dear neighbor lady whose fence I painted every year. The boots and shirt eventually wore out; the wool breeches, waist-size 28 inches, brought laughs; they hang still in a cedar-lined closet, a frequent reminder of the happiest days of my life—and participating in what we thought were original college pranks.

Skunk-Hunting, Dorm-Flooding, and Proctor-Hazing

Mischief at Mont Alto was abundant. Only a few of the 65 students who matriculated in that early Second World War year had autos, and, for those who did, gas rationing stickers were in short supply. Consequently trips to the girls' colleges in Chambersburg and Gettysburg were infrequent; fun was found nearby. Anyway the girls at Wilson Women's College in Chambersburg were snooty toward us woodsy types.

Among the freshmen foresters was a demolition expert. One farmer's outhouse—that I know about—blew to smithereens in the moonlight of an icy-cold late November Saturday night. Professor Cope had heard about it by breakfast time. Few words were said. He simply showed us the best white pine lumber for the seats for the two-holer; eastern hemlock 2x4s for studs, and some shakes (could they have been American chestnut?) for siding and roof. The school's little peckerwood mill supplied the demand, sawed by previous year's students. We would refill the wood box later in the year. Before the day was done, the man of the soil had a new privy over a newly dug pit with seats hand-sanded to a smooth finish. No splinters would stab the backside of a visitor to "this little home." James Whitcomb Riley caught the scene poetically in *The Passing of the Outhouse*:

When memory recalls the past and moves to smiles or tears,
A weather-beaten object looms right through the mist of years:
Behind the house and barn it stood, a half a mile or more,
And hurrying feet a path had made, straight to its swinging
door.

* * *

And still I marvel at the craft that cut those holes so true,
The baby hole, and the slender hole that fitted sister Sue.
That dear old country landmark—I've tramped around a bit,
And in the lap of luxury, my lot has been to sit.

Some other tales to tell. One dealt with limburger cheese supposedly borrowed from the kitchen larder for a night snack. Instead, the culprit(s) placed it on the hot-water radiator in a dorm room. The cheese could not be scrubbed off; plumbers were called, the radiator removed, and one of the pair, not the object of the culprit(s), moved into my room. Mutt Sparks, my first roomy, had vacated the school in a longing for the love of a young lady in New Jersey. Mutt earned his name because he was the tallest student and because he shared a room with the runt of the class. My new roommate, Richard Clark, became a cherished friend. We bunked together after the war and have continuously kept in touch. He went on to study for the Episcopal ministry at Cambridge, returning to forestry as a second career after earning a Master of Science degree.

Then there occurred the water fight between the upper and lower floors of Conklin Hall. The stairway looked like Niagara as the second-floor fellows utilized the firehose. Mr. Cope arrived; mops were distributed and floors scrubbed dry before anyone retired for the night. (Conklin, restored to its original picturesque beauty and with a stain-glass window, has been preserved as a student center for the present-day branch campus of the Penn State-Mont Alto Commonwealth College.)

The night before final exams were to begin for the Fall term, some wise guy suggested we move a coal car from its position on the rail siding under a coal chute near the only entrance by paved road to the campus. Studies for exams satisfied, the fun began. A prankster released the coal car's mechanical brake. The heavy load gave way slowly to the slight downward grade of the track, and then was braked by the "conductor" when squarely in front of the entrance gate and fully straddling the snow-covered and icy pavement. Faculty entry was blocked.

Not until the school secretary Samantha Heefner and faculty members who lived in nearby towns (three lived on campus) arrived to give final exams was any of the resident staff aware of the devilish stunt of the preceding night. Instructors and secretary slid on the ice and slushed through the snow in their hike to the classroom building and Wiestling Hall. The beloved father figure, Professor Cope, again showed his skill at leadership; no exams until the coal car was back under the chute. That's on the incline! How could so few manage the task? Enrollment by this time was down to about 45: draft boards and disillusionment with work in the woods took their toll. Nevertheless, with crowbars (sometimes called prybars and pinchbars) and chucks and wedges, we inched the heavy steel car, little by little—but only to the side of the road. "Prof" was lenient. Exams could now begin.

Skunk-hunting at night, led by Dutchman Bob Burkholder, Jake Longwell, and Bob Early, also provided diversion from study; many the pair of Keds and trousers left buried in the ground behind the old stone dorm. The malodorous secretion ejected when the animal is startled won't wash from clothes, no matter the amount of bleach appropriated from the chem lab applied to the fabric. Years later Bill Haig reminisced about mice trapping—in Gene Delatre's dorm room.

John Kent, a wiry young bachelor, served as English instructor and dorm proctor. His room on the ground floor also served as a place for finding both consolation and individual improvement in writing skills (to him I owe so much!), as with each student he'd go over the essay or composition assigned for the week. We were to write about forestry, but not about forestry solely: the narrative was to show a connection to the larger world, to one's faith, future, or fortune. I titled one of my essays, dealing with science and faith, "Where were you, Job?" Mr. Kent also taught English Bible, meeting in the Science Building on Sunday afternoons. Many attended the class out of respect for Mr. Kent; some because "Prof" Cope, strong in the Lutheran catechism of his German ancestry, would appreciate our wisdom in attending. (Mr. Cope still said, "Outen da lights" when electrical energy was to be saved.) I carried my new *Scofield Reference Bible*, a gift from my older sister Hope, an oft-published poet, when I left home. She had reason to know I'd need its instruction.

It was a shameful act to pull on Mr. Kent when a character—we never learned whom—tied a dead rattler around his door knob before he arrived home after a date with the younger cook late one night. Like the rest of us, Mr. Kent went to war; and though I looked for him following the conflict, I never learned his fate. Prof. Cope wrote me, shortly after Kent's induction, that he was a corporal in the Military Police.

About Mr. Kent and the cook, I should relate that he was the only faculty member who ate breakfast with the students. As dishwasher, I knew his coffee was prepared separately. Soon I found hidden the jar of a supposedly anti-aphrodisiacal compound used to spike the coffee drunk by the "scholars." (We thought it was saltpeter, but I find no evidence in the medical literature that this could be so.)

Apples ripe, and no one to pick them in the fall of 1942. Orchardmen enticed Mr. Cope to enlist us in the effort. For several days we climbed, picked, and toted, all the while over-eating bellies-full of the delicious fruit until weariness drove us to our bunks.

No beer busts at Mont Alto. Mr. Cope saw to that. A rumor, perhaps enhanced by the faculty, told us why he was abstinent. He had been a district ranger on the Coconino (or was it the Kaibab) National Forest, nearby the Grand Canyon. In Cope's time it was the Cook National Forest. His crew had been on a fireline for days. Exhausted when the blaze in the grassy stands of old-growth ponderosa pines was surrounded, the firefighters celebrated with a keg, possibly home brew provided by a local moonshiner. The fire was only controlled, not out. In the dry, windy night on the high plateau, the embers came back to life. Now, the fire's own heat generated wind that hastened its spread. No longer a surface fire running along the ground, the flames crowned in the tall yellow boles with resinous foliage. The ranger was responsible. (Harold Geiger quite recently squealed that hard cider left in jars on Conklin's window sills "cured" nicely.)

The "Scholars"

Scholars we were at Mont Alto. We took classes in English, botany, dendrology, algebra and trigonometry, chemistry, engineering drawing, and physical education during a 4-day week. We spent another full day each week in the woods or tree nursery: planting, lifting, transplanting, harvesting, thinning, scaling logs, measuring tree heights and diameters, running lines, and laying out sampling plots. We devoted Saturday mornings to ROTC, sporting 1903 Springfield rifles for close-order drill. We called the Reserve Officer Training Corps Rode-a-see (now Rot-zee). During this course, a recruiter from the Army came to visit. He said that if we joined the Enlisted Reserve Corps, we'd be able to finish college, go in the service as second lieutenants, and exchange the "cowardly 3" initial number on our draft cards for the bravado of a "1". From then till April, 1946, I was 13186849: it's engraved on every bone and fiber of my body, like a genetically coded chromosome. Oh, the recruiter did mention we'd be called up if the war heated up and we were needed. It did, and we were, though we completed the freshman year.

Conklin Hall, the dorm built by students in 1909-11 with rich red sandstone hewn from a nearby hill and named for the second state Commissioner of Forestry (Mont Alto archives).

Professor Alvin Grove, the only faculty doctorate, taught botany, and taught it thoroughly. Near semester's end he noted a student's textbook still in its bookstore jacket. The professor picked up the book and upon opening it heard the crackle of a never "cracked" heavy text. Not only Ned, but everyone in the class now had to outline the whole text. That was a good way to get a good grade on the final. I suppose we all did. To every student's question, Dr. Grove returned with, "How's cum?" It was his trademark, and we all mimicked him.

Early 1940s faculty: l-r. George Harvey, Leon Skinner, H. Norton Cope, Henry Chisman, Rogers (?), Marx Moller (?), ROTC Instructor (?), Secretary Semantha Heefner, Bill Pfeiffer, David Kribs, Otto "Shorty" Pfleuger, and Jim Bethel. Skinner, Rogers, Kribs, and Bethel had left prior to Fall term 1942. Kribs taught botany at State College and Bethel eventually became dean of the School of Forestry, University of Washington (Mont Alto Archives).

Then there was dapper Henry Chisman, MF from Duke. He could pass for Clark Gable, mustache and all, as the librarian obviously would agree. With Mr. Chisman, we had a great field instructor; the woods came alive. Happy Hank, as the students affectionately called him, headed up the civilian defense effort, giving weekly instruction and drills that qualified us as home-guard wardens—as if Hitler's hordes could find, or care about, us hidden in these hills. We sorrowed for Mr. Chisman, as his wife was hospitalized with terminal tuberculosis just five miles up the mountain road at the state sanitorium. The hospital was founded, as was the old Pennsylvania Forest Academy, by Dr. Joseph Rothrock.

Otto W. Pfleuger taught dendrology. He had been head of the Philippine Forestry School, but poor health forced him back to his beloved Penn's Woods. We called him "Shorty" at his insistence. I suppose he was 5'2". (At 5'4", if I can look a man in the eye, I subconsciously say, "What a runt that fellow is.") At Thanksgiving dinner, Shorty sat on an unabridged dictionary at the head of our table; his wife, who had headed the Philippine Home Economics program, was our hostess at the other end. We were a family and especially well-mannered on that occasion.

Forest School campus from the classroom building. Conklin, the stone dorm, is in the background. Wiestling, in the foreground, housed offices and dining room. (It was earlier the ironmaster's residence and named for a Civil War colonel.) Double Cottage, in the center, housed the cook's apartment and infirmary (Mont Alto Archives).

All dendrology was taught in the forest and arboretum. Anything resembling lecture material was expected to be self-taught from Harlow and Harrar's heavy tome. One didn't carry this text to the field. For winter characteristics needed to identify deciduous trees, we toted Bill Harlow's *Key to the Twigs*, a paperback. During the December final exam, Shorty took us to a 20-inch dbh tree, the lowest branch of which was more than 15 feet above the ground. Oh, none of us correctly named the large deciduous tree *Sassafras albidum*, for the species' stems seldom exceed 3 or 4 inches dbh in Pennsylvania. As small trees, we knew well the sassafras of the Lauraceae family, for we often boiled tea from its root bark while hiking the trails of the Caledonia and other nearby forests. We wrote the common and scientific family, genus, and species name on little slips of paper as we did each week for selected trees, and signed our names. Peeking at another's paper would gain a stern stare from Shorty. I never dared do it again!

The arboretum was a real asset, especially so for those who'd go west. Most of us yearned at that stage in our studies to head to the tall timbers of the Sierra-Nevada, the Coast or the Cascade ranges. We'd know, if we were so fortunate as to find a job in the West, the identifying characteristics of many of the principal conifers growing in those regions. The arboretum, established from seeds, dates to 1905.

Professor William Pfeiffer, because of low enrollments—when he wasn't tying flies for trout fishermen—spent his work day at administration, freeing Director Cope to spend more time with his students in the field. Cope and Chisman alternated teaching Introduction to Forestry. For that course, using appropriate formulae, we each constructed a personalized Biltmore stick, based upon the distance from the individual's eye to his fist. No one else could accurately use my stick, for no one else had so short an arm. I attached a hinge so it would fold into two parts. We used the Biltmore for determining diameters at breast-height. On the reverse side we carved and inked a hypsometer scale by which we measured tree heights and the numbers of merchantable logs in the trees of the second-growth forests.

A few more words about Mr. Cope, leader extraordinaire.

He served with the famous forest engineer battalions of the American Expeditionary Force in Europe in World War I. Students remembered him long after graduation for his Socratic method of instruction: "If you can't answer the kid's question, ask him one." Prof. Cope granted merits for proficiency in handling oneself and tools while engaged in woods work, for making nature collections, and for accuracy in technical subjects like compass use and cruising. My scout collections earned some of these merits.

The physical education instructor was later famous as a champion at tying trout flies. His association with forestry, although on the faculty of the physical education department, also garnered some infamy about the School in the press: forestry studies are not truly academic. Just fun and play was implied in the angling accounts that ran in the nation's papers. In 1994, the *National Geographic* called "George Harvey a legend who caught big brown trout at night with a huge, hairy fly he invented." The colorful browns—sides stippled red and white, red streaks on tails, and with bellies of gold—challenged anglers among the student body.

Harvey also lectured us on venereal diseases. By the end of the term, we thought the school had an epidemic with which to deal. Everyone had some of the symptoms described in the instructor's orations: legs bowed forward, eyes that bug out, sore on a lip, pimples, sore throat, body rash, moist toes, etc. Yep, the coach scared us.

Algebra and trig came easily for me for I'd used the same two texts in high school. However, Mark Moller, a civil engineer, covered both subjects in a single semester. We'd need them for mensuration and surveying.

The Penn State Forest School inspired and motivated me. My transcript records 6 As and 2 Bs the first semester. The faded copy, all hand lettered, called them 3s and 2s, respectively. I made the Dean's List. Later, if you're patient, you'll note the significance of this achievement. (C earned 1 grade point; B, 2 grade points; A, 3 grade points; D earned none; F was a -1; and a -2 gave recognition to cheaters and plagiarizers.)

Reunion

I mentioned earlier the Episcopal Emmanuel Chapel, erected in 1854, at the entrance to the campus. Irregularly a clergyman held services there for us and a few surrounding farmers. Some students went to the churches in the town of Mont Alto, a pleasant walk. Indeed, the bulletin for the 1941-42 year read, "All students are encouraged to attend regularly the services of one of the churches in the village of Mont Alto...."

My privilege it was, in 1986, as a Presbyterian minister, to lead the worship service for the first Mont Alto Super (all classes) Reunion. Retirees and their wives packed the little sanctuary, singing lustily to the accompaniment of one forester's wife's professional ability at the keys and pedals of the beautiful organ, now with electrified bellows.

Later I learned from Maurice Goddard how much the service meant to him. Especially as we sang William Merrill's rousing hymn,

> Rise up, O men of God!
> Have done with lesser things;
> Give heart and soul and mind and strength
> To serve the King of kings.

It was in this chapel that the son of "Pop" Goddard—as he was addressed while serving briefly as director of the Mont Alto School following WWII—was baptized. From the stone church, too, the infant son was buried. Professor Goddard, a Maine native awarded an honorary doctorate, went on to become Director of the School of Forestry at Penn State and subsequently the Secretary of Pennsylvania's Department of Forests and Waters (later the Department of the Environment), serving under many administrations of both political parties. A generation of students knows him as "Gramps."

For the occasion of the Super Reunion, a pulpit Bible was procured from Penn State's Archives. It was said to be the Book from which John Brown, the Messianic Calvinist, preached on his way to Harper's Ferry, where the Shenandoah River joins the Potomac in its run southward. Some claim it was the abolitionist's last message on his way to an immortal destiny resulting from the raid on the armory astride the Shenandoah. There along the River, the Appalachian Trail now threads its way from Mount Katahdin in Maine to the peak northeast of Brasstown Bald near Amicalola Falls in Georgia. (Another account has Brown, using an alias "Isaac Smith," only attending services.) A marker at the School entrance reads:

John Brown Raid
Captain John Cook (sic) Cooke, one of Brown's followers,
was captured near here on October 25, 1859, nine days after
the raid on Harper's Ferry. He was hanged December 16,
two weeks after John Brown.

End of an Institution

Staunchly and alone, Professor Pfeiffer held the fort when other foresters on the faculty and most of the students left Mont Alto to further the fledgling North Carolina forestry program at Raleigh in 1929. N.C. State then could have a graduating class of 4-year students in its first year of operation.

It happened this way. As the competing Department of Forestry at Penn State in Centre County began to gain strength, it usurped the Forest Academy's turf. The Academy (as Mont Alto was then called) enabled Pennsylvania to be the first state to train its own foresters. Gifford Pinchot, the nation's first chief forester and subsequently (1920-1922) Pennsylvania's Commissioner of Forestry, was elected governor in 1923. He served until 1927 (and again from 1931-1935), opposing during his first term the existence of the second school, even though it dated to 1906. Mont Alto seemed to be his favorite child. The Pennsylvania State College budget, with the forester in the governor's chair, now only allowed for an agricultural extension forester. When Pinchot left office at the end of his first term, his successor not only restored the funds for the State College department in the School of Agriculture, but added insult to injury at Mont Alto. The Penn State Forest School now would be for freshman instruction only. Sophomore and upper-level courses would be taught at State College (now University Park). Director Julius V. Hofman left to be NC State's first dean, taking with him from Mont Alto in 1929 most of the faculty, the sophomores, juniors, and seniors. (The Hofman State Forest in North Carolina's Pocosins honors the former Pennsylvanian.) Through the Super Reunion mentioned earlier, the few living among the 243 graduates (over the school's 28-year life) of the pre-'29 school preferred to remain aloof from those of us who would follow. It was our loss. When none lives to carry on their traditions, the last survivor will drink the commemorative bottle of spirits. Gratefully, Henry Clepper, class of 1921, has left the record in numerous memoirs in books and in *American Forests* magazine.

The Wrestling Match

This chapter opened with mention of a missed mattress. We'll close it on that note. Just after the coal car episode and with some final exams taken, we dragged mattresses from the empty beds to the hall for wrestling matches. That would break the tedium of studying for the morrow's tests. Somehow, I failed to recall Coach Harvey's admonition that, when you fall, make sure your adversary is on the bottom; and if you must fall, don't miss the mattress. The challenged, or was it the challenger's, name is lost to memory, but we both heard the bone snap.

To the infirmary in the upper story of Double Cottage, and adjacent to the cook's quarters, Mont Alto's sole medical man was called. I think he used 50 yards of 4-inch wide adhesive tape, criss-crossing my chest and back, to set the collar bone. Remaining finals were privately administered; the cook graciously served as nurse. Dick Clark packed my Scout knapsack and footlocker and Mr. Kent took me to Chambersburg to catch the bus to Washington. I think then I learned the good doctor was a veterinarian, but claimed by all to be the best medic in this rural countryside. X-rays at Walter Reed Army Hospital, where my older brother, Hobart, now an Army Air Corps captain and fortuitously on furlough, got me treated, proved the claimants correct: the vet set the fractured bone perfectly.

When the year ended, Phi Eta Sigma, a national men's honor society tapped me (1 of 33 in 1943 for Penn State) for membership. As we'll learn later, Miss Magruder would be surprised. I think I'd found my niche.

Winter scene of Double Cottage in the early 1940s (by Tom Neff, courtesy of Mont Alto archives).

Alas, Penn State administrators at State College closed Mont Alto, the little school nested among the eastern white pines, the eastern hemlocks, and a vast number of deciduous species. The war required sacrifices. Spring semester would find us looking for sleeping quarters in State College. For some, a sad Christmas would follow the end of Fall term, many of the remnant 40 called to serve their country before the new term began. Some left forestry discouraged by the turn of events. We were closely knit, loved the Forest School, and envied the professors. Maybe someday we could teach at the college founded by Joseph T. Rockrock (medical doctor turned forest conservationist) in 1903 and guided through its first years by George Wirt—who only a year earlier had completed forestry training at the Biltmore Forest School.

(Elizabeth H. Thomas has meticulously chronicled *A History of the Pennsylvania State Forest School, 1903-1929*, published in 1985 by the Pennsylvania State Forest Academy/School Founders Society.)

Old-growth red spruce on a knoll not far from South Mountain. Note the abundant reproduction coming in under the shade-tolerant parent trees (USFS photo by E.S. Shipp, 1927).

2

THE SCROLL UNROLLS

Ginkgo, planted in 1762 in London's Kew Garden, and the author's daughter, Jean. Stems of these dioecious stink-bomb trees lined streets of the nation's capital.

The corner of Fifteenth and A streets in the nation's capital city, a mile from the domed building of stately columns and pilasters in the Corinthian style of architecture, seems an unlikely place to begin a journey that would lead into forests throughout the world. As I let the scroll unroll, perhaps dear reader, you may note the connection.

At the time of my birth, my father, a captain in the District of Columbia National Guard and captain of the DCNG rifle team was at Camp Perry, Ohio, for the annual competition. The team, not knowing what to call me, sent a silver-plated loving cup bearing the inscription, "D.C.R.T./To Perry Walker/1924." Perry would have been my name, save for a drunk next door named Perry G. In that year "Silent" Calvin Coolidge was president, the hometown Senators defeated the New York Giants to win the World Series, and Congress granted full citizenship to American Indians.

My recall goes only to the mid-1920s, the "Roaring Twenties," with its high-spirited flappers and high-rolling economy that followed the First World War. At age 3, we'd sit on the curb and call the kids across the street "chocolate drops," and they'd reply, "soda crackers." The imaginary strip down the middle of the pavement was not crossed by the kids alone. Black children, towed by parents, could enter the alleys of our neighborhood, shouting, "Lady, yo' got any ole braid?" And Mom would give them bread, and more. In summers they also sold good watermelons in the alley behind the houses, hawking them as "good to the rind." Where they grew the giant juicy fruit I never knew. Perhaps their expertise came naturally: *Citrullus vulgaris* is native to tropical Africa. Our house was near enough to Union Station that Mom often fed the hobos who came to the back door. The "mark" of a soft touch on the fence remained elusive to us. But the tramps would know to which door to go.

We "farmed" grapes. Mom planted the vine and Pop erected the trellis over the 6x8-foot plot of soil in the back yard. The vine was a producer. From its fruit Mom annually canned the year's supply of jelly.

Those were Jim Crow days in the District of Columbia, even though Lincoln's North won the war. Segregated schools, streetcar seats, and drinking fountains were the rule. So, too, restrooms, lunchrooms, churches, and parks. It was *de facto*, not *de jure*, segregation. A fact, not a law, but treated like a law.

Parks, of which the city abounds, were easy to segregate. No one used them. National Capitol Park Police made certain of that. Don't step on the grass! So kids played ball in the alleys or streets. Principals locked school yards at 3 o'clock sharp. We found other places to play, like the city dump, where now stands the huge RFK stadium. There I learned to smoke cinnamon cigarettes, the only kind Mr. and Mrs. Steinberg at the corner store would sell my "rich" buddy, Jack. They cost a penny each. There, too, the big boys taught us about sex. Indeed, all we'd ever need to know, and it was accurate. Why professional teachers must carry out the task remains a mystery to me.

While not a church-going family, Mom sent us to Sunday School. Has anyone else been expelled from such an institution? Hiding behind the kitchen stove, I heard Mom say to the visitors from Epworth Methodist Episcopal Church, South, "Oh, that couldn't be my boy." Perhaps I was 6 by then. Later, when Uncle George, a highly respected Methodist clergyman from a tall-steepled church in Pennsylvania, came to visit and discovered his sister's two youngest progeny had not been baptized, he proceeded to remedy the matter. So I, along with my younger brother, Mark, were baptized (maybe they called it christening). My Baptist friends still can't comprehend being baptized in (with) a *Wheaties* cereal bowl. (Such a "chalice" came in every package of "The Breakfast of Champions.") Chuckling about it now, I think the inoculation by sprinkling "took," even if the sacred rite was performed in the kitchen.

Genetic War

We digress to explain the cultural wars raging in the

genes of my chromosomes. Pop, born Hobart T. Walker, in New Tazewell, Tennessee, and reared in a Kentucky Mountain cove through which flowed Wallins Creek in Harlan County, left the back woods to join the Marines in the First World War. I thought he was joking when he gave as his reason for enlisting "to get a horse blanket and a pair of shoes" until I read Catherine Marshall's *Christy* about life in the Southern Appalachians at the turn-of-the-century. Wood and coal were plentiful resources in these hills: the latter was mined from a hole in the yard for cooking and heating the little stone house on the side of a steep slope that I relocated in the 1960s. Visiting but once as a child, I recall my grandmother smoking a pipe. Her son, a tall, erect, athletic hillbilly who could take on anybody became a sergeant and, following "the war to end all wars," commanded a company of 89th Division infantrymen in the National Guard. In that task his troops help put down riots in the nation's capital. One involved the First World War "bonus marchers" camping in a park along the Anacostia River during the Great Depression. In this 1932 task, my father served immediately under Major Dwight Eisenhower who, in turn, was subordinate to Colonel Douglas MacArthur. World War vets were called upon to put down their comrades in arms who, many homeless and hungry, wanted the money promised by Congress, to be granted fifteen years later, for service to their country. (Congress relented and gave the money. My father, with a bunch of cronies, blew his bonus on a fishing trip to Cuba.) Because Pop wore boots that hugged his legs to the knees with his military uniform, I told my first-grade teacher he was a bootlegger. The term, overheard in that time and place, had just come into my vocabulary. I'd learn its true meaning.

In contrast, Mom, born Laura Pearl Johnson of a large progressive farm family in the fertile Susquehanna Valley of Pennsylvania and principally of German stock (hence my middle name Colton), was strict about orderliness. How these two, the rough and ready southern mountaineer and the cultured northern lass came together I attribute to my mother's leaving the rural one-room schoolhouse for "the excitement" of working for the Veterans Administration and my father's assignment to the Marine Barracks just south of the Capitol. From this union came a son, the forester, constantly torn between doing things the easy, careless way and "if it's worth doing, it's worth doing right." I compromised on, "If it's worth doing, it's worth doing, and let's get the job done." Perfection is for more deliberate types.

These were the years affected by the twentieth amendment to the Constitution, forbidding the sale of alcoholic beverages, and its rescission with the twenty-first amendment in 1933. I recall Mom being so distraught over booze in our neighborhood that she dragged us kids house-to-house getting signatures to prevent the opening of a liquor store across the street. She won. So it seemed consistent that an essay by my mother, written for a class at Bloomsburg (Pennsylvania) Normal School in about 1904, was entitled, "Reverence."

On Bending Twigs and Leaning Trees

Sometime in these early years we visited friends in the country. I stepped in a yellow-jackets' nest, lived through the agony of multiple stings and, Mom later said, complained about the absence of sidewalks and having my shoes caked in mud. From that nearly singular family venture, my later wanting to be a forester amused and amazed her. But then she would quote Alexander Pope: "As the twig is bent, the tree inclines." And I could only ponder the significance of the line of the 18th century English poet.

I learned early to be willing to place less than first. In one summer, the school playground was open for inner city kids. I think I was in second grade when I won the silver medal for coming in second. There were only two of us in the race.

At Principal Diedel's Kingman Elementary School in the Great Depression years, we mid-city kids learned to garden. Workers plowed the hard, cinder-covered ground and we planted and weeded; but with the shortage of money was the shortage of rain that gave rise to the Dust Bowl even in the East. Our crops seldom reached harvest size.

About age 9, my buddy Jack's dad gave him a copy of The Cub Scout Handbook. We set up a den in his basement and, without any further advice, lived the life of the Cubs.

We knew of no cub packs or den leaders. Jack, who became a popular saxophonist in a dance band, and I remained cubs until we turned 12, but that's a tale for a later time.

To earn money in those pre-teen days, we'd work up a route to collect neighbors' old papers and magazines from musty basements. The junk shop at 15th and H streets paid 8¢ a hundred pounds. We could bind with twine 10¢ worth on a little red wagon, and with that, following the conventional Saturday night bath, splurge on a 5¢ Eskimo Pie. The rest went into a savings account started at my birth when someone (oh, I'd love to know whom) gave me a $5 gold piece. Mom, the ever-obedient citizen, didn't read the fine print when Franklin D. Roosevelt took the nation off the gold standard: she dutifully swapped it for a paper silver certificate. The law, unrealized by her, provided for continuing possession of gold pieces with sentimental value. What would a collector pay for that coin now?

In those years, too, I got a dime each year for painting the yard bench of an arthritic neighbor. She never forgot my fence-mending chores, for in her will she left me a hand-crafted solid mahogany hinged box, I judge *circa* 1860, and an ivory egg that hung from her husband's watch chain. Hollowed out and threaded, when unscrewed it reveals a minutely carved chicken, also of ivory.

We followed the lamplighter around the block and up the alley as night fell. He rode a bike, stopping at the posts to insert a lighted wick on a pole into an opening at the base of the lamp. The street lights burned consumer gas, manufactured from coal at the gashouse beside the Potomac River, now the site of the famous Watergate hotel and office building. The spent coal, becoming cinders, was spread on the schoolyards to cover the soil and thus avoid traipsing mud into the buildings. But, ouch, when you fell on those sharp-edged, gray-colored, light-weight, fist-sized objects, blood gushed from the cuts and abrasions. I think *natural* gas began to flow through the pipes to the lights and houses during the Second World War.

The narrow alley, too tight and the bends too much for even the large cars of the 1930s, served as a playground. There, we followed the ice man as he delivered blocks to kitchen window boxes during the warm months. He'd give us little slivers to suck on while we made summer "snowballs" for alley battles from the manure dropped by the horse pulling his wagon. The moist fiber served as binding for the molded missile. (Pharaoh and Moses both knew about that.) Garbage was collected, the slops carried away for farmers' hogs, and milk delivered in reusable glass bottles by the same kind of conveyance.

For kids without playgrounds, bicycling provided pleasure and an escape from the inner city. The problem was that we four kids had but one bike and that belonged to my older brother, who, much taller than I, got angry when I lowered the seat. So I got saddle-sore and endured thrush, like a diapered baby, after every ride. One Sunday, Jack and I rode to the Washington-Hoover Airport, where now stands the Pentagon, a long trek for 11-year-olds. The age can be verified, for we were in the radio hut aside the runway when news came across the wireless radio that the great entertainer Will Rogers and his pilot, Wiley Post, had been killed in an airplane crash in Alaska. That occurred in 1935.

Trollies

Without a family car, we got around on trolley cars, also called streetcars, that ran on tracks imbedded in the pavement down the middle of the streets all over town. In between the rails was a 2-inch wide trench into which was extended the steel connection to the source of the current. The "juice" was safely several feet below street level.

In outlying districts, an overhead "trolley" was the source of the current. That wire hung from wooden poles, paralleling the tracks. At some point, the motorman would stop the car, get out and crank up the underground gizmo and release the rod that would spring up to connect with the overhead line. Returning to town, *vice versa.*

The streetcar's electric motors gave off ozone, a pungent-smelling noxious odor that would cause vomiting, especially for those not acclimated. Kids would run for windows that would open, though most would be stuck shut. It was especially nauseating on cold winter days. In summer, the company rolled out its open-air rigs. These were fun to ride. No sides, just a step that ran the length of the car—perhaps 40 or 50 feet—on either side. The conductor walked the step, noting who got on at each stop, and collected the tokens. Tokens were dime-size, and full of holes. They cost 30¢ for 4. School kids, however, got books of tickets that cost 3¢ each (i.e. for ride).

The car seats were of hard straw "wicker," without springs or cushions. Floors and sides were of wood. Motormen worked the windshield wipers manually.

I think no one was ever turned away because the trolley was full. The conductor would just push folks down the aisle, sardine-in-a-can style. I was always too short to reach the overhead hangers for standing passengers.

Printer's Devil

At about age 14, I went to work as a printer's devil at a hole-in-the-wall job press in the basement of a shabby hotel at the corner of 13th and New York Avenue, N.W., two blocks or so from the White House. At that age, I knew the hotel was a house of ill-repute. The opening wage rate (for the printer's devil!) was 10¢ an hour. The printer's devil cleans type, cleans the ink from presses, picks up pied (spilled) type and returns it to the proper font cases. Printers used simple machinery, changed little then from when Gutenberg invented moveable type in the mid-1400s. In the earliest days of offset printing, I hand set in that shop every street name for a map of the City of Washington, and made a proof sheet from which the customer cut and pasted the names for his map. L.B. Prince Map Company captured the field. I have a copy of the map to show for it. Mr. Prince became a millionaire.

We used benzene to remove ink. We practically bathed in it, not then knowing the chemical is cancer-causing and withdrawn from general use.

When the printshop owner went to war, he owed me $5. I now sit at the old oak desk he gave me to settle the debt. I took it apart and pulled it across town in three red wagon loads. It took a year to bleach out the ink and build a center drawer, which still doesn't fit.

Stink-bomb Trees

Major Pierre L'Enfant, the French engineer hired by President Washington to lay out the nation's capital city, must have been having a bout with the bottle when he mapped Ames Place, a one-block-long narrow street mid

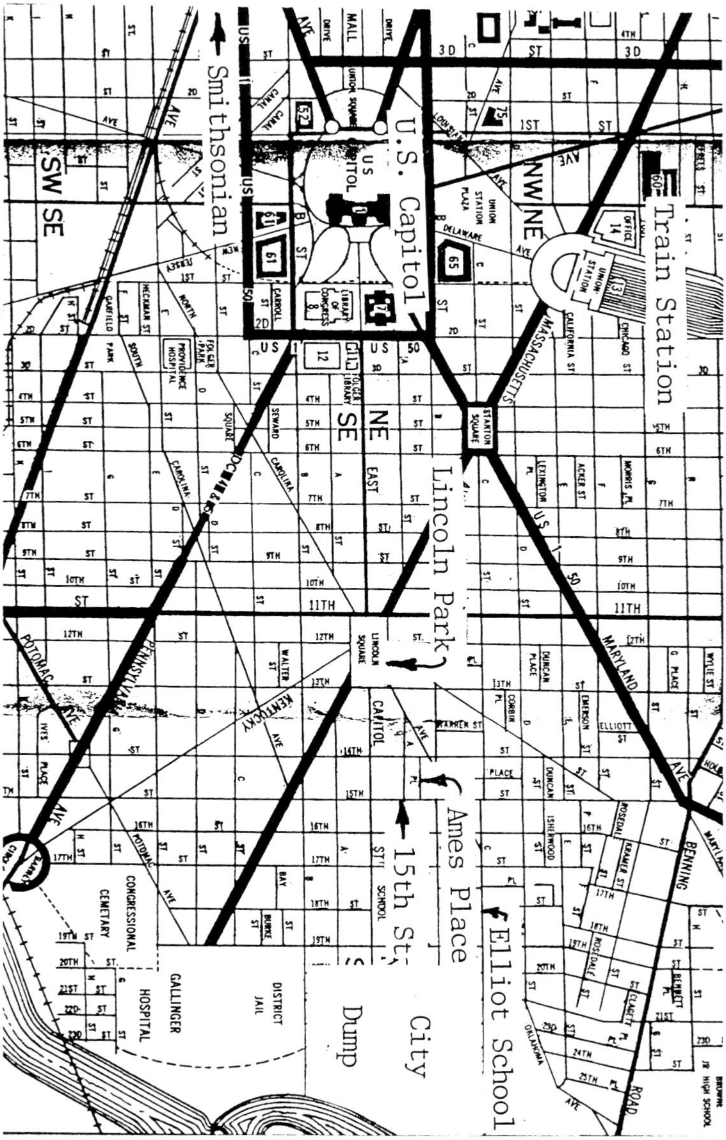

Small portion of map of Washington, for which the author set in movable type every street name—the method differing little from that employed by Guttenberg in the 1400s. North →

way between A and B streets. Urban planners of the early years of this century planted stink-bomb trees on both sides of the lane. *Ginkgo biloba* fruit squashed underfoot on the pavement gives off a sickening smell like that of rancid butter. Only diligent washing of hands and clothes eradicates the foul aroma, so tenacious is the odor of butyric acid. Probably the species' ability to quickly provide shade and its tolerance to smoke and dust encouraged city managers to plant the ginkgos, an oriental botanical oddity, in the compact soils where the air is greatly polluted. Few trees, it turns out, are as insensitive to dioxides of nitrogen and sulfur and of smoke and dust as is *G. biloba.* Drought and floods seldom affect it. Producing shade quickly, ginkgo resists insect infestation and fungal infection. And the fan-shaped leaves of the tree describe it more decorously as the maidenhair tree.

We used the plum-size fruit for initiation rites. Every new kid had to eat one, unless the newcomer was brawnier than the citizens of longer tenure. No way could a fellow retain the mushy fruit, whether green or matured to yellowish-orange with the progress of the season.

If only the city's gardeners had known *G. biloba* is dioecious (only male or female flowers are borne on a single tree), they could have planted only male trees. But forgive them, for sexual orientation of a tree of this species may take 50 years.

Few rules we learned from our parents. They instructed by example; we knew what was expected of us. I do recall three table instructions: no elbows on the table, no singing at the table, and the Bible says "not to lick your knife" (which text I haven't located).

From this environment, escape was sought. The way out was through Scouting—even in a concrete-jungle setting—and, from Scouting, to a forestry career.

3

"ON MY HONOR"

Hiking the Appalachian Trail in Virginia, 1941.

Jack and I joined Troop 20 when we turned 12. The shoulder patch designated us the Cardinals. I blew the whole $8 savings account at the Perpetual Building Association for the uniform. Charlie Callison, a young bachelor employed at the Navy Yard on Water Street, was scoutmaster. We met in the Congregational Church across the street from Lincoln Park where, as previously noted, the park police made sure we didn't play ball. Sometimes heckled, we'd carry our uniforms to meetings in bags. Charlie's mother always had the cookie jar filled, for we were frequently at his house learning knots, first aid, and compass and pacing—all subjects serving me well as a forester. Charlie, like the rest of us in the neighborhood, lived in a house 18 feet wide on a lot 18 feet wide, a dozen such row houses of identical architecture lining the street. To go from front to back door via the outside, we went around the block and up the alley. In that alley, I witnessed a shoot-out. The policeman missed the culprit: He got away by running in someone's back door, locking it behind himself, and exiting the front to the street.

Charlie owned a new 1936 Ford V-8, 4-door convertible. Its doors opened dangerously backwards. With canvas roof stored, he'd stack a dozen boys in the car. Returning from a Court of Honor one night, he took us through the winding and hilly roads of the Soldiers' Home several miles north of the Capitol. It was a roller-coaster ride. A cop stopped Charlie for speeding and reckless driving. Beside the road the officer required the scoutmaster to form us into a line, give the scout sign, and to recite the oath and law. No further conversation, we rode home quietly and slowly.

Perhaps two weeks later, with impunity, we sang to the tune of "Casey Jones was an Engineer:"

Henry Ford had a Model A;
  He tried it out on the new highway.
Got it up to 60 and he lost his nerve,
  Couldn't slow it down when he cruised around the curve.

Passed a Packard and a motorcycle, too.
  Speedometer setting on 72
When he stuck his head out the window
  And gave a loud shout,
"I've got it in second
  And I can't get it out."

So, if you're going to heaven
  I'll give you a tip,
A Model T will make the trip.
But if you're going to Hades,
  Get a Model A,
'Cause it will outrun the devil
  And you're bound to get away. (Anonymous)

Author (front, second from left) awarded Eagle badge by the Smithsonian's Dr. Bartsch, 1938. Rev. Lynn C. Drake, Council Scout Executive, is at left.

I rose to Eagle rank in Troop 20, as did many others. The Cardinals were not the kind of kids to appreciate preaching about doing a "good turn daily." Rather, we learned about nature, especially at the Smithsonian Museum. There, the curator of minerals took us behind the show cases and, labeling envelopes and bags he supplied, would drop in a small rock or mineral and tell us something about its origin or characteristics: Alum, for instance, will pucker your lips. I still have samples with the Institution's numbers inked on them. The curator's usual inebriation may have encouraged his generosity.

Dr. Paul Bartsch, president of the local council of the BSA, taught us bird study, although he was curator of mollusks at the Smithsonian. The tall, courtly gentleman met us before daybreak at the Virginia end of the Lincoln Memorial Bridge. There, at the entrance to Roaches Run Bird Sanctuary, we parked our bikes and he passed out the binoculars. Plank and earthen trails led through the swamp, atop which the Washington National Airport's runways now rest. A variety of habitats enabled us to do the field work for bird study merit badge, while back at the Smithsonian we practiced ornithological taxonomy with stuffed examples and professionally prepared mounts.

We earned swimming and life saving merit badges at the YMCA, to the side of the White House on G street. Older readers may recall one of President Johnson's cronies got "caught" there. While struggling with the weights, my appendix ruptured and, following removal, abscess set in. (Dr. Cole promised to put a zipper on my belly for convenience.) During the long recovery, Mrs. Waldron's seventh-grade class brought me Jules Verne's *Twenty Thousand Leagues Under the Sea* and *The Mysterious Island*. Those books I treasure because they got me started reading. Until then, the funnies were hard work. I still have the 37 letters Mrs. Waldron requested my Elliot Jr. High classmates to write. One especially I cherish: It is addressed to "Dear Little Fat Boy." "Crisco," for fat-in-the-can, was more descriptive!

Elliot was more disciplined than Army basic training. Signs posted in every room reminded that "Elliot is a work shop, not a playhouse" and "Silence is a mark of good breeding." To chew gum was a "federal" offense, as in

Singapore. We had good shop instruction: print, woodwork, and metal.

A Camp for Self-Sufficient Scouts

In the summers of those years we roughed it at Camp Woodrow Wilson at Burnt Mills, a few miles from Silver Springs, Maryland, and the end of the Georgia Avenue streetcar line. From there with knapsacks on our backs we could hike or catch a ride to the camp. The cost to participate amounted to 25¢ per week. The camp had some 8-man cabins at $1.25 per bunk per week, a first-aid kit, and a commissary. Spending whole summers there, we bartered with farmers for chickens and vegetables.

A Camp Wilson Boy Scout 8-bunk cabin and patrol (author, far left).

We rose to reveille, raised the flag at the 6- or 7-acre parade ground to the bugler's "To the Colors," cooked our own meals over open wood fires, stood inspection daily, and swam in the Northwest Branch of the Anacostia River just below the water filtration plant that dumped its foamy yellowish scum into our swimming hole. Some scouts luxuriated at mealtimes with Coleman stoves fueled with white gas. My older brother, the inventive kind, pitched our tent by a waist-high oak stump upon which he mounted a wood-burning stove improvised from scrap steel he found around the place. We claimed that site year after year. To eat hot food, we cut our own firewood. Destined to be a forester, I still carry on a knee the scars of the bite of a two-man crosscut saw. After we lowered the flag to the bugler's Tattoo and Taps had sounded, new kids were subjected to snipe hunts, fictitious manhunts for escaped criminals, and collapsed tents.

The several-acre campsite is yet identifiable, although now overgrown with brush and surrounded by high-rise apartment buildings.

A Baptist minister of a church in Silver Spring, the Rev. Mr. Loftus, was camp director, but the day-to-day operations were the responsibility of a building contractor named Jordan, one of Mr. Loftus' deacons. Both were expatriates from Texas in those Great Depression years. And, of course, with the Lone Star State's English, Jordan was pronounced Jerden.

Mr. Jordan led chapel, at the bugler's command, on a rock pinnacle overlooking the Northwest Branch. I think we sang, "Fairest Lord Jesus" and "Love Lifted Me" every Sunday. Without accompaniment, the Texan had all of us lustily singing *forte*.

Mr. Mac's Scouts

A girl destroyed Troop 20. Charlie married her. At about that time our family moved to Twenty-first and N streets, a mile north-northwest of the White House. I must find another troop. Troop 72 was of a different style. Meeting in the basement of the Church of the Pilgrims, two blocks from home, bowling alleys and a gym were available for our use. We could play ball on the front lawn. Scoutmaster Lacy P. McColloch ran a tight ship: he required us to wear full uniforms. He, himself, dressed like Lord Baden-Powell, the British founder of the world-wide scouting movement. Mr. Mac was a botanist with the research arm of the U.S. Department of Agriculture. While his experiments dealt with getting tomatoes to consumers free of blemish, his hobby was making us kids into botanical taxonomists. The troop's library, well-stocked with natural history volumes, was a great help. But it was our leader who, without an automobile, led us to the end of streetcar lines and, from there, on various trails (to Cabin John, to the abandoned Chesapeake and Ohio Canal tow path along the Potomac River, or to a farmer's land) to teach us plant names and characteristics as we collected specimens in makeshift plant presses for later mounting. Mice ate my stored grass collection; but two loose-leaf volumes, in which forbs are displayed on paperboard and covered with cellophane, continue to this day to show flower colors and parts.

Early saxifrage (*Saxifraga virginiensis*) from the Scout collection. The plate below the mount reads, "Collected at Cabin John, Md., April 21, 1940. A perennial growing on exposed rocks and dry hillsides. Named from saxon, a rock, and frage, to break, referring to habit of certain species of growing upon (fractured) rocks."

Each patrol had its own room for meetings and the troop had a shop for crafts. Mr. Mac built an imitation fireplace in one corner of the assembly room. In this he mounted an electric fan in front of red and yellow lights so that, during each Friday night's closing ceremony, the rotating fan made the lights appear like a *bona fide* campfire as we said the Great Scoutmaster's Prayer and the bugler sounded Taps. Prominently displayed on a wall of the Assembly Room was George Washington's orders to his troops: Profanity would not be tolerated in the Revolutionary Army nor in BSA Troop 72.

Mr. Mac had a personal collection of oaks, having found four rare species of *Quercus* as shrubs in crevices in outcrops of gneiss along flood plains.

Troop 72 library in the Church of the Pilgrims.

The availability of the troop's library encouraged us to read, especially in the natural sciences, wholesome literature like James Fenimore Cooper's *Last of the Mohicans*, and of course *Boys Life*. Perhaps there I first took a real interest in books. Perhaps that experience led me into writing as a pastime.

For the dedication of *Ecology* and *Our Forests*, I wrote, "To Lacy P. McColloch, botanist and scoutmaster, who inspired a youngster . . . to observe the things of nature and the nature of things."

The minister of the Southern Presbyterian Church of the Pilgrims, the Rev. Dr. Andrew Reid Bird, could have been the New Moses. With a full silver mane parted in the middle, the preacher was a commanding figure. Having been scoutmaster of this troop for many years, he obviously felt privileged to visit briefly during meetings.

By the time he stopped me on a stairwell during a break in one meeting, I'd come regularly to Sunday School at the Church and participated in the youth activities on Sunday afternoons before the evening worship. Unlimited ice cream, supplied by the wife of Brigadier General Frank J. McSherry for the weekly gathering, provided the attraction. (Elizabeth McSherry was the daughter of the Bostonian Shoe Company founder; the general was Paul V. McNutt's military attaché in Franklin Roosevelt's War Manpower Commission.)

'Twas on one of the clergyman's visits that he stopped to say, "Laurence, I want you to learn all the verses of a hymn." And I soon could recite

> All hail the power of Jesus' Name,
> Let angels prostrate fall,
> Bring forth the royal diadem,
> And crown Him Lord of all.

Early on Mr. Mac quickly recognized my problem: Dues were 10¢ a week. At Troop 20, the charge was a nickel. So he issued me a pass key and appointed me custodian to keep the quarters clean in exchange for the required stipend. The pass key also gave access to the tall tower of the white limestone gothic edifice nearby the Rock Creek Parkway. Crawling up behind the huge organ pipes in the balcony and up the belfrey's ladder, we blew bugles to wake the dead—out of the hearing of our beloved leader, of course.

The scout council owned Camp Theodore Roosevelt on Chesapeake Bay for, I thought, "rich" kids. Boys paid about $10 to attend for a week. Cooks provided meals in screened dining halls. Mr. Mac finagled a scholarship for me. It was no great favor: As a freeloader I felt obligated to earn

Taps was sounded and the Great Scoutmaster's Prayer said to the flicker of the artificial light in the fireplace as the meetings ended.

merit badges for the bronze, gold, and silver palms that attach to the ribbon of the Eagle medal. Forget advancement, I say: "Send a kid to camp to have fun."

Scout camps provided good training for World War II army life. Twenty-five-mile forced marches for an infantryman, carrying a full field pack, were made easier by the earlier experiences. So, too, the camaraderie and discipline enabled the natural progression from boyhood to manhood.

From US Forest Service tasks on a national forest ranger district, to an experimental forest as a scientist, and on to academia, scouting continued to be an inspiration. This was evidenced in the School of Forestry at Stephen F. Austin State University. So many students were Eagles that the School's unofficial motto was, "From scouting to forestry, a natural progression." Those students nominated me for the Distinguished Eagle Scout award, presented in 1975.

National Park Service Influence

Living nearby Rock Creek Park encouraged contact with National Park Service naturalists who regularly conducted nature walks on the trails. A kid would be drawn to the life of a park ranger-naturalist by the appearance of these botanists, ornithologists, and zoologists, always decked out in sharp green uniforms and wearing campaign hats. And we'll learn this kid was.

In the parks around the city we never found Indian arrowheads or other native-American artifacts. However, musket lead and small bore cannon balls, even before metal detectors, could be found at the sites of the War of 1812 skirmishes in nearby Maryland.

Rangers in the Shenandoah National Park also stimulated interest. We'd go by the office of the Potomac Appalachian Trail Club on H street, lay down 25¢ per person per night for the use of a shelter on the trail within the Park, and pick up a key. With a scout backpack, poncho, and one of Pop's U.S. Marine Corps blankets each, and a few light-weight foods, we'd go by Trailways or Greyhound a hundred miles or so to the Appalachian Trail and lesser foot trails crossing any one of several bus routes. We'd hike north or south and, when it was time to head back toward the city, flag down another bus on another Virginia road.

Adirondack Club cabin in the Shenandoah National Park (l. to r. - Joe White, Mark Walker, and author, 1947).

Our bulky *Guide to the Paths in the Blue Ridge* was loose-leaf, so we carried only the maps required for the anticipated journey. Locked cabins, marked on the maps, provided emergency food, dry wood, complete first-aid kits, and shelter at (for us) 2-day hiking intervals. My buddies and I slept in Adirondack lean-tos, also spaced at these intervals on the Trail, on alternate nights. Firewood and food were always replenished, even if it meant going to a mountaineer's cabin in a cove in the valley below Buzzard Rocks or Pocosin Mission to buy the food or, as a last resort and only in the summertime, to purchase it from a park concession at Big Meadows or Panorama.

My kid brother, younger by five years, made the trek with me over a Thanksgiving weekend (as I recall). We got lost in a blinding rain that obliterated the trail in the Limberlost to the east of White Oak Canyon Falls and above the cove the old maps called Big Tom. Well after dark, we located the trail and stumbled upon the Adirondack shelter that was our destination. Mark, by then, was suffering from a frightening chill, every bone of his body shivering like a quaking aspen leaf. What should I do? Go for help? Where? There'd be no traffic on Skyline Drive, several miles distant, at this hour. And if I gave instructions about our location to a passing motorist, could a park ranger ever find us before daylight?

Fortunately, the lean-to was well supplied with dry wood. Building a fire all across the 12-foot opening, I heated football size rocks in the coals, used these hot rocks to steam-dry the Marine blankets, wrapped the 11-year-old in the blankets, and used more hot rocks to dry our clothes and to sheathe him with warmth. By sunrise on a beautiful clear day he was raring to go, fully recovered. We spent that day replenishing the fuel supply, getting a few winks, and heading for U.S. 211. Mark made one more trip, a foot-crunching expedition with me (and a buddy named Leon Testa we'll mention again). The kid brother preferred the life of a musician. As a sergeant and chief flutist in the U.S. Marine Band, he accompanied Margaret Truman as the president's daughter sang in White House recitals. (The Library of Congress has provided tapes of his solos.)

First Publication

"Red Cedar: Good Whittling Wood," a story with photos of a jewelry box, book ends, letter opener, etc., carved in Troop 72's shop from a *Juniperus virginiana* log, appeared in *Popular Homecraft*. The publisher never paid, though a writers' agent tried to collect. Good experience!

The jewelry box whittled from an eastern redcedar log.

National Guard Duty

"On my honor I will do my best" and "A Scout is Trustworthy." One of the more notable lapses in the oath and law occurred when the DC National Guard recruiter found me. Three months short of 16, I signed on when learning the 260th Coast Artillery (Anti-Aircraft) Battalion would spend a chunk of the summer of 1940 camping *en route* to, and along, the scenic St. Lawrence River near the Thousand Islands and Plattsburg, New York. Europe was at war. Though still not involved with troops, the US displayed military strength. Wearing the uniform of the US Army seemed an appropriate progression from scouting. The officer swearing me in and who knew my father, a former 29th Division NG company commander, did ask if Pop knew about my enlistment. Not yet 5'4", the medics provided a waiver on the height requirement. (The 29th, known as the "Blue and the Gray" for the location of its units along the Mason-Dixon line that separated the old North from the old South, led the charge at Normandy on D-Day in WWII.)

The 260th camped each evening in small-town fairgrounds where we set up our equipment. Often the high school band would join with the Coast Artillery band for a joint concert. In a small Pennsylvania town, Fannie Landback played the horn; we corresponded until she married while I was in my freshman year in forestry school. I think her dad didn't appreciate a soldier walking his daughter home. How could he know the soldier was probably no older than his Fannie.

The outfit had seen service on the Mexican border (it would return there, as we'll see later) and in France in WWI. Hence the fleur-de-lis and a prickly pear on its shield. The battalion's motto: *Ferio, Tego*, I strike, I defend.

The 260th spent some weekends of the Fall of 1940 at town squares in the Shenandoah apple-growing country of Virginia. There we set up tents and positioned the big roving carbon-arc lights to flood the evening sky. High-school girls came by. To one, I asked, "When do you roll up the sidewalks in this hickville?" Warrant officer Joseph Berst, my CO, heard the smart-aleck crack, called me aside and offered some fatherly advice about courtesy. It's the small towns I eventually learned to treasure and to which I desired to escape. The guardsmen, almost all of whom were fathers as well as civilian soldiers, made sure we gullible kids behaved.

Before long I was a PFC-equivalent in the Ack-Ack unit, bringing my pay to a couple dollars for a weekend drill. I had turned 16 when the call came to be mustered into full-time service "for a year" in January 1941. We unpolished the trucks from olive to olive drab and loaded them on flat cars for the ride by train to Camp (now Fort) Bliss, Texas, on the Mexican border. Disappointment soon followed our excitement as names were called and honorable discharges presented to the high school boys. My document, as a reason, mentions only "to complete education."

Letters in 1941 from my CO, Warrant Officer Berst, at the El Paso camp addressed me as Dear Shorty or Dear Mascot. Mr. (as warrant officers are properly addressed) Berst said the fellows took over a nurses' home at the largest hospital in the city. They planted cactus on Bugle Street, while the CO directed the choir at the Lutheran Church.

Before the callup year ended, the Japanese struck Pearl Harbor. I never learned a single word following the war about the destiny of those fine men. A military history says the Army renamed the outfit an anti-aircraft artillery gun battalion and inactivated the unit at Camp Phillips, Kansas, in May 1944.

I trust, dear reader, that you've noted that this chapter, like others, is really about mentors. So here I introduce lines a former student clipped from an uncopyrighted *Wall Street Journal* ad. It was a thoughtful gesture.

Do You Remember
Who Gave You
Your First Break?
Someone saw
something in you once.
That's partly why you are
where you are today.
It could have been
a thoughtful parent,
a perceptive teacher,
a demanding drill sergeant,
an appreciative employer,

or just a friend.

***

Someday you might
get a nice letter.
It could be one of
the most gratifying
messages you ever read.

I hope some of my mentors read this chapter

Troop 72's Scoutmaster Lacy P. McColloch, USDA botanist, who inspired us to "do our best to do our duty to God and country."

4

MISS MAGRUDER'S DISAPPOINTMENT

Remnant eastern white pine in a New England woods, a reminder of the author's gardening job following high school. This lone pine, a famous landmark of the Androscoggin Valley, mysteriously disappeared in 1956 (US Forest Service).

I dare not show my high school report cards to my kids. I recall "A"s in academic subjects only in Miss Magruder's biology classes. And that could be attributed to the botanical collections made as a scout in Mr. McColloch's Troop 72. The mounted specimens impressed her.

Western High, erected in 1890 in a large estate called The Cedars on the edge of Georgetown, was an academic school. Army buses brought children from Ft. Myer and Ft. Belvoir in Virginia for quality education. Dr. Elmer Newton, a tall goateed medical doctor, was principal. (He began teaching years before as a substitute for an ailing chemistry instructor, staying on to retire with our 1942 graduating class. We all sang "Elmer's Tune," a popular song of the times, in his honor.) No foolishness here; boys and girls used separate entrances. (One may still see the gender words chiseled in stone above the porticoes of what now is The Duke Ellington High School for the Performing Arts.) Western in those years claimed more graduates who also succeeded at West Point's US Military Academy and the US Naval Academy at Annapolis than any other high school in the nation. The wildlife biology twins, Frank and John Craighead, famous for their photographic exploits, preceded me by a few years. Edward White, the astronaut killed in the Cape Canaveral fire, was in the class of '48.

In those depression years, PhDs happily accepted high school teaching assignments. An English, chemistry, and physics instructor at Western High were each accorded the title.

Cadet Band

Most of the boys didn't take physical education or shop. The School, however, like all District of Columbia high schools, encouraged enrollment in the high-school cadet corps. The boys carried out precise close-order drill with old Spanish-American War 30-calibre rifles. Each spring the eight high schools participated in a competitive drill for coveted medals and prizes. Lt. George Hawkins, business instructor and Army reservist, guided the cadets.

Western High School's 4th Regiment Band, 1942. The captain here walks with the judges, just behind the drum major (photo by Christina Testa).

As I liked a lot of kids in our 4th Regiment band, I joined it, though I had no musical knowledge or ability. “Captain” Donald McAdoo, the band teacher, was a great encourager. Eventually, I played the alto horn or the bass tuba, depending upon whether the oomph or the paa was most needed for particular selections. When someone asked why one would "pick-a-low" instrument, the stock answer was "To play low-and-grin, by Wagner." Once when returning to the bus following a parade down Pennsylvania Avenue, passersby all grinned. I thought the comedy was the little guy with the big horn. Rather, a buzzard had aimed its full load square in the shiny brass bell of the big bass tuba.

My musical inadequacy rewarded me with the band captaincy in our senior year. The captain only marches to the side, wears a Sam Browne belt, carries a saber (the quartermaster issued a short model), and gives commands. He directs the instrumentalists in the concert mode. Musicians in that little 28-piece collection of brass, reeds,

Captain of the 2nd place band in the city-wide competition.

and drums couldn't be spared for a useless officer and the band have any chance at winning the city-wide competition. We came in second, beating out larger and more melodious Souza types because of our spit and polish. Brass buttons and bells glistened in the afternoon sun as we marched sharply to the cadence of *The Washington Post*.

Fond memories emerge in recalling friendships made in the band. Leon Testa, one of the better musicians in the school, regardless of the instrument chosen, preceded me as captain in the class of '41. He set me up to be his successor, I think, because I liked his little sister. I loved her so much I lost my junior high school diploma, which I stuck in my rear trouser pocket so my hands could be free to carry her violin home from graduation. No one turned in the document. The crummiest trick I played on Christina was to let her try to teach me to dance for the senior prom, and then I invited Marian. Chris, whose mother taught me manners, became a US Navy nurse and married the executive officer of the White House yacht, *The Williamsburg*, moored at the Washington Navy Yard. That made Chris a companion on an occasion to Margaret Truman.

Washington Redskins Player

In 1941 I played for the Washington Redskins. That's right, with George Preston Marshall's football team and slinging Sammy Baugh, the Texas Christian University "great." What position? Oh, bass horn in the band. The yearn to travel again showed itself. When we weren't at a home game at the 7th Street Griffith Stadium, not as large as some high school stadia today, we made the trips to put on our two half-time shows. On the field we spelled USA or HELLO.

We could also form USA.

At Philadelphia, I paused to buy a candy bar after the game. Looking up, I saw the rest of the bandsmen were out of sight. Spotting a rear platform of a train marked Redskin Special, I hopped aboard. The train soon moved out to Baltimore, while I, the life of the party clad in Indian feathers and dark red blouse and trousers trimmed in leather, was well cared for en route to the wrong city. Baltimoreans kindly chauffeured me home. Honesty now compels me to report that I sold for cash the two complimentary tickets each bandsman got for each game.

Marshall (laundryman, not the military general) showed up in fur-lined boots for Thursday night band practices. His presence aggravated the director, John D'Andelet. John's patience paid off: The 100-piece band was celebrated at Washington's Willard Hotel. The seating chart for the banquet, to which all of Washington's and football's brass attended, shows me seated by some "big" guys. The extravagant Marshall was the first professional team owner to hire a marching band and to have his own team song, "Hail to the Redskins." The musical boosting was an asset that continued into the nineties

We leave the 'Skins with these historical notes. The Redskins had treed a bear, beating the Chicago team 7-3 early in the season. Now, in the East-West playoffs, the Bears were to beat Washington 73 to 0. In the 4th quarter, we bandsmen in Indian garb warmed up at the sideline: We were ready for the coach to send us in.

'Twas on December 7, 1941, at the conclusion of a home game that everyone was asked via loud speakers to remain for an important announcement. The news awaited the final tally of the game. "The day of infamy," as Franklin Roosevelt called the Japanese bombing of Pearl Harbor, heralded the beginning of war. Statesmen and military personnel fled the stadium to take their posts throughout the city's government offices.

Career Counseling

Miss Magruder, the biology teacher, doubled as career counselor. Toward the end of the senior year we visited about my future. As her seemingly diligent student sitting across the table told of his hopes to be a forester, her expression changed noticeably. Studying my grade card, she simply said, "Laurence, I'll see if I can find you a place in the CCC." And to the Civilian Conservation Corps I would have shipped, had not the war reduced enrollment applications in forestry schools so severely that The Pennsylvania State College cancelled entrance exams and, with the transfer of Roosevelt's Tree Army to the fighting army, the CCC exited the country's forests and parks.

I digress here to tell of Mrs. Dembitz, newly arrived from *der Fuehrer's* Germany. She taught us first-year

Latin. No wonder Hitler's hordes were so effective, if the elderly woman's disciplined instruction was an example. No foolishness. Words like *alba, nigra, rubrum, lutea, borealis, glabra, aquatica,* and *sempervirens* came alive later in Prof. Pfleuger's dendrology course. I continue to hope for a Latin scholar to team with a taxonomist to provide a glossary and pronunciation rules for a forester's lexicon. I learned *distichum* dis-ta-cum; some colleagues say dis-tich-um. Which is correct?

Cape Cod Capers

With Miss Magruder's discouraging appraisal allayed by an encouraging word from Mont Alto's Professor Cope, and while I was still too young for military service, Mrs. Elizabeth Jones McSherry, our Sunday School teacher and a general's wife, proposed a novel idea. The Jones family compound, on Chappaquoit Island off the coast of Massachusetts' Cape Cod, would be short of domestic help during the 1942 summer. The general's Japanese servants had been involuntarily moved by President Franklin Roosevelt's executive order to a detention center, necessitating the need for Mrs. McSherry to find a maid, a cook, a butler, and a gardener. Marion, daughter of a milkman and a student at a Boston college, took on the dusting chores; while Mim, graduate of Fanny Farmer's Cooking School, handled the kitchen tasks. My high school friend Leon, now in pre-seminary studies in preparation for ministry, signed on as butler; while I agreed to maintain the extensive lawns (with a hand-pushed, non-motorized squirrel-cage lawnmower common in those years) and retrieve the mail daily upon arrival of the train at West Falmouth on the mainland. My chores included feeding five pounds of horse meat daily to a huge black Great Dane. The matriarch named her dog Candace to commemorate the beautiful Ethiopian queen of biblical fame.

Leon, with witty charm the girls all loved, promptly pre-empted Maid Marion, a most perfectly featured lassie. Mim, ten years my senior, was like an older sister whose friendship would continue many decades after she married the manager of the Jones' estate in Boston.

It had to happen! The general and Roosevelt's war

production chief, Paul V. McNutt, arrived for a weekend when Mim was ill. The gardener was drafted to cook, and directions given that each of the gentlemen was to receive four eggs sunnyside up and unbroken. I think the tally totaled 20 hen ovaries before the butler delivered the requested number to the dining room.

To retrieve the mail I varied the means. I could ride a bike over a wondering byway and cross the wooden one-lane bridge connecting the little spit of land on Buzzards' Bay to the town of West Falmouth. (Seamen, unqualified in ornithology, misnamed the inlet: the birds were herons.) Or I could use the rowboat. Often I'd take the water-tight pouch and swim across the bay. Gasoline was rationed by the government and jealously protected.

Once Mrs. McSherry sent me with the McSherrys' son, Mike (who became a noted Vermont physician), to see a doctor in Boston. We had reservations in the Bradford Hotel (today the "Festered Fleabite") near the train station. Upon entering and noting the ornateness of the lobby, I considered it appropriate to ask the price. Told something like $10, I said, "Mike, let's go to the Y (I observed it near the station); we can save some money." We swam, played ball, and had a great time; but what a scolding I got when Mike bragged on the great time he'd had at the Y. We stayed at the hotel during future visits to physicians. (These were my first nights in a hotel. When the man in the red coat extended his hand, I shook it.)

The general's wife also taught me some table manners: In a restaurant, one doesn't stack the dishes for the waitress, even if the poor girl has to crawl over the table to retrieve them.

For the summer's work the gardener received $50.

Benny and the Bomb

Over by Georgetown University was another city dump. Benny O., the son of an Army ordnance colonel, used it for testing his basement-built rockets and explosives. When, on occasion, he got back to the school building after lunch, he'd warn us to watch and, in the distance, we'd note in the air the success of Benny's venture. It was when he unleashed a new kind of stink bomb in a hall locker that

Dr. Newton had no other choice. Benny transferred to Woodrow Wilson High. The school's alums, thanks to the persistence of Lorraine Segrist Donohue, class of '43, elected me to Western-Ellington's Honor Roll in 1994. Now, what would Miss Magruder say to that?

Escape

TV commentators frequently refer to Foggy Bottom, a few blocks from our home. Speakers and hearers assume the title relates to the muddleheaded State department folks who labor there. It may, but its origin probably dates to when Washington hired L'Enfant to design the capital city. Today's lower Constitution Avenue, nearby the mouth of a covered stream that feeds into the Potomac River near the Watergate Hotel, was once a river swale subject to settling fog. When it rains, water flows from the curb drains *up* onto Constitution Avenue, not the other way around.

Where the Watergate Hotel and Office Building of Nixonian fame now stands was the Christian Heirich Brewing Company. Amendment 20 of the U.S. Constitution closed it down. During that time, however, the Women's Christian Temperance Union, as though to mock the beer brewers, managed to erect a monument on a little plot of public land where several streets come together immediately in front of the brewery. The statue of William Jennings Bryan, the staunch prohibitionist, golden-tongued orator, and several-times candidate for the US presidency on the Democrat ticket, reminds one of more orderly times.

From this environment, as the patient reader has previously learned, I'd escape.

Don't worry if your job is small,
And your rewards are few.
Remember that the mighty oak
Was once a nut, like you.
Anon.

A stately live oak, the species used for knees and keels by colonial tall-ship builders (USFS photo).

5

THROUGH THE SCHWARTZWALD TO BERCHTESGADEN

European forest of Scots pine planted on an overgrazed moor.

As we completed spring semester, 1943, on the campus at State College, all of us awaited the call up of the enlisted reserves. First stop along the way to Germany's Black Forest and Hitler's hangout in the Austrian Alps, places playing significant roles in mapping this forester's future, was Fort Meade, Maryland. A supply sergeant tossed to me the government-issue clothes and a Gideon at the 76th Division chapel gave each of us a pocket-size, brown-covered New Testament. The train's whistle soon sounded and the wheels began to roll to a destination unknown to us lowly privates.

Camp (James Walker—no relation, honest) Fannin was a collection of run-down cotton farms in East Texas. Company A's 4th Platoon, 67th Battalion, would be my home from June to September. From our parade ground we made 25-mile forced marches through ankle-deep sand to Big Sandy, carrying full field packs (including overcoats and bedrolls) and wearing gas masks in the July and August heat. While the "branch immaterial" training center was named for a rum-running, slave-trading Confederate Civil War general on the lam from the US, sympathetic folks in the nearby town of Tyler, where nary a drop of rain fell that summer, exclaimed that the GIs were "out at camp, fannin'." All of us had prickly heat, the rash of red pimples all over the body accompanied by intense itching and tingling.

En route to Big Sandy, we hiked the main (only?) street of Winona, cheered by townsmen at their store-front, awning-covered checkerboards as we plowed deeper furrows in existing ruts. (Today, a paved road scenically winds its way between a state hospital, once the site of the camp's infirmary, and I-20 that presently cuts through the old army base. Pine and broadleaf forests seeded in upon abandonment of the infertile, coarse-textured beach-like soils.) As the littlest guy with a last name beginning near the end of the alphabet, I seemed to always end up in the 4th Platoon of a company of the 3rd Regiment. That meant I brought up the rear on these forced marches. If a soldier passed out, you could leave him by the roadside for the medics to pick up, but not his M-1. The rear-rankers dragged into camp toting three or four 12-pound rifles. And the rear rank never got a 10-minute break. It took that long for commands to reach us. "Take a 10-minute break's over" became one command.

To remind us college kids that good soldiers gripe, the commandant hung signs in the barracks that shouted, "Quitcherbeliakin', We're at War." Fannin didn't even have a stockade for malcontents.

At Tyler, where the Presbyterian Church entertained GIs and fed us Sunday dinners in the members' homes, I met the girl of my dreams, a high school senior. She laughed at my corny jokes. Her mother also cooked delicious meals, introducing me to okra, butter beans, and slabs of meat that weren't chopped up in stew. Alas, the Scottish redhead married an Air Corps bombardier.

Back to College

Prepared with infantry basic training for wherever needed, we were physically tough. The army now stashed us in various institutions of higher learning until further notice. The whole camp, apart from the cadre, consisted of youths with a year of collegiate study. I shipped to Washington and Jefferson College in western Pennsylvania. A 3-day pass allowed a brief visit home, coincidentally while my older brother, the captain, was there. We always fought; but now I, with broad shoulders and muscular arms developed on Fannin's obstacle course, was confident I could lick the skinny flyboy, though he was a half-foot taller. 'Twas not to be. Maybe a crack about my lowly rank ticked me off. As at age 4 and 14, he pinned me down spread eagle until I said, "I'm sorry." To this day, regardless of the transgression, I find it easy to say "I'm sorry," wondering to myself the sincerity of the expression.

We pranked like college kids at W&J, filling government condoms with gas from the jets at the fraternity house where we bunked. Tying the opening and attaching a string that served as a wick, we lighted a gross of them simultaneously and let them sail in the night. Washington, Pa., was aglow in the darkness of a moonless sky.

We marched six abreast, wearing overcoats, before dawn to a downtown hotel for breakfast. Along the way, in the dark, some wise guys unbuttoned the ornamental overcoat belts of the rank of soldiers immediately ahead, rebuttoning the woolen strap to the coat of the soldier to the right or left. At the command "Fall Out," buttons flew as the students pulled themselves apart. While one rank did this to the fellows to their front, the six men to their rear gave them the same treatment.

Hot-foot pranks should have earned us all a stay in the stockade. The college's library tables were so huge, a fellow could crawl under the massive oak planks from one end to the other, plant a paper match between the sole and the upper of a GI boot, ignite the plain end, and be back at his seat before the business end of the match let the wearer of the shoe realize he'd soon have a blister. That sport ended when the platoon's lieutenant, garbed in enlisted men's shoes and trousers, was erroneously identified by the joker to receive the hot-foot treatment.

Penn State, the transcript reads, awarded me 12 credits for the math, science, and other courses taught at W&J. At the conclusion of the term, I was offered the opportunity to go to a medical school in Boston: the Army was running out of battlefield medics. Not long after war's end, my buddy, Sheely, who accepted the Army's proposal, had a thriving general practice in the mountains of Kentucky, his home state.

My father wrote me a letter that reached me at this station, suggesting I should study hard. His first letter, in contrast, sent while I worked as a gardener on Cape Cod, condemned the idea of going to college as a waste of time. He was confident I could get on as a printer, his life's work. Leon, the butler, reminded me years later that I wept when I read that first note.

From W&J, I shipped to Camp Claiborne, Louisiana, in the midst of second-growth longleaf pine stands in the Kisatchie National Forest. The 84th Infantry Division, the

Illinois National Guard "Railsplitters," was packing its gear for the European Theatre. Learning a Corps of Engineers' forestry battalion was training nearby, I tried to get a transfer to serve with the unit made famous in the First World War by Colonels Henry Graves and William Greeley, both serving as chiefs of the US Forest Service. The forest engineers moved sawmills to near the front to harvest timbers for bridges and other materiel required in the conduct of war. No, the 84th couldn't spare this lowly private to the loggers; but in the middle of the night a couple days later they did spare me to the Signal Corps as a dah-de-dah Morse Code radio operator.

Having shared a bedroom with the youngest ham operator in the nation, my brother whose QSL read W3GXI, and going to sleep every night for years to the tune of dah-de-dah-dit-dah-dah-de-dah, I considered it no joy to transfer to the 65th Signal Battalion, code named Wig Wag. (How's that for originality?)

The Limey Boat

For two weeks we wandered the Atlantic Ocean in *HMT, the Esperance Bay* as part of a convoy from New York, via the African coast, to Liverpool, England. Privates were in the bottom hold (or was it hole?), called steerage for the shipment of cattle. We slept three deep: pallets under the tables, pallets on the tables, and hammocks above the tables. We were ordered not to take off our shoes, lest Hitler's subs discover our convoy and we'd need to quickly go over the side into lifeboats. Except once. Midway on the journey we took showers in salt water. Guards watched as we scrubbed with GI brushes, for salt in the sores and, especially, athletes' feet from not bathing and not removing shoes, was a torture that should have been saved for the enemy. Or were the English sailors our enemy?

Before we embarked, the Limeys had bought all but a few candy bars and bottles of soda pop available for the American soldiers. After a few days at sea, the ship's store was empty and the black marketing of nickel items for a dollar began.

Sergeants marched us single file to the deck one hour each day on a rotation basis. We hugged the old ship's rails, disgorged what food we hadn't previously relinquished, took a few breaths of salt air, and then were shoved back to the foul-smelling crowded hold of vomit and BO. And to think I volunteered!

Captain Blanding, a West Pointer and our CO, unwilling to believe the complaints dispatched to him by a non-commissioned officer, brought his bedroll to bunk with us to satisfy himself that the bellyaching was unjustified. He, apologizing, didn't stay the night.

Rumor reached us that the first-sergeant passed out on the non-commissioned officers' deck. We never saw him after that. In delirium, he repeated over-and-over, "Monty needs five men Get Monty five men." It was appropriate. Mess sergeant Montillio, in civilian life a Waldorf pastry baker, was continually demanding that the top-kick conscript men for his kitchen. Indeed, back at Camp Bowie, Texas, where we staged for a short while, we stood for hours in formation waiting for a general to present Monty a medal for having "the kitchen with the least number of flies of any in the Corps' area." Five men can swat a lot of flies.

Inspection was a daily event, requiring a fresh shave. For water one could choose the saline solution, straight from the Atlantic, in the ship's heads; the single canteen of fresh water issued daily at the mid-afternoon meal, or the "coffee" spilled into your canteen cup at the mid-morning mess (a biblical term for food). Coffee was the lathering choice for the required tonsorial treatment.

Stars and Stripes reported the dilapidated old merchant ship sunk on its next voyage. No Americans on board, we cheered.

On to the Seine

From Liverpool, on England's west coast, a train carried us to Downington, near the port of Southampton, from which we sailed in the night on a Landing Ship, Tank (LST) to the mouth of France's Seine River.

Two obstacles deterred our entering the fray with the Germans somewhere near the coast of France and along the Seine River. The first of these was at an encampment near Salisbury in southern England. We had just arrived in the night by train from Liverpool, zig-zagged all across the island country to bivouac on what we soon learned was a farm. The guard on duty called out the whole platoon to confront a company of silent intruders that cautiously advanced under the cover of the cloudy, moonless night in the fog of the coast region. We had no ammo for the M-1 rifles, so we fixed bayonets and proceeded with the attack. Had we been issued bullets for the 30-caliber weapons, a farmer would have billed the US Government for a lot of cattle. Only when the animals mooed and munched did the adrenaline in a bunch of neophyte soldiers return to its glands.

We filed on board the LST for the night trip across the English Channel from Southampton two days after Christmas. The landing ship, tank, suddenly slugged the shore, the bow quickly opening in order to allow a ramp to lower to the river's edge. Up this steel roadbed trucks and tanks were expected to file onto the ground along the Seine River's shore. The official account reads the landing near Rouen. One 2-1/2-ton truck blocked the movement. It was the one to which I was assigned. The lieutenant yelled at me to drive it. I'd never driven a deuce-and-a-half, as the 6x6s were called. There was no choice. After several stalls and knowing nothing about double-clutching, I finally got the heavy, loaded rig to solid ground. For the next 3 hours I drove, without lights, in "grandma" gear, the extra low of the transmission. Occasionally a flagman would yell at me to speed it up; but I didn't know the gear pattern, and if I stopped that truck, would it ever start again? "Feeling" the road in blackout driving, I proceeded, never seeing the truck ahead until we rendezvoused

All this time, the truck driver, a big bull of a man from the hills of Kentucky, was riding with the munitions in the bed of the truck. After this I became his assistant truck driver. Which meant I rode shotgun, my head stuck out of the roof of the truck on the passenger side with a 30-caliber machine gun, while he rode with a "grease gun," a light-weight automatic, on his lap. Once he got careless: the bullet from his gun went through the roof of the truck. We patched it with a bolt. The big fellow could change a truck tire!

Sergeant Guion was real "GI." Everything must be spit and polish in order for him to gain another "rocker," the lower stripes on a sergeant's chevron. We were to have an inspection. A general was to check us out. The snow was deep and the thermometer dipping. Frozen feet were our reward for obeying the sergeant. Ill-equipped, we hadn't been issued Army snow-paks or arctic shoes. We spent the night hunkered down beneath some trucks for protection, still waiting for the general's inspection.

Moving On

Catching up to the front late one night we were ordered to set up our radio in the upper story of a farmhouse, enabling us to extend an antenna from a window for better reception and transmission. With the arrival of first light and a look out of a window of the house, we discovered we were at a graves-registration site. The quartermaster unit used dump trucks to bring in the frozen bodies from the battle raging in the distance. Americans were treated gently, the soldiers assigned the task caressing the bodies of their countrymen and patting the heads of the fallen when unbuttoning coats in the search for dogtags. The enemy's troops, in contrast, were dumped from the trucks by hydraulic cranks, skulls cracking on the ice-covered ground. Faces were slapped and bayonets used to rip clothing in search for the metal disks that would bear the name, rank, serial number, religion, and blood type of the slain soldiers.

The Quartermaster Corps' personnel stuffed the bodies in mattress covers (we all carried our own caskets), tied one tag to a toe and the other to the cotton tick. Our radio crew gladly left that scene, the "undertakers" seemingly no longer belonging to this world, even if it meant providing communications for the infantrymen not far to the east. Why I do not know, but suddenly, upon first witnessing this massive human destruction, we laughed hysterically. TV perhaps has made us immune to such morbidity; perhaps it is the screen's distance from the scene that steels our emotions to such violence.

Fortunately, there's a comedian in every outfit. He, alone, keeps the others sane. We were moving eastward in a command car, an open-air-taxi type with 4-wheel drive, canvas roof, and no doors. Our radio gear was stashed in the rear of the 4-passenger unit. When we came upon a German soldier *sans* head lying across the narrow road, the driver stopped to allow removal of the obstacle. The fellow riding shotgun got out, shouting, "Wake up, wake up, you're on guard." Only one who has had this call every 8 hours for weeks on end can appreciate the humor in it. Then the American recognized the enemy as a Hitler Storm Trooper, wearer of the coveted SS ring. Our buddy couldn't disengage the ring from the swollen digit, so with his bayonet, he took the finger.

Our crew seemed to be sent this way and that, toward Belgium, where the Battle of the Bulge was yet to take the lives of thousands of college students. Then, circumventing Paris, the message said to go south to provide coded radio communications for a French corps.

I wrote a sweetie back home that the place in which we were holed up reminded me of "how saintly you are." The censor missed it; she responded, "You must be at St. Marie." (Aw, I knew the mail wouldn't go out until this battle was history.) All along the way were destroyed forests: tanks, fire, and, later, harvests by the forest engineers saw to that.

Five of us served together in Europe, always on detached service with this division, that corps, or the French army. Company headquarters was foreign to us, and we to them. That was apparent when promotions were dispensed. (I have Army papers certifying I was a T-4 high-speed radio operator, but never received the sergeant's pay nor the stripe beyond that of a T-5, an "honorary PFC," with corporal's pay.)

While with the French, our SEGABA machine was stolen. We couldn't code and decode Morse messages in five-letter scrambled word groups. Captain Blanding received a battlefield demotion, destroying his career, for the infraction. French Moroccans, the enlisted men of that nation's army, dumped the truck, with the machine in it, into a ditch.

Crossing the Rhine

The Rhineland battle began as the Battle of the Bulge ended. Lt. Gen. Jacob Devers intended to eliminate the Germans in the Colmar Pocket west of the Rhine between the Swiss border and the Saar River. Colmar was our biggest battle. During the shelling we found temporary

The brass and bronze artifacts from the Colmar museum.

safety bivouacking one night in the basement of a huge bombed-out museum. Coming off guard duty for which the five of us rotated at 4-hour intervals (guard, sleep, radio) when able, I discovered in the wee-hour darkness that one of my cohorts had stuffed a cadaver skeleton in my bedroll and, on its head, placed a German helmet. I didn't get a chance to get even, the museum was shelled from across the Rhine within minutes. In my escape, I grabbed a pocket full of bronze "trophies" from a shattered showcase. (In 1989 I tried without success to return these "souvenirs" to their proper exhibits.) Later we learned that, just the night before, the Germans used this building for a command post, their radar readily picked us up, cannons zeroing in to scatter the Americans and Frenchmen.

Colmar surrendered February 3, and a week later, safe from German shelling, we crossed the Rhine with the US Seventh and the French First armies near Mannheim, not far north of the Swiss boundary. The engineers had laid a pontoon bridge, named for 7th Army commander Alexander Patch. The original span, several hundred feet to the south, was destroyed by the Germans in a futile attempt to keep the Allies to the west of the Rhine. Soon we camped in the *Schwartzwald*, Germany's famous Black Forest.

"Hans" and "Fritz" greeted us as we set up camp under the resinous conifers of the *Schwartzwald* and awaited orders. The German lads took us to their frightened father, the *forstmeister*. It's a crazy world: soon we were drinking tea and eating *der mutter's* cookies in spite of directives forbidding fraternizing. We left behind some C rations. Even the European beech trees, resembling and related to American beech of the genus *Fagus*, were not cut in this forest, near-sacred to Germanic peoples. (When walnut was no longer available, Hitler's troops carried rifles the stocks of which were carved from *F. sylvatica*. Some lexicons trace "beech" and "book" to the same Anglo-Saxon root, the early writings of Germanic peoples inscribed upon tablets of this wood.) On those smooth-barked stems, although tempted, we carved no notes to posterity.

Just as we were breaking through the "impregnable" Siegfried battlements, leaflets dropped from the sky. Designed to break morale, they read,

> Patton, your general said three days ago:
> "It will cost tons of dog-tags to break through
> the Siegfried Line and to cross the river Rhine."
> Now you know!
> . . .
> two pieces of tin-plate with a number on them—
> to be brushed aside when they have served
> their purpose . . .
> That's all you are worth to him!

We traveled swiftly through Bavarian towns and cities. Regrouping beside a railroad, Al, I think it was, spotted a hand-pumped track-inspector's railcar and suggested a ride. We went eastward, perhaps some 10 miles, to the outskirts of a town when wisdom finally prevailed and we returned to the assembly point. The next day we "took" the town we had visited the evening before.

We raced from town to town, seldom stopping for long. The map issued us at war's end shows the 65th Signal Battalion a part of the XXI Corps (its emblem an acorn in a 4-leaf clover), to which we were somehow attached, cutting a wide sweep from the Seine River to the Austrian border.

Sad News

About this time, we went to the rear for showers, portable units erected by the black Quartermaster Corpsman, and for clean fatigues and underwear. While waiting in a chow line, I opened the abundance of mail that caught up with me. The first was from the captain brother, date-lined Cunming, China, telling of his thrills flying the "Hump." So unusual the missive's conclusion, I folded and pocketed the letter. The last lines read, "Remember the things Mom taught us, and 'if we live in the spirit, let us also walk in the spirit.' Gal. 5:16."

Later, still in the line with mess kit and canteen cup, I opened a letter from my father. Out fell a clipping from the Washington *Star*: Hobart's plane crashed while flying the China-Burma route. Soon the platoon leader, not knowing of the mail already received, sought me out with an American Red Cross cable that related the same message. My initial reaction: disbelief. I thought "Hobo" was playing a prank. I suffered for Momma, for the hard-working brother was properly and for good reason the "apple of her eye." My sergeant neatly lettered the Biblical admonition inside the front cover of the Scofield Bible carried in my duffel bag, the canvas satchel awaiting us at the showers. A stain-glass window in Washington's Church of the Pilgrims memorializes the airman. President Truman attended his reinterment, from Cunming's above-ground tombs, in Arlington National Cemetery.

The author's brother, an Air Corps captain, in front of a hut in Cunming, China.

En route to *Munchen* (Munich), we detoured to open the gates of Dachau, 20 miles north of the city. Handed a Flit gun, an infantry sergeant said, "No one leaves without opening shirt and trousers and receiving a generous dose of chemical." I did as ordered, realizing years later that the spray was DDT, the top-secret discovery of a Swiss chemist. Dichloro-diphenyl-trichloroethane, actually invented by a German scientist in the 19th century but then of no known value, was needed to replace the Chinese-produced insecticide distilled from the Oriental chrysanthemum flower. The Japanese blockaded its export after Pearl Harbor. The Swiss researcher, we're told, related his discovery to the American Embassy in Bern, after which President Franklin Roosevelt assigned existing laboratories the task of DDT's manufacture. Until then the Rickettsia carried by the rat-infested flea that migrates to humans living in trenches and foxholes was responsible for the winning and losing of Europe's wars. History credits with too much military brilliance the "Napoleons" of various periods. The microscopic organism, the source of typhus, warrants acknowledgment.

At a concentration camp near Muhlhausen in early 1945.

Only recently did I learn that 2448 Protestant and Catholic clergy who had opposed Hitler were among those imprisoned at Dachau. One thousand and eight were yet alive among the thousands of other political prisoners when we liberated the death camp. Chaplain Edward L.R. Elson, a Presbyterian who led worship for our unit and who later was Chaplain of the U.S. Senate, joined us at the wretched site of the heinous crimes.

Wig Wag Under Cotterkey

I now go into detail about Wig Wag because the U.S. Army Military History Institute at Carlisle Barracks has no history of the 65th Signals. The Institute has only a pamphlet that lists, with all other units, its three campaigns (Ardennes-Alsace, Central Europe, and Rhineland) and a notation that the unit received the Meritorious Unit Citation.

A 1966 historian's account explains an argument between The War Department ("a soldier first, a technician second") and the Signal Corps' insistence that its "units need not be highly proficient in front-line combat tactics." The article also explained why we were shipped out four months prior to completion of scheduled training and were quite unprepared for battle: "...requests came from the ETO asking that the unit (the 65th Signal Battalion) be shipped immediately for Ninth Army Operations."

My notes, a unit chronology provided by the History Institute, and a map issued by the 21st Corps of its route across Europe, enabled the writing of the following account.

The XXI Corps, coded Cotterkey, literally ran in circles from Phelsbourg to Baccarat to Morhange, then back to the east of Phelsbourg *en route* to St. Marie. Skidding on icy roads, we joined the fray and, as directed, set up communications on top of a hill. The enemy figured we would; only the zero temperature and the three feet of snow kept us from triggering the land mines that, we soon discovered, had been planted beneath where we parked the command car. The Corps' map shows we again traveled in circles, the headquarters taking a direct hit at Colmar, and the rig peppered with shrapnel as we made a "strategic" withdrawal to Logelbach and returned to Ribeauville, from which we left eight days earlier, to reorganize. I think Major General Frank Milburn had a liking for Morhange, for again we circled back to the French town, this time for showers and dry underwear. On to St. Avold to bivouac by a lake before the drive to Bitche, for which Cotterkey got a new name, now the Sons of Bitche.

Next day, rather night, it was at Pirmasens, miles from nowhere, that we spent stuck in a ditch just to the east of the Siegefried Line, the concrete wall supposedly impenetrable. Some guy's gun went off accidentally, for a moment scaring the heck out of us: what Kraut would have found us in this ditch?

We could scarcely keep up with the infantry to provide communications until approaching Neustat where a 5-day battle cleaned out the enemy. Soon we discovered the Gestapo headquarters masquerading as a school supply warehouse. I have a slide rule to prove it. Crossing the Rhine River at Worms on the German side, the Corps raced on past Lambertheim, where we rode the RR handcar, and to Erbach where our crew "bivouacked" in style in a still-standing fine German home. By the second of April the Americans had taken Hardheim and Wig Wag's 5-man radio team had taken over a farmer's barn, now enjoying fresh milk straight from the udder.

But the war wasn't over. Stymied for nine days at Tauberbischofsheim while continuously strafed, "the five" were assigned another home where the proper occupants had left behind real eggs and jelly, and where we holed up and enjoyed a bed and a bath until the enemy ran out of ammunition or patience. At Ochsenfurt we again were delayed by resistance fighters and "bed-check Charlie" who found our camouflaged radio rig in a stone quarry. A quick swim in a crystal-clear mountain stream near Rothenburg revived us for the chase to Aalen, on the way seeing thousands of horses left dead in the Nazi retreat. Two days later, at Dillingen, we released Indian POWs from the African campaign. What a joyous time for them!

Yet "Jerry's" planes kept busy, and kept us busy, trying to destroy the only bridge across the Danube River. Pausing at Augsburg, we helped to set up 7th Army Headquarters, camping on the zoo grounds. Yep, you guessed it: some prankster opened the fox pens. Here, too, we captured two "admirals," complete with medals on their tunics; they were street-car conductors. Skirting the lake at Schondorf, the five radiomen of the 21st Corps took over a rabbit farm at Weilheim and adopted a billy goat for a mascot. While the main thrust of the Corps opened the concentration camps at Bad Tolz, we five, as previously noted, were north of Munich opening the gates of Dachau, releasing the thousands of Jews and "disloyal" Germans in striped garb. By then we quit taking prisoners; there were too many. On the 6th of May, Hitler's Army Group G surrendered at Degerndorf. Shortly thereafter the famous 101st Airborn Division dropped into Berchtesgaden to find Adolph Hitler supposedly a suicide.

Another detailed itinerary, supplied by the Army's history institute, traces the 65th Signal Battalion from Rouen and on to Nancy, Gerbeviller, Morhange, St. Marie, Colmar, Ribeauville, Morhange, St. Avold, and Bitche in France, and to the German towns of Pirmasens, Nuestadt, Lampertheim, Erbach, Hardheim, Tauberbischofsheim, Ochsenfurt, Rothenburg, Crailsheim, Alen, Dillengen, Augsburg, Oberschondort, Weilheim, Bad Tolz, Degendorf, Schwabisch Gmund, Leipzig, Mulhaussen, Schwabish Gmund, Hohenheim, Mohringen, Bad Orb, and Bad Mergentheim.

General Order 243 awarded the 65th the Meritorious Unit Plaque "for superior performance...in...exceptionally difficult tasks...." We called the gold open-wreath insignia, worn on the sleeve, the "toilet seat."

War's End

As we approached the Austrian border, our destination was Hitler's hangout near Berchtesgaden in the mountains along the Austrian border. As the war wound down, we took prisoners, stragglers in the Alpine Mountains, tossing their weapons into the nearest creek or lake. They were too heavy and too many to tote. We each retained one sidearm as a souvenir. Lugers were preferred; I liked the Mauser, still a convenient pistol.

Following a day in the forest of Bavaria, and assuming the war was, for all intents and purposes, over, we sat around a campfire. From somewhere a shot rang out; we hit the ground assuming a sniper on the loose. Battlefield sawbones at Battalion Aid sewed me up and sent me with other wounded via ambulance on an all-night ride to a tent hospital (M*A*S*H.-style), the 27th Evac. Hosp., near Munich. A week there, receiving 48 shots of penicillin, one every three hours, fit me for a C-47 Hosp. Evac. plane ride to an evacuation hospital at Reims, France. No room available in the one to which I was flown, I was soon on a hospital train, sleeping between white sheets, until arriving at Lyons' 36th Gen. Hosp. Patched up and sent by train to the 2nd replacement depot (reppel deppel) at Worms, this time by 3rd class (bullet holes made a leaky roof), I learned through the grapevine that our gang had coated its rifles in cosmoline for shipment to the Pacific Theatre. I wasn't about to be left behind.

Word spread that if you could get from the reppel depple to your unit on your own and had your CO notify the reppel depple of your arrival, you would not be AWOL. So after two days I left the camp with the mail pouch, hitching rides with military trucks to Stuttgart and Augsburg, finally finding my buddies at Gmund. A great reunion took place. A review board determined that friendly fire got me; a soldier, also sitting around a campfire nearby, was showing off his German revolver.

That wound, dear reader, made me a disabled vet and, thus, eligible for Public Law 16, somewhat superior to the GI Bill for college studies. PL-16 and the GI Bill together got me through all but one semester of my undergraduate and graduate education.

Shipping to the Pacific was an unfounded rumor, at least for now. We moved quickly to Leipzig, not knowing that President Roosevelt (deceased less than a month when the war ended in May 1985) had given this strategic university city to Stalin's Russia at the infamous Potsdam session. Five-letter code groups filled the airways. Not until we sailed from Le Havre for home a year later did we know our business in the old city made famous by Bach and his organ recitals at the Lutheran church and where he served as cantor at the nearby *Thomasschule*. We were evacuating scientists before the "Rooskies" arrived. Later, the press reported, Werner VonBraun, the rocket wizard, was among those evacuees. Stalin's soldiers wanted our watches (they were Bulova GI), our food, and anything else we'd swap them. With strong-arm tactics they muscled us, barged into a *foto* shop, and insisted on a collective photograph, the prints made pronto by the scared photographer. Within two hours of the arrival of Russian troops, we packed our duffel and made fast tracks for Karlsruhe.

The Russian "comrades" insisted on this photo, promptly developed and printed. The American in the front row, left, along with the author in a tie, soon left Leipzig as more "Rooskies" rolled in.

Occupation Troops

We had opportunities to travel during the months we must remain in the Army of Occupation. Several weeks in the woods of England, Scotland, and Wales introduced me to stands of Scot's pine on their own turf, as well as Douglas-fir and lodgepole pine introduced from North America. But the highlight was to be the three months at Nancy, the French forestry academy attended by Gifford Pinchot. Alas, upon arrival by train, we learned the dorm was full and that we would entrain for Paris, there to study French language and civilization at the Sorbonne, part of the *Université de Paris*. In the French capital, we lived in a confiscated Japanese dormitory, learned the French language, and traveled the country to visit historical sites and, under a bit of pressure, to put GI gasoline in the GI buses to see the woodlands of the lowlands, the *Foret d'Othe*, the Palace of Fountainbleau built by Louis IX in the forests of Bière, and the broadleaf forests of Lorraine, near Nancy, where I suppose Pinchot learned mensuration.

Dick Kerr and author in the Occupation Army in Germany's Bavaria.

The instructor struggled with us restless soldiers. On one field trip, he kept tugging at a turtle-neck sweater that covered his tie. Finally, in exasperation, he removed the tie, exclaiming as it came apart in three pieces, "It's ze only one I've got, but it's real silk." While we couldn't replace his tie, for the Christmas of 1945 he received two GI duffel bags of GI goodies that included blankets, overcoats, and GI C and D rations. The captain, commandant of the class, assigned another low-ranker and me to deliver the goods, *via* Paris' metro, to the professor's flat, a two-room apartment. We assumed he would barter the gifts for needed family food.

Shortly thereafter we sat for final exams. The professor

announced that "20 would be ze perfect grade, 16 ze passing grade." As I noted the 14 on my returned paper, the professor declared, "Due to ze circumstances, 12 will be ze passing grade." Penn State gave me 3 credits for the course; but more importantly I was now ready to take any university's doctoral French qualifying test. However, I had no dream at that time of being anything other than a "dirt forester."

One doesn't forget the ravages of war: a lone crucifix with a Christ 20 feet above the rubble in the center of a village is a reminder that it was Hitler who said, "We shall erase their cities." Now Chaplain Elson stood behind remnant pulpits, great and ornate, in gothic churches to tell "the old, old story." Sometimes glockenspiel chimes still sounded from town halls as in Esslinger on Neckar. And we Allies had spared much of *Schwaebisch Gmund* in the romantic valley of Rems.

40 & 8

Almost a year after the armistice, we piled like logs into a wooden boxcar left over from the First World War. On the sides were the words *40 hommes et 8 chevaux*. Eight horses didn't share our hay-carpeted floor; but 40 men, eager to get to the dock at Le Havre, France, willingly endured the cold and the meager rations for the five-day rail trip from Stuttgart, Germany. A pot-bellied stove, its vent protruding from the sliding door, provided sufficient heat, added to that of our bodies, to survive. We wondered aloud, "Who won the war?" The conquered often fared better than we did.

A few days at the wharf and we were ready to walk the gang plank to the deck of one of Henry Kaiser's Liberty ships. At sea, I stood guard on the midnight to 4 a.m. watch on the WACS and nurses quarter-deck. One night as the sergeant of the guard posted me, Military Police were completing an investigation; some gal was selling herself for $5.00 a peek. The money was confiscated for the ship's charity fund. As an irony, a chaplain was berthed in a cabin on that deck.

We cheered as we passed the Grand Lady in New York's harbor. The Army then quarantined us at Fort Dix before dispatching us to discharge centers. I went by train to Fort

Meade, there received a new uniform, and headed by bus to Washington. Leaving the Greyhound at the city's shoddy bus station on K street, I was rudely halted by a burly homosexual. Carrying the heavy duffel bag filled with souvenirs I scurried the dozen blocks to home. I then renewed my pledge to flee this town as I fled the deviant.

Climbing the 16 steps to the front door, the reflected light of the street lamp on the gold star in the window told me this would not be a happy homecoming.

On furlough during the Occupation a soldier got a chance to take in the sights. Acid rain was allowed to consume monuments, in contrast to the careful management of the continent's forest.

Tanks and troops avoided wetlands like these in the low countries. Nor, during the European war, did they dry out to become subject to pyrrhic conflagrations. After the war, the author found such conifer swamps in northern Europe drained and converted to high-yield forests.

6

A NITTANY LION

Xi Sigma Pi's red pine plantation at Stone Valley twenty years after pledges to the honor society, including the author, planted the stand. Dr. Robert Shipman, with whom I worked while in the Forest Service and for whom Penn State's Shipman Forest is named, provided the photo.

To utilize Public Law 16, better than the GI Bill for disabled veterans, the government required applicants to take career aptitude tests. The two-day battery of quizzes at American University placed me in a category for social work. When I returned to meet with the career counselor and to learn of this decision, my prompt response was "No thanks, I'm going to be a forester." More discussion. PL-16, if one satisfied the career qualifier, could pay for an education through the terminal degree. Never mind, the government could keep its money. Left alone while the counselors huddled to discuss this stubborn veteran, they decided that, somehow, forestry deals enough with people to be considered social work. Indeed it does, but of that I was then unaware. The bureaucrats certified my training through the Master's degree, at that time the terminal degree in forestry.

I pondered why the tests set me up for social work. I think it was the loaded questions, to which I'd answer what I thought was the right thing to do, not what I'd really do. Example: If you chanced upon an old woman lying in the gutter, you'd (a) pass her by, (b) help her up, or (c) stomp on her. Yes, I passed by many an old man or woman sleeping on the vents that release warm air to the streets as heated air moves from a central plant to the government's buildings in the nation's capital. But I would have checked (b) on the quiz.

Western Wandering

Now, I must wait out the balance of the summer until Penn State's fall term begins. A government veterans welfare program provided an incentive not to work. I "joined" the "52-20 Club." Title V of the Serviceman's Readjustment Act of 1944, signed by President Roosevelt shortly before his death and more popularly known as the

GI Bill of Rights, authorized the charity. If on one day each week the former serviceman went by a Veteran's Administration office and signed a form that he was willing to work, and found no task to do, he'd receive $20 for that week. This could go on for 52 weeks. So Dick Keer, an army radio operator buddy from Philadelphia, and I decided to see the USA. We'd covered much of western Europe while in the Army of Occupation. Dick and I began the tour with a swim in the Atlantic Ocean off New Jersey's coast, hitchhiked to the Pacific for a dip in its surf, and visited a little of northern Mexico. Dressed in GI khaki, *sans* hats (so we were not masquerading as soldiers), the trip took two months. We could depend on making 25 miles an hour, day or night, and it was safe. We rode with blacks who needed white people to buy their restaurant takeout food in southern Illinois and with long-distance truck drivers who needed someone with whom to talk to keep them awake. Along the way we climbed Pikes Peak, which Lieutenant Zebulon Pike noted, as he surveyed its base, never would be climbed by man. (Mapmakers are not sure that what we now call the mountain given Pike's name is the one he described for the War Department's files in Washington.)

Once we got a ride in a moving van in which we slept on sofas in exchange for helping the driver unload the cargo. Dick knew the driver had money with which to hire locals for the heavy work. We settled for a steak dinner in Jackson Hole, Wyoming, just south of the Tetons.

At Yellowstone National Park, we opened cans of food and set them in a hot spring to heat. The first night we slept on couches in the lobby of the big lodge until the night watchman tossed us out. Kindly, he showed us the furnace room where the concrete floor served as a bed until light. Hearing about our situation, a park ranger stashed us with mattresses in a storage room for several subsequent nights. Mattresses make good blankets. As no one would give us a ride inside the Park, we were about to give up and head west when a couple stopped who had "traveled with us" along the road, noting our progress, all the way from Ohio. 'Til now their car was loaded. Relieved of its luggage at the lodge, they now had room for us, and showed us the beautiful falls, geysers, and hot springs. The couple was German, we knew their homeland, and Dick, fascinated by the Teutonic language, became a professional linguist.

Dick and I heard the noontime recital by the Mormon choir in Salt Lake City, then swam in the saline sea. Ouch! Salt in athletes feet abbreviated the dip in the Great Salt Lake. From there a rancher responded to our plea, dropping us off somewhere near the Bonneville Flats, on US 40. Lips swollen by the desert sun and hours later severely chilled by the high-country late night air, we flagged a Greyhound. The bus must have been exceeding the century mark, for it required at least a quarter mile to stop. The driver patiently waited. Winded as we climbed aboard, we paid our way to Reno, the sole occasion during the whole trip to pay for a ride. Arriving at the gambling town early in the morning, we found little evidence anyone had gone to bed the night before. Even kids were tossing dice and pulling the levers of one-armed bandits.

We sold blood at a San Francisco YMCA for $25; it seemed like easy cash. We backtracked to the Sierra-Nevada to roam the Sequoia country and camped in Yosemite National Park. An army friend of some wealth showed us Los Angeles—when it was still orchards, wide-open spaces, and pure, hazeless air and sky. I'm glad I got to see it then for, on my next visit to the City of Angels, two decades later, solid and liquid particulates in the atmosphere resulted in temperature inversions—the upside down climate associated with smog. Gases from vehicular exhausts in the 1960s so stung my eyes that I thought I needed to visit an ophthalmologist until a newspaper headline warned that, while the pollution was abnormally high, there was no need to panic.

Crossing the Hoover Dam that brakes the Colorado River, we were so awed by its engineering grandeur that we thanked our movie-star chauffeur for the ride from Glamor City and chose benches on which to snooze to wait out the dawn. A thoughtful guard threw a GI overcoat over each of us sometime in the frigid night. We arrived, too, at Carlsbad Caverns after the cave gates had closed. Benches there also served as bunks until a prowling skunk disturbed our dreams under the starlit sky.

For the return east Dick, not well, went by bus, leaving

me to wander a bit. My route took me through Tyler, nearby the now-abandoned Camp Fannin. I called on the redheaded lass who married the bombardier. She was now a young war widow.

Having dutifully made ourselves available for work each week at a VA "52-20 Club" office, Dick and I had a pack of government checks awaiting our return home. They'd come in handy. Some say I should be ashamed to tell this story, but Congress passed Title V specifically to keep the millions of returning vets from cramming the jobless rolls.

The Campus

With credits earned in the Army Specialized Training Corps (ASTP) at Washington and Jefferson College, 3 credits from Paris's Sorbonne, 6 as ROTC equivalents, and 6 for correspondence courses taken during the occupation, I could graduate in two more years. To do this required some heavy loads in the college lying in the Nittany Valley of central Pennsylvania. I was game to try.

One of the correspondence courses taken overseas was economics, the other business law. I blew the final examination for the law course. The generous instructor, somewhere in the States, gave me another chance, first explaining that I should not, as a judge, decide cases on what I think is right and fair, but rather upon the law. Ever since, I've held that business law is simply "a legal way to skank your buddy."

At Penn State, we inked speech outlines: to inform, to convince, etc. The first of these I called "The Death of a Giant." It was about the bigtree (*Sequoia gigantea*); another, I wince, was to convince the audience of the necessity for a government program to regulate timber cutting. It opens with the account of an entire Texas town of 2000 people sold to a wrecking company. The outline says no more of the facts: who, what, where, when, how, and why.

The speech professor, a slight fellow, would stand at the rear of the lecture room, frequently interrupting a student's oration to shout, "Turn the rascals out." Older readers remember the shouted phrase with reference to politicians. I suppose the instructor thought us too timid in getting our ideas across to fellow collegians.

Just as we had worked in the Mont Alto State Forest's 23,000 acres of northern hardwoods, which had been under management for 25 years, now we'd discover the low-grade broadleaf trees of the Barrens. Many of our fires were set in this 12,000 acres of scrub. We loaded the trucks for Stone Valley, a 3300-acre tract, acquired, I believe, from the Soil Conservation Service as a Land Utilization Project. The LU lands, rundown farms with worn out soils, were purchased a dozen years earlier during the Great Depression. The feds moved the farmers to more fertile sites.

H. Artur Meyer, a Swiss, taught mensuration, utilizing his personal woodlot for much of our instruction. In class one day he asked for a slide rule to speed up a calculation. One was handed him with multiple scales on both sides, eliciting the comment, "Ach, ist too confusing. Who hass a slide rule?" Now, a 6-inch stick was passed to the professor. Again, "Iss not accurate. Now who hass a slide rule?" From the rear of the class I passed along a "souvenir" borrowed from an SS (storm trooper) headquarters masquerading as a school supply depot at Schwabish-Gmünd. "Ach," Dr. Meyer said, "iss like de one I used as a boy."

Dr. Meyer's "Management Without Rotation" appeared in a 1943 issue of the *Journal of Forestry*. In the paper he compares the volume distribution of unevenaged forests of his native Switzerland and that of a virgin beech-birch-maple-hemlock forest of northern Pennsylvania. We could recite the paper's details. Five of us seniors (Walter Anderson, T. A. Friend, Jr., L. N. Gauthey, W. C. Keller) made a farm woodlot plan for the professor's Tussey property: The chestnut oak (*Quercus montana*) and red oak (*Q. borealis*) stand needed a sanitation treatment, some Christmas trees were ripe for harvest, and a field should be planted to white pine (*P. strobus*).

Professor Henry Chisman moved with us, as did Professor Cope, from Mont Alto to State College. Happy Hank earned considerable officer rank in the Navy, taking off only the gold braid in converting his dress blues to civilian attire. He also changed his ways of dealing with students, now handling silvics like a military class. That is, until in an officious tone, he called on the most decorated soldier in the class, one who made Lt. Colonel in the Pacific. "George" (his name escapes me), stood, saluted, gave the correct answer, said "sir," and sat down. The professor, embarrassed, promptly stomped out of the classroom. Never again did he wear the officer uniform; always from thence he was his old self, admired by all—especially the secretaries and librarians.

Peers talked me into testing Dr. J. P. Kelly, the plant taxonomy professor. From my war-time collection in Germany, I pulled a European forget-me-not and included it with our Centre County mounts. Dr. Kelly promptly knew we'd tried to pull his leg. "This *Myosotis scorpioides,*" he said, "grows in the Rhine River bottomlands of Germany."

Some of us not raised on a farm took a physical education course in equusations. We figured the horseback riding experience would be useful on the job. A White Russian, dressed in the nattiest riding habit you can imagine, a swagger stick in one hand and a bullhorn in the other, led us in the exercises. When one fellow found himself under the belly of the horse, holding on to the trotting steed for dear life, the Colonel's bullhorn barks, "You on Brownie! Who gave you permission to dismount?" Later, my horse passed the final exam and I flunked. No, it was the other way around. I went over the hurdle; the horse did not.

The soils professor, Dr. Anthony Richer, taught me as much about college lecturing as he did about the "medium in which plants grow." Each of his soils lectures had a well-worded philosophy story. Like this one: A sidewalk superintendent comes upon a construction job. He asks an artisan working with trowel, mortar, and stone, "What are you doing, Mister?" The workman answered, "I'm laying stone." Nearby was another man doing the same task. He, too, was asked, "What are you building?" With a little more enthusiasm than the first mason, he answered, "I'm building a wall." Further along was a third laborer handling the same job. To the same question, he answered, "I'm building a cathedral." The soils course inspired me to do graduate work in forest soils.

Now, Professor Edwards appears. He's the oldest member of the faculty, teaching logging and lumbering as he had learned it at century's turn. Oh, how I wonder about my own instruction, and think of Bobby Burns' Scottish rhyme,

"O wad some Pow'r the giftie gie us
To see oursels as others see us!"

Extra-curricular

Fraternity life wasn't for me, but no bunk could be found at a cheaper rate. Motive aside, I pledged Tau Phi Delta, and still wear proudly the small gold tree pin in my lapel. TPD, founded in 1924 as a club for non-fraternity foresters, was solely for forestry students. Prof. J. A. Ferguson, forestry department head and advisor, had written in 1925: "If a fraternity instills in its members a spirit of loyalty to God, to country, and to fellow students and to self, it is worthwhile."

Tau Phi Delta foresters in 1943. Jean Chovet, in the second row with pocket handkerchief, gave the supreme sacrifice in WWII. The author: top, right end.

Dr. Wallace White, wood technology professor, served as advisor to the chapter, one of three in the United States. (Other chapters were established at the universities of

Minnesota and Washington.) Once initiated, I settled in at the old 4-story brick and stucco house with the wrap-around porch on Fairmont street, a mile from the campus. My pre-war friend Dick Clark joined me, and we remained roommates until graduation. A cook named Brownie continually angered the brothers by diluting the catsup so as not to overextend the budget.

A deal enabled me to fire the TPD furnace in winter for my room while I washed dishes at a wealthy Jewish fraternity around the corner in exchange for the best meals anyone—student or connoisseur—could desire. They called their large house the Pink Elephant.

Soon Dr. William Bramble put me to work counting seedlings and other chores. A crew I recruited spent a Saturday counting newly germinated maple seedlings in a northern hardwood stand, plots having previously been established and treated in various ways. When Dr. Bramble saw the data, he, in his quiet manner, suggested that he and I go back to look at the plots. My crew had spent the day tallying nannyberry shrub seedlings. In defense of our error, both *Viburnum lentago* and maples of the genus *Acer* exhibit opposite foliage.

The State's Department of Forests and Waters first called on forestry students at State College, as at Mont Alto, to fight wildfires. The tanker truck, stored at a private garage a couple of blocks from campus, became my responsibility. After a forest fire was out, we'd refill the tank, cruise down sorority row, release the water, refill the tank, and store the truck. Dr. White, a Phi Beta Kappa from his undergraduate years, always dangled the coveted key from a chain across the chest of his vest. On fire duty, he was the time-keeper, recording each student's coming and going. Always the vest and the PBK ornament, even on the fireline, accompanied his boots and riding breeches.

Summer Camp Calamity

Professor H (we'd best leave it there) was so humble he was proud of his humility. He held the doors for the students. Unfortunately, he was summer camp director for several score students, all veterans, at Blue Jay, near Marienville, on the Allegheny National Forest. We lived in a CCC-type facility built in the 1930s by the federal government for rehabilitating older men, those not qualified by age or behavior for the exemplary Civilian Conservation Corps. A physical education instructor was dispatched from the college to stand us in line for roll call and mail call, not a life-style relished by a bunch of ex-GIs who had tramped the jungles of the South Pacific and endured the skirmishes of North Africa and western Europe.

We'd had enough when the director announced that he had invited every forester in the state to be our guests for a steak dinner on the last night of camp. Already, five of us had signed a letter to the editor of the campus newspaper, complaining not so much about the meals but that we were paying $14.50 a week for two cheese sandwiches and an apple for lunch each day. Why no variety? The director said he liked cheese sandwiches.

So, after supper, I, as instructed by the director, got a crew together and loaded the stake body truck with two loads of wood for a campfire around which Victor Beede, department head, would lead the singing of *Aluoette* for students and visitors. (Prof. Beede had worked in a French Canadian logging camp.)

The director approved with great praise the pile of wood (it must have equated to 6 cords) and dismissed me to drive the faculty (except the director) and students into Marienville to attend a movie. The truck was jammed for the five-mile jaunt over rocks and ruts. The faculty took seats in the theater. In the darkness, however, the students collared me and left. The faculty, both engineering and forestry, were stranded in the little village's movie house. Unknown to me (honest), the students had serious business in mind back at Blue Jay. I didn't fret; I'd return for the teachers when the movie ended.

Back at Blue Jay, the students hid the fire gong and the hose nozzle, turned off the water at the well, doused the huge stack of wood with gasoline, and torched it. The director, drawn in his undershorts from his bunk by the commotion, was taunted with imaginable epithets on how best to extinguish the blaze. When one senior (we were a mixture, due to the war, of rising juniors and seniors) dumped a bucket of water through a window on what he thought was the director's bed and learned it was the camp desk, he returned with another bucketful for the bed. In the glare of the fire's light, the director called by name a student for help. The poor fellow, upon moving to assist, was tackled and dragged back by his colleagues. The heat singed barracks' paint 50 feet away. As the fire burned itself out, I returned for the teachers, some of whom had already taken lodging at the local inn. (As they got the same treatment as the students, and some of them, too, were vets, I still wonder if they were in on the mischief.)

When the department head arrived from the campus at State College and learned of the incident, matters got worse. Both sides uttered ugly words. Punishment would be by grading. The five of us who signed the letter which was given by the newspaper editor to Prof. Beede, to see if such a situation could really be true, received an F (-1) in forestry. I earned an A (3) in engineering. The Fs were changed to Ds (0), so we wouldn't return the next summer. That is the only D in my college work.

A year later, in preparation for graduate study at Yale, I asked Professor Beede if he would write a letter of commendation. He agreed to do so; and then, in his very officious and dignified manner, asked, "Walker, from whom else will you request a letter?" I answered, "Prof. Cope; he and I always got along well together." "Walker," the department head said, "I think you should ask Professor H. for a reference. After all, he, too, is a Yale forester." (Drs. Bramble and White were also Yalies.) To this, "After last summer, I don't think the H (a nickname students gave Mr. H.) would give me a good reference." "Quite the contrary, Walker. Professor H. told me that you were the only lad who failed summer camp and didn't cuss him out." Years later, returning to campus, I chanced upon the director. His greeting from under the bushy red eyebrows that waved like dried pine needles in the wind: "I don't remember your name, but I remember you're the only student I failed who didn't cuss me out." The professor, I'm confident, meant well.

Courses at summer camp included applications of silviculture, more dendrology, lumbering, and engineering surveying. The latter involved land-boundary marking, theodolite recording, and logging road layout. A thorough understanding of courthouse records was required.

A virgin white pine (*Pinus strobus*) and eastern hemlock (*Tsuga canadensis*) stand (soon to be blown down in a storm) grew nearby on the Allegheny National Forest at Hearts Content. We tallied younger stands of many ages and especially appreciated the coppice wild black cherry (*Prunus serotina*) reproduction. Here the optimum of the four factors of site—climatic, physiographic, biotic, and edaphic (soils)—come together to support the finest stands of this valuable furniture wood. Restrain the deer from consuming the sprouts and the forester has a continously regenerated stand, rotation after rotation.

Summer camp and campus instruction through my 1948 graduation indicated that the thrust for Nittany Lion foresters was to train us for state and federal employment. Apart from Mr. Edward's lumbering courses, there seemed little enthusiasm for encouraging industrial vocations.

By graduation time I had passed the forester Civil Service examination. It was a long subject-matter test. I considered myself well-qualified for whatever task to which I'd be assigned. I learned. Oh, how I learned how much I yet needed to know.

IVCF

I became associated with Inter-Varsity Christian Fellowship, a student-led organization on campus, and served as its president in my senior year. One of our assignments involved conducting the worship service once each month at the Rockview State Penitentiary. Nearby was the State Tree Nursery where we students studied seeding and planting. Thus Penn State and the State Pen were interconnected. How would the reader like to prepare a message for prisoners, including a college faculty member's father, for a Mother's Day delivery? Lincoln's famous words, "Everything I am or ever hope to be I owe to my mother," hardly would be appropriate.

As president of the Penn State chapter, I was expected to attend a summer training session at Campus-in-the-Woods on Fairview Island in the Lake of Bays nearby Ontario's Algonquin Provincial Park. So I hitchhicked from the college's Blue Jay forestry camp in the Allegheny National Forest, arriving a few days late for the 4-week

Penn State's Inter-Varsity Christian Fellowship Conference at Camp Hatetoleaveit (Hate to leave it), 1947.

session.

I knew better than to carry a weapon into Canada, so I left my Mauser with a fellow student to tote back to State College. 'Twas a good thing, for just south of Buffalo, two police cars and a motorcycle officer converged on me at an intersection soon after I got out of a car, ordered me to drop my bag, and began a search of self and duffel. Ah, the officers discovered I could read, so they didn't get their man. Invited to hear a replay of the culprit's description, I agreed that it seemed like me. Assured by the policemen that I'd be halted at every corner, they relayed me in an assortment of patrol cars to the bridge over the Niagara River. A bench, this one overlooking the Falls, once again served as a berth for the night.

At that conference, I sat under the world-renowned theologian, Carl F. H. Henry. I also had a chance to wander in the cutover and regenerating stands of white spruce (*Picea glauca*) and black spruce (*P. mariana*), always hoping I'd see a spruce grouse. The "fool hen" allows a person to come within striking distance before it flies, although this characteristic is more widely recognized for the western Franklin's grouse. Because it can be easily

Dr. Gordon VanWylen, Penn State instructor who later became dean of engineering at The University of Michigan and president of Hope College, with the author. Gordon tried to talk me into enrolling in Princeton Theological Seminary.

knocked down and then killed with a stick, lost woodsmen credit the bird with saving their lives. In this glaciated lowland country, balsam fir (*Abies balsamea*) forms pure stands. If resin from the minute mammary-appearing blisters of the "she-balsam" was being collected for glass cement, I did not see such an enterprise underway in those Northwoods.

"Friends of the Trees"

For career performance, none of the three most important courses, in my opinion, of my undergraduate studies were cataloged *FOR*. Along with the soils and speech classes, noted above, technical writing was the most helpful. No one ever failed as a forester for poor forestry training; it is communication skills that often separate us foresters into categories involving success. As the late Stephen Spurr, a forester and president of the University of Texas, stated in concluding an address before the forestry students and the whole faculty of Stephen F. Austin State University: "I meet in my work many people smarter than I, but no one better educated."

My peers elected me to the College of Agriculture Student Council. One didn't "run" for such an office. The honor was in being asked by your fellow foresters to represent them. The faculty and student members of Xi Sigma Pi, the forestry honor society, asked me to join them in 1947. In the latter case, initiation required that each pledge, following a solemn candlelight ceremony, retreat to a solitary place (my monastery cell was a men's room), there to consider the meaning of the phrase, "Honorable friends of the trees." After an eternity I was escorted back to the fore of the assembled members in the candlelit room to deliver my thoughts on the subject. I spoke to the brothers of Solomon's famous proverb, "Wealth makes friends," though I think the wise king said it sarcastically. This was followed by the admonition, also from Solomon's repertoire, "If a man would have friends, he must show himself friendly. And there is a friend that sticketh closer than a brother." Finally, I recited some lines from Emerson's Waldeinsamkeit:

I do not count the hours I spend
In wandering by the sea;
The forest is my loyal friend,
Like God, it useth me.

From the campus at State College, we would visit stands of wild black cherry. These second-growth stands, about 100 years old, probably were established immediately following a blow-down (USDA photo).

7

JUNIOR FORESTER ON THE SABINE

Ranger Ivan J. Nicholas, in uniform, in a 40-year-old "selectively cut" pine stand established on an old field (USFS photo by Todd).

The Civil Service Commission announcement in the Spring of 1948 called for some 30 junior foresters (P-1 grade, the entry-level professional) for the US Forest Service. Penn State graduates alone could fill the need. Other federal agencies, state organizations, and industries recruited a few. I had an advantage: disabled veterans not only got 10 points added to their CSC examination scores, their names went to the top of the list.

We spent much of the last term at State College boning up for the exam. The faculty produced a file of old exams from which we studied. It was a subject matter test, not a round-peg-in-a-square-hole psycho quiz. This example sticks with me: In a forest tree nursery you observe here and there, among the longleaf pine grass-stage seedlings, stems that are making height growth. What species are their parents? And which species is the paternal? The test was given to the entire graduating class at the State College Post Office.

I got a good grade, received the 10 extra points, and was listed near the top of the roster. A CSC-assigned physician signed off on my physical and mental condition. (He also tried to get me to give up forestry and go to medical school. The country was still short of doctors; but why, twice, should I be encouraged in that direction?)

Lacking confidence in receiving a Forest Service appointment, I had already gone, by mail, directly to the superintendent of the Shenandoah National Park, bragged on my knowing every trail in the Park, and received an assignment as a ranger. I also informed him that, should the Forest Service opportunity come along, I'd want to leave the Virginia mountains. He agreed, and I bought a uniform. Soon, however, I became a Junior Forester on the Sabine National Forest in East Texas. The southern openings were my choice because I figured everyone else, from every northern school and the several southern colleges, along with all the graduates of western institutions, would request the West. This reflects, I think, some misguided instruction in silviculture. In the South were the opportunities to watch trees grow, to manage forests for goods and services, and to participate in the greatest "green revolution" the world will ever witness. But no one told us so.

Sabine National Forest portal, 1940s (USFS photo).

On To Texas

I checked as my first choice the Texas National Forest for another, and better, reason. It was the redhead who laughed at my jokes when I met her while stationed at Camp Fannin five years before. The war widow still laughed at my jokes when I "chanced" upon her during the cross-country hitch-hiking tour following my Army stint. In those two years since discharge, I put on a lot of miles, all *via* the thumb, between State College or DC and Tyler or Wheaton College in Illinois. The bombardier's GI pension paid for her escape from the family farm. I skipped my graduation to be present for Anne's at Wheaton. Seeing her and the Sinclair family, all with railroad passes, off by train following commencement, I raised a thumb for Tyler, Texas, arriving only a day later than the train. Scared as a kitten stranded in the top of a tall tree and thinking of all the freedoms I was voluntarily surrendering, the preacher had to repeat the vow question. "Well, do you?" A fellow with Yankee blood in his veins, a woods worker, and maybe a Republican, it was a bit much for Mr. Sinclair to handle. However, we borrowed his car and honeymooned paddling a canoe among the broadly buttressed baldcypresses and the brightly hued floating water hyacinths in the black backwater bayous of Caddo State Park. (Years later I was asked to work out the ecology of the vegetation of a stream that empties into the earthquake-created lake.)

Assigned to the Sabine

The letter said to report to the Supervisor of the Texas National Forests in Lufkin. I got off the bus mid-morning in late June, wearing my only suit, a woolen gray, to make a good impression. Directions given at the bus station sent me south on US 59. My heart sank when, soaking wet and toting my suitcase from downtown, I had been directed to the *Texas* Forest Service. A kindly forester drove me to the west side of the city to the National Forest supervisor's office in an old CCC warehouse on State 103. From there, on that day, a staff forester drove me to San Augustine where I'd settle in on the Sabine National Forest, one of the four forests that comprised the Texas National Forests (now the National Forests in Texas).

A GI would already know about Tex Ritter, the guitar strummer, who made the Sabine Forest area famous with his song about railroading through these woods on the H, E&WT:

On the East and West line
Old East Texas a-sure is fine.
Drop me off just anywhere,
Tenaha, Timpson, Bobo, and Blair.

Tenaha, the name now given to the northern half of the Sabine as a separate Ranger District (Yellowpine is the south end), and Timpson continue as busy little towns. The highway department still posts a sign for Bobo, while Blair can be located to the east in the woods. "Tenaha, Timpson, Bobo, and Blair" by the Second World War was a major chant for gamblers. Risking cash payroll disbursements, the ritual chanted over a pair of ivory cubes with black dots called for a 10 on the dice.

Ranger Ivan J. Nicholas and I were the only professionals for the 180,000-acre ranger district (which was synonymous with the National Forest. Ranger "Nick" served in that capacity from 1946 to 1955. But John E. Johnson, a local experienced man (LEM) who began working for the Forest Service in the CCC days, was indispensable. Though not a professional forester, he had assistant ranger rank and salary. His father before him rode horseback as a fire patrolman in these woods. And from Johnny and Ranger Nick I learned more forestry than in all my undergraduate studies. Only then did I begin to learn what the work of a forester and the tasks of the professional are all about. From them I learned much silviculture that could be passed along as a forestry school teacher to the next generation. *Axes, Oxen, and Men,* my history of the Southern Pine Lumber Company, is dedicated to Ranger Nicholas.

The Sabine River, separating Texas from Louisiana, and for which the forest was named, had once been an important artery of commerce. *The Neches Bell*, a river steamboat according to a local keeper of historical accounts, carried goods from Beaumont on the Gulf of Mexico to Logansport at the upper end of the Forest. Cotton flowed south, the production of that important agricultural commodity accompanied by the cut-out-and-get-out logging practices being the reason Logansport is no longer a port. Silting in of the river and its tributaries occurred quickly and worsened until the Toledo Bend Reservoir was filled with water in 1965-66.

Silting in of the nation's waterways encouraged Congress to pass the Weeks Law in 1911, enabling the federal government to purchase land to protect the navigability of navigable streams. Some mountain forests in the East and in Arkansas were bought with this authority for national forests. But it was the Clarke-McNary Act of 1924, an amendment to the Weeks Law, that encouraged the acquisition of cutover land in order to assure an adequate timber supply for America's future.

National Forest maps still show the location of *Camp* Brittain, a family "village" of canvas tents to the south of Haslam, a company town. At Haslam, W.R. Pickering built in 1913 what was claimed to be the second largest sawmill in the country. When he cut out in 1936, Pickering took the cash from his sale of land to the government and headed to California's northern timberlands. Left behind was a Mrs. Garrett, widow of a foreman, to look after matters. Twenty years later she reminded me that I had sold her government crossties.

The Sabine was added to the roster of national forests when the government purchased the land from several lumber-manufacturing companies in 1935-1936 for about $2.80 per acre, property taxes prepaid. Timberland acquired covered parts of four counties. Mostly mixed stands of loblolly pine (*Pinus taeda*) and shortleaf pine (*P. echinata*) and bottomland hardwoods and southern baldcypress (*Taxodium distichum*), the Sabine also included stands of longleaf pine (*P. palustris*) at its southern end.

The CCC established an extensive plantation on both private industry (by contract) and government lands that included slash pine (*P. caribaea* in that day; now designated *P. elliottii*). One locates on maps the vast Moore Plantation extending from Pineland to Yellowpine. The first planting on the Sabine, perhaps in the Texas National Forests, the trees from a Louisiana nursery were ferried over the River at Pendleton. No loblolly pines were available the first year for this project.

Early on the job, the aforementioned Johnny, 15 years my senior, attempted, I think, to have some fun getting the new kid lost in a logging operation in a river bottom. Leaving me for a reason unknown, he told me to wait. I assumed he was "going behind a tree." After a couple hours there and day's end approaching, I walked out, probably a good two miles, to his pickup. This neophyte didn't intend to spend the night in that moccasin-infested maze of river-bottom skid-road trails, all of which look alike following a harvest. I had not yet encountered alligators. Somehow, I walked straight for the truck as a bee heads for its hive. An hour later Johnny arrived at his pickup. He never said a word. I wonder, was it he who was lost in the woods?

I suppose all rookie JF's were asked to check the nuts and bolts on a fire tower. The 110-foot tall "erector set" model at Dreka had no stairs, just an open 1-foot wide ladder mounted to the exterior of the galvanized iron tower. Later, wire screen was wrapped around the ladder, but not before an elderly look-out had a heart attack on the job. Lowering him was a scary task.

Snake Scare

Working alone, before the days of snake leggings and hard hats, I used the official marking ax to blaze and stamp the butts of hardwood trees for a small timber sale. "US" was molded into the hammer head of the short-handled ax. Hearing a swishing sound, I looked down; there at my knee was the open mouth of a moccasin. Instantly I knew why herpetologists call them cottonmouths: inside the jaw is a pure white ball of cotton-appearing flesh. I don't understand the physics of the next maneuver, but I went straight up, then sideways 10 feet before gravity brought me down.

Stupidly angered, I cut a fork stick, trapped the reptile, removed its head, and quit the woods for the day. Adrenaline shock soon followed as I realized that had that calf-size pit-viper sunk those two extended ivories into my leg, the buzzards would have enjoyed my eyeballs before Ranger Nick could have found me in that soggy bayou.

I don't recall this event, but Oscar Pate, a logging superintendent, insists it is true. He was having a

The South's JFs in 1949. Jack McElroy, 3rd from right, middle row, became supervisor of the Texas National Forests. From left to right front - William P. Thompson, Naval Stores Conservation Program; Leroy K. Kelley, Texas National Forests; Mark C. Hopson, Flood Control Project, Missisippi; Author; Farris E. Dendy, Mississippi National Forests; Dudley W. Rinehart, Ozark (Arkansas) National Forest; James J. Cann, Mississippi National Forests; and Charles A. Ott, Texas National Forests. Second row, left to right - Clinton K. Sykes, Mississippi National Forests; Alfred P. Mustain, Alabama National Forests; John W.C. Eaton, Kisatchie (La.) National Forest; Donald E. Bylsma, Alabama National Forests; Jack E. McElroy, Ozark (Ark.) National Forest; Carl G. Heczko, Mississippi National Forests; and Homer B. Starks, Florida National Forests. Back row, left to right - Lionel R. Johnson, Mississippi National Forests; P.F.W. Prater, Assistant Regional Forester in charge of personnel; Merlin J. Dixon, Kisatchie (La.) National Forest; Judson L. Hudson, Naval Stores Conservation Program; H.C. Eriksson, Regional Training Officer; J. Herbert Stone, Regional Forester; C.A. Connaughton, Director, Southern Forest Experiment Station, New Orleans; and Patterson B. Int-Hout, Mississippi National Forests (USFS photo).

discussion with a tenderfoot forester about a cutting operation when he backed the young forester to a stump and invited him to sit down. The stump was a freshly cut pine. If that occurred, the trousers remain glued to that resinous "seat" and the backside of the neophyte had a turpentine burn.

Silvicultural Practices

Ranger Nick always made the district's quota for timber sold and harvested. As I recall, we sold enough stumpage in the 1949-50 fiscal year to pay for the forest, bought in 1935-36. Purchase of the national forests under the authority of the Weeks and Clarke-McNary acts has to be one of the best deals the government ever made. These southern forests would continue to be good investments if the radical environmentalists would allow foresters to manage this renewable natural resource.

After protection from wildfire and timber trespass (theft), our principal effort went toward thinning and timber stand improvement (TSI). Pine thinnings went to sawmills and pulping plants. TSI involved marking hardwoods that were of commercial value for hardwood mills that marketed them for bridge timbers, crossties, and other purposes utilizing low-quality stems. TSI also involved utilizing the herbicides ammate and 2,4,5-T, the former in cups chopped into tree bases and the latter in frills made with axes at waist height.

In the mid-1980s, a summer section of graduate students needed a 3-credit course. The one we put together had us going to the old atlases of Sabine practices (thinnings, especially) and then finding the sites on the ground where the effects of these practices could be ascertained three and a half decades later. Here's what stood out for those students: Thinnings of loblolly and shortleaf pines, resembling what forester antagonists call selection harvests, just don't work. No seedlings or saplings could be found in the understories.

On the Sabine in Shelby County grew the national champion water oak, so designated by the American Forestry Association. AFA records show it to have a circumference at breast height of 253 inches, a height of 77 feet, and crown spread that averaged 104 feet.

Arrowhead hunters find these woods a good place to find Indian artifacts. They appear following trash-moving, gully-washing storms.

The Bootlegger's Vehicle

Ranger Nick assigned tasks. Among my first was to check out a hardwood timber sale in the Sabine (meaning cypress) River bottom, a broad flood plain of rich alluvial soil. The vehicle issued me—along with a badge (don't wear it where it can be seen), a passkey, a marking hammer, and a government driver's license—was a 1940 Plymouth sedan. We called the revenuer-confiscated bootlegger's car the Taxi, for you could still see the word under the thin coat of Forest Service green paint. Hidden inside the doors' interior panels were shelves for stashing narrow flasks.

When rain came to the bayous, I called it a day, hiked out to the Taxi, and headed for the ranger station, an office above a grocery store in San Augustine, some 50 miles distant. I soon slid off the ungraded ungraveled dirt road and into a ditch.

The FS was too poor to own come-alongs or any other equipment useful for dragging a car from a ditch to the "road." (I had never driven, even in the Army, on an unpaved road; when pure silt mud's involved, it's an art.) I hiked to a peckerwood mill operated by two brothers living in adjacent homes. I learned later I "yoe-hoed" at the house of the wrong brother, awakening him from sleep. One would have graciously helped me; the one I selected growled, "Get the tractor from the shed and bring it back."

I'd never been on a tractor saddle, but I did have a flashlight with which to study the gear chart plate. The tractor and I were on our way. By midnight, I had dragged the Taxi to the ridge of the road and returned the John Deere. Another two-hour's hike back to the bayou's edge would have me on my way to a warm, dry bed. So I thought. Cautiously advancing the old car, it quit after a few hundred feet. The gas gauge read empty. But I had filled the tank that morning! Had I pulled loose a fuel line? Checking with the flashlight answered that question negatively. Ah, here's the problem: the bootlegger's gas tank had been peppered with buckshot. Apparently soap, which gasoline won't cut, had been used to seal the holes, and the soap dissolved while the car sat in the water-filled ditch.

To Shelbyville I then trudged through the mud, arriving just as Mr. Strong was opening the country store.

The kindly shopkeeper filled a five-gallon can with gas, left the merchandise in the care of his wife, and on his tractor we went for the car. Another hour later, filling the tank at Strong's store, I was on my way to San Augustine, tho' the fuel gauge by then again read empty.

The FS quickly located a pickup truck for the city-bred Yankee forester to drive. Months later, I learned FS auditors turned hand springs when they received the bill for the gas, refusing to pay. The FS contracted with Texaco, while Strong's Store sold Gulf. Solution to this problem involved a federal institution in Kansas that manufactured highest-quality paint brushes. As Leavenworth's products were not on inventory in the ranger district's warehouse, Mr. Strong was happy with the barter exchange.

Heck, I couldn't even drive on hard-surfaced road, especially in western Tennessee following a long drought that left a layer of loess silt on the blacktop, accompanied, just then, by a few drops of rain. As I passed a car, I also passed a sign, blocked from my view, which said "slippery when wet." Anne was just beginning to tell me about the warning when it happened. We landed upside down with the hood resting against a powerline pole. The farmer charged Liberty Mutual $10 for each of four fence posts sliced by the 1947 Chevy we bought as we left New Haven in 1949. The car we passed was driven by a local physician.

I tell you this story because the doctor called it a miracle, and I agree it was. Janet, then six months old, had just been shown off to her grandparents in DC. She lay on top of a crib mattress that covered the back seat and leveled by suitcases on the floor. Now, as the car skidded on its roof, one would think, the roof would be the only thing between the baby and the ground. Call it physics if you wish, but the mattress rolled with the car, the infant protected by the mattress from the crumbled roof as it slid along the ground. Janet received but a nick on the cheek from a splinter of broken glass.

I still hold the record for spending more days stuck in the mud than any forester ever assigned to the Sabine.

(Much of the trails were soon paved.) Driving through slick silt is an art. Young Castle inherited the art. While he couldn't read and write and, therefore, couldn't get a driver's license, he could manipulate a pickup through every obstacle. On rainy days, Castle met me at the end of the pavement in Patroon (the village name meaning an indentured servant, probably named by a Dutch family who settled there after a sojourn in New York) and returned me to solid ground when the day's work was done.

Stealing Posties

Ranger Nick asked me to check out an alleged timber theft in Compartment 22. I began the search at one end of the tract, traversing at 10-chain intervals until hearing the *which which* sound of axes hewing a log. The cadence was so perfect it could have been timed by the drummer of a Glenn Miller band. Parting the haw and titi understory brush, I looked upon two large men, their muscular arms the size of my waist. Dressed only in undershorts and with beads of sweat rolling off their shiny black skin, they shaped with precision, using 14-pound broad axes, railroad crossties from the post oaks (*Quercus stellata*) of the upland flat. They worked from opposite ends of the log and on opposite sides, slicing with razor-sharp axes, to shape the ties, the bark and wood on the same side of the timber as their naked legs.

Unable to whistle and not wanting to spook these fellows, I sang (probably a hymn!) as I approached. We chatted a while; they'd later haul these timbers by horse and wagon all the way to Haslem, at the north end of the forest, to sell to Mrs. Garrett, the widow, already mentioned.

I carried a pad of free-use permits to issue for government trees. The tie-hackers got one, authorizing them to cut every "postie" (as both the trees and their products are locally called) on the Sabine district. In other compartments a timber-stand improvement crew busily girdled trees of this species if their presence impeded the growth of pines.

Learning to Plow

Here's where I learned to plow. Not with a tractor, but really learning to plow. As all the local workers talked daily of the "forty" they'd work that evening, it occurred to me that I should know something of the effort expended in the task. Ah, there's an old black farmer behind a mule in yonder field. I parked the FS pickup and walked to him. Few fences restrained people or pigs in those days. I promptly realized he was scared. The shield on a vehicle meant the law. And when I asked him to teach me to plow, he had greater reason to fear. No one in their right mind would want to learn to plow. The old man was probably born with his hands holding the reins of a harness or the handles of a plow. Since he was still hesitant, I offered him a dollar. Now his conviction was affirmed; no one would pay to plow; you get paid for plowing.

Eventually he gave in. I'd go 50 feet and look behind to see how well I'd done. The old man would be bending over, slapping his thighs in a hilarious laugh. Then when I'd look forward the mule would be six furrows over to the left. I'd straighten up the line and, so help me Hannah, I couldn't help it, I'd look back again to see the old man groaning with laughter. And again, that quickly, I'd find myself many furrows to the right.

The plowman earned his dollar, redoing my mischief. And I learned a new Biblical lesson from Luke's gospel that "Anyone who puts his hand to the plow and looks back is in a whole heap of trouble" (paraphrased).

Syrup making on farms within the Forest's purchase unit boundary was a common sight in the fall of the year. Folks cut the cane by hand and pressed out the juice with mule power, the animals circling the press a thousand times to squeeze out the last ounce. They cooked the syrup in vats fired with wood; some they hid away to ferment to a "beer."

A Friendly Town

In San Augustine, a county seat of some 1500 people, this kid from the concrete jungle of the inner city found a home. While notoriety is written in the town's history and a Texas ranger ("One riot, one ranger") assigned there was our neighbor, no one locked doors. A quiet village! Yes. On the other hand, Mrs. Payne, a widow from whom we rented a room until better accommodations were available, showed us a picture of the community's hanging tree from which five negroes hung. She also operated from her back porch a bank for blacks, who couldn't get loans elsewhere, and from whom she obtained usurious interest rates. Anne had been forewarned by the black sharecropper on her parents' Smith County farm of "mean San Augustine." This naïve white newcomer didn't find it so; I gave the school commencement address. Longwinded I was, as I read the talk over now. (The regional forester in Atlanta learned of the invitation and wrote to commend the effort. The FS needed friends.)

County Judge R.N. Stripling became a friend. His son was counsel to the Dies Congressional Committee on Un-American Activities during the Truman years. The son, Robert, coauthored with sports-writer Bob Considine *The Red Plot Against America.* The judge gave me a personally autographed copy as I walked past his 1900-vintage drugstore. (Later, the inside well was uncovered.)

San Augustine's Memorial Presbyterian Church, organized in 1838 (Esther Matthews and Winnie and Thelma Nicholson photo).

The town people, and especially the members of the Memorial Presbyterian Church in the old village, were good to us. Without a minister, the manse was vacant. The church folks moved us in, the nicest home in which either of us had ever lived. In exchange, as a layman, I led

Sunday services, morning and evening, continuing to do so until the congregation called the Rev. J. Lester Spencer, a demobilized Army chaplain who had served the nearby POW camp. On a more recent occasion the rural newspaper announced a forthcoming service during which I would speak. The article noted that I had "pastured" the church. "That meant," someone said, "you turned us out to grass."

Local War

San Augustine and surrounding counties from the earliest days of the Republic of Texas were peopled with Scotch-Irish whose tempers could flare with little provocation. The Neutral Zone, a No-Man's Land lying immediately to the west of the Sabine River and extending toward the setting sun for 50 miles, was infested with outlaws who took advantage of the set-aside strip of turf while Spain and the United States argued over the new nation freed from Santa Anna's Mexico by Sam Houston at San Jacinto. Too dangerous for settlers, it remained the private domain of these rascals until the Regulators decided to bring order. Against their effort stood the Moderators who insisted that the Regulators were over-zealous in their attempt to corral the roughnecks and selective of whom they would regulate. The region's War Between the Regulators and Moderators cost about 200 lives, according to a Shelby County history. Various family feuds continued for decades, even until the Second World War.

The timber-marking crew, during times spent with them, told and retold the Shelby County Courthouse skirmish story. The county seat was at the little hamlet of Shelbyville at the southern extremity of the county, newly surveyed and established by the legislature. Now it sat within the national forest purchase unit boundary. The county judge (chairman of the county commissioners' court, an elected body principally responsible for road maintenance) preferred the courthouse to be in the new community of Center in the middle of the territory. According to an unofficial county history, the judge conscripted some blacks who, in those immediate post-Civil War days of 1866, loaded in the night a two-wheel ox cart with all the records filed at Shelbyville, cut through the woods (there was no direct road), passing only three houses in the wilderness, and hid the documents in the new settlement. When the courthouse was erected at Center, the judge, fearful of the Shelbyites, built into the structure for his personal use a hidden trapdoor escape route.

How glad I was to have been a citizen of San Augustine when, years later as dean of the School of Forestry, I met up with a state representative at the Austin airport. As Rep. Steve Burgess and I chatted, a member of the state Railroad Commission arrived on the scene. Dressed in the highest-fashion cowboy attire, from 10-gallon Stetson to one-of-a-kind silver-embossed western boots, the commissioner asked a lot of questions about the school and the university. Suddenly, the 6-foot-plus politician leaned way over, pushed his finger into my chest, and blurted into my face, "Wayer yoa frum?" Just as quickly, Steve interrupted to exclaim, "Oh, Larry Walker, He's from San Augustine!" The high-fashion cowboy's response, "That's ok then. I thought yoa was one of these (expletives deleted) Yankees who come down here to tell us how to run our schools and colleges." For many reasons, we'd found a wonderful home in San Augustine.

Public Relations

Ranger Nick had a speech defect which most people attributed, because of his name, to a Russian accent. That meant I did the public relations work, gave the school and civic club talks, and originated a forestry noon-time Wednesday broadcast over KDET in Center soon after Jack Bell, its program director, first opened the mike in February 1949 (Jack's still involved, always dressed in the latest cowboy fashions, a principal participant in Texas Outdoor Writers Association events.)

I talked over the low-wattage AM Station for 15 minutes. FM had not come along and KDET was so weak that you could hardly pick it up in San Augustine, 35 miles to the South.

I began each message, "Friends of the Trees," on one occasion quoting the prophet Joel: "Oh, Lord, to thee will I cry for fire hath devoured the pastures of the wilderness, and the flame hath burned all the trees of the field." Fires were our big problem, some 40 recorded on the north end of the Sabine during a two-month period. I pointed out that the $2000 suppression cost could have been used to gravel roads. To gain interest, I mentioned our people by name: Lamar Duncan on Dreka Tower and Albert Jones at Chambers Hills lookout. I noted that Homer Kay ran the Timber Stand Improvement crew which had to be reassigned to the tougher fire suppression task when Lamar's and Albert's alidades triangulated on a fire. In these homilies I promoted such things as San Augustine's Forest Festival, put together by the math teacher, Mrs. Harlow Johnson, Johnny's wife.

I would intersperse a little forest history—ghost towns that resulted from cut-out-and-get-out practices, free-market enterprise, something about photosynthesis, and receipts to the counties, in lieu of taxes, from timber sales:

1937 =	$ 18
1940 =	300
1948 =	15,000
1949 =	33,000

The economy was picking up.

Timber as a renewable resource, the need for thinning, and how land-owners should care for their own woodlands were subjects of these radio visits. That we now had 19 men employed on TSI crews was deemed an important topic, as was the gravelling of certain roads. Gravelled roads, I learned, were paved with clay and a few small stones. FS engineers certified to the volume of stone in the contractor's loads, but I saw no reason for it. They remained slippery mud slides to me.

As these were the days of the Korean War and President Truman's proclamation on timber production and price controls, I employed patriotic themes. This was important, for southern pulp and paper plants had run "at full capacity from Victory in Japan Day to the present." I find little in these chats to quote verbatim for the present reader.

Ranger (oh, he was proud of that title, losing it in a promotion, because he was the best timberman around, to the supervisor's staff on the Kisatchie NF in Louisiana) Nick thought I should have a uniform for these PR tasks. I still had a fancy Park Service belt, shirt, and trousers, but I had sold the Smokey Bear hat and coat. My kid brother's Marine Corps' green blouse would be close enough, provided I could replace the brass buttons with official FS leather nubs. So I wrote the Fecheimer Company in Ohio, the FS supplier who, so tearful of my plight, sent a whole set *gratis*. Anne altered the Marine style to that of the FS and affixed the buttons. (Dear reader: JFs earned $2666 per annum. That came to $100 in each of the 26 every-two-weeks pay periods each year. The FS didn't provide uniforms.)

We carried badges, but didn't wear them. Nick's predecessor had been shot at; apparently the scare was justified. We were not policemen, but wildland managers, in spite of the Agriculture Secretary's directive that we report all whiskey stills found along our creeks.

The author, JF on the Sabine.

Contrarywise, prior to beginning work (cruising, timber stand improvement (TSI), timber-marking) in a compartment, we made the rounds of nearby country stores, talking loudly of where we'd be working as we drank 5¢ soda pop. Local citizens understood: We'd see where the stills had been. We also wouldn't have to fight incendiaries that weekend!

Wildfire Solution

Ranger Nick sized up the fire situation. It wasn't that our people were setting them; but the payroll showed our people were fighting them. They earned 35¢ an hour fighting fire. Wages for road, bridge, and trail crews, TSI workers, and timber markers amounted to 40¢. A fellow could put in 40 hours at 40¢ and earn some extra dollars on weekends at 35¢. The ranger called everyone to a meeting at the Dreka fire tower work center to announce a new rule: When you've worked your 40 hours, that's it for the week. If you chase smoke and hoe lines for 10 hours on Sunday at 35¢, then your 40¢ an hour work week will end Thursday afternoon. If a Friday night fire or a set on Saturday occurs, Nick said he would bring in crews from the Angelina district. No accusation was made. The men caught on; their "friends" abruptly ceased setting the woods on fire.

After putting out a fire in the night, we got lost trying to find our way out of the woods. One-by-one, the five-man crew in the moonless night slid down a steep bank and into a creek that, according to our compass and recollection, shouldn't have been there. We built a fire, dried out, and awaited dawn.

To save money, I'd take the call by a crude radio from the firetower men who had determined a fire's location by triangulation and then I'd try to find it. When spotted, I'd return to the truck and, by radio, request the number of fighters required, drafting them from the nearest crew, regardless of its present task. One can wander, even with compass and pacing and maps, an hour or more in dense smoke before finally coming upon the flame.

We all feared an assistant regional forester responsible for fire prevention and control. He'd show up unannounced from Atlanta, inspect fire tool caches placed at various locales and write a report which we got six months later. By then, we couldn't defend ourselves, and he knew it. All across the South, I learned later, he was not well received.

Calling on a family we suspected of burning the woods, I was invited into the shack built of scrap lumber, the ground beneath visible through cracks in the floor. We conversed briefly, but for this I was totally unprepared: A child, still in diapers and not yet able to walk, was sitting on the floor smoking a cigarette. We classified the fire as a "smoker-caused incendiary," attributed to one of the older boys, maybe 8 or 10 years old.

On another occasion, the crew and I called upon a black man to discuss a fire we had just corralled nearby his place. I suppose it was instinct that caused me to greet him with a handshake. Neither he nor the crew—all white men "because we didn't know how to work blacks"—knew what to make of this violation of a social norm in the woods of Deep East Texas.

Early Prescribed Fire

Ranger Nick carried out one of the South's earliest prescribed fires in the Moore Plantation. Its purpose was to remove hazardous fuel and to improve the range herbage for cattle and wildlife. CCC boys planted the Moore, now a wildlife management area, including in the mix slash pine, the natural range of which does not extend west of the Mississippi River. We intended the fire to run against a south wind. The fire refused to run. About mid-afternoon Nick sent us with our backfire drip torches to a FS road, earlier a tram bed, on the tract's south boundary, there to set a fire that would run with the wind. It did. It swept clean the soil of herbaceous material, pine straw, and litter until... Until the head and tail fires met! There a brief firestorm occurred, the holocaust consuming everything in about a 5-acre area. The controlled burn, apart from this exception, did not injure the planted pines. Fire running with the wind, under prescribed conditions, is a cool fire.

Fifteen years later, searching for sites to show students upon my going to work at Stephen F. Austin State College, but having long forgotten the prescribed burn, I chanced upon charred soil in a large barren zone. To the ranger station I went to learn what had happened here. Out of the compartment file folder fell a piece of rotting, worm-eaten yellow-dog, the cheap newsprint-quality paper on which we hand-wrote memos to the file. I had signed this one in

1948, soon after arriving on the job, calling trees in the one-day field examination *loblolly* pines. But they were slash pine. I think, as I look back, that I reasoned well: The Sabine isn't within the natural range of *P. caribaea* and I had never seen the species. To tally the saplings as loblolly pine was logical. Oh, the barren area was the site of the 1949 firestorm noted above.

Entrance to the Moore Plantation. CCC "boys" planted slash pine to the right of the road and shortleaf pine to the left in 1936 and 1937 (USFS photo, by Alva Williams, 1940) .

CCC Work

The Civilian Conservation Corps enrollees erected the facilities used by the East Texas Fox Hunters Association for their meetings at Boles Field, beginning in 1926. They continued to meet annually at the site, astride a Pickering Company tram road from Haslam, after the land was purchased for national forest. According to a Shelby County history, red foxes were imported and in time became established in the forest. The annual chase lasted 8 hours. Walker foxhounds and redbones predominated; bluetick hounds didn't seem to participate. A national cemetery for fox hounds was laid out in 1941 on government land.

Portion of Boles Fox Bowl during a 1940s annual meeting of the East Texas Fox Hunters Association (USFS photo).

Hundreds of folks came for "the thrill. not the kill, of the chase," setting up primitive camps, using the amphitheater, called Bowles (Boles) Bowl for gospel singing, and being entertained by the cowboy evangelist, B.B. Crim. If Crim couldn't get converts at the end of his fiery sermons, he'd tell the audience where it could "go;" he was "going coon hunting." Crim could be vulgar, telling women in his meetings to "Cross your legs. Now, the gates of Hell are closed." But the Baptists brought him back, year after year.

The CCC boys left their mark on the Sabine. The Patroon (puh troon) camp housed 240 enrollees and officers in 6-man tents and later cabins. They erected the Dreka fire tower, converted tram roads to graded roads and, according to photographs, boxed the roots and soil of six 8-inch diameter pines for shipment and planting at the State Fair grounds in Dallas. Did the trees survive? At Redhills Lake, the "tree monkeys" and "bush apes" raised a natural dam on a branch that flowed from a spring to form the finest cold-water lake in East Texas.

Boles (Bowles) Bowl on the Sabine (USFS photo).

Many of our laborers and subprofessional workers had been in the CCC. Of the $30 enrollees earned each month, $25 was sent home (this compared to the $21 army privates earned). The Sabine hosted three other camps: Milam, Zavalla, and Redhills.

The earliest camps were a tent city and the Exhibit Hall at the town of Center's fairground. Eventually six 40-man barracks were erected at each camp. Johnny Johnson managed a "colored" camp at Bannister. For camps of both races, he'd spend two nights teaching the enrollees to sign their names rather than making an X. Into the 1950s we still needed to provide this instruction for local laborers.

During the Second World War, CCC camps in and near the Sabine housed prisoners of war. Some 200 POWs salvaged trees following two severe ice storms in the 1943-44 winter. (The butt two logs of trees injured in those freezes, but not then harvested, today remain as evidence of the severity of the storms. They crook at their height at the time of the storm, a lateral branch assuming dominance to replace the apical stem.) A FS crewman, recalling the period, told me how, having used the German-invented chain saw in Europe, these men of the

African campaign carried out their tasks with dispatch, always finishing by noon. The Geneva Convention hadn't figured on power saws when establishing maximum daily cutting requirements for POWs.

Special-Use Permits

Once a year we made the rounds of permittees, those folks farming a few acres of industry land since before FS acquisition. To avoid hassles and to keep friends, we issued special-use permits at minimal or no cost. In approaching a permittee's shack we'd stand back and loudly call out HULLO, lest a vicious dog find your rump tasty, as one did of mine. Finally, someone would appear, see the "gov'ment" shield on the side panel of the pickup, and show fear. Government always meant trouble for these people and, as a Snuffy Smith cartoon of the period illustrated, there was no distinction between the Treasury department's revenuers and the Agriculture department's foresters. So Snuffy says to Aunt Louisie as the tree-badged fellow wearing the Smoky Bear hat walks up the trail, "Shoot 'em. Works for the same outfit, don't he?"

When the HULLO is answered, you learn that no Jake lives there. Then, when told you want to give him a free-use permit so he can continue to plant, chop, and pick his cotton, the resident responds, "Oh, *that* Jake. He's in the back room." Many of the Jakes legally could have taken possession under squatters' rights laws: Fence and farm the land for seven years and its yours.

Schools and churches received special-use permits. We even provided free stumpage for them. They got local peckerwood mills to convert the logs into rough-sawn lumber for the buildings.

Land-Line Controversy

One old-timer and his wife inherited the East Texas mean streak. The controversy involved a line being surveyed that separated their property from that of recently purchased government lands. The objection to the work of the surveyors was so strong that the old man ran off the survey party with a shotgun while the Missus, at an appropriate distance, cradled in her arms what appeared to be a 30-06 rifle. Even the US Marshall, sent in from

Beaumont, was greeted uncivilly as he stood by the Jacob staff upon which was mounted the Forester's compass. As I recall, final settlement of the dispute took place in a federal district courtroom in Beaumont. Old witness trees, cut at the base and carried to the courtroom, convinced the judge that the elderly folks did own the land. FS surveyors were trespassers. But such opinions were not unusual, for jurists and juries in these quarters generally took the side of those opposing the federal government's intrusion into the lives of citizens. They still do.

Ticks

Ticks, those pesky vermin transported by cattle and wildlife, agitated this forester and his wife. The 8-legged arachnids come in three sizes: "seed" ticks (larvae), smaller than the head of a straight pin; "yearlings (nymphs)," about 1/8 inch long, and adults, 1/4 to 1/2 inch, depending upon how full are the bellies of blood or of eggs. The South's wood ticks (*Dermacentor*), unlike New England's deer ticks (*Ixodes*) that spread Lyme's disease, serve as vectors for Rickettsias, close kin to the typhus organism. Rocky Mountain Spotted Fever ticks also spread a Rickettsia. Most disease is spread in the nymph stage of these "hard" ticks with life cycles spread over two years.

Brushing against tall grass or herbaceous weeds knocks them off their roosts, seemingly by the thousands if they are seed ticks just released from the female adults' belly. The infants fan out quickly to cover your clothes, finding tunnels through garments to one's skin, then digging in for blood. A mulberry branch, the rough texture of its foliage velcro-like, comes in handy for swatting them from clothes. Yearlings do a better job of burrowing into a human's hide, but the fully grown adults, following the last moult, viciously mine for blood. Heat from a match or a lighted cigarette makes them back out. Pulling them out or digging them loose leaves the pincer jaws under the surface of the skin, often resulting in infection. In those days, we didn't have imported fire ants to hold ticks in check (according to some reports) nor many deer as their hosts.

Ticks must be removed before entering the house. On the back porch, and disrobed at the end of the day, I had

Anne inspect me for ticks. They are destroyed only by being torn apart with the fingernails. Squashing won't do, nor will drowning or discarding. We missed a few, but Lyme's disease had not yet been discovered nor had Rocky Mountain Spotted Fever been diagnosed in East Texas.

Not so bad are the chiggers, the notorious redbugs that burrow under the skin. While an alcohol rub helps soothe the itch, bib overall farmers standing on the town square just scratch wherever it itches. The scene can be hilarious.

I mentioned above the lack of deer in the woods. In three years working the Sabine, I never saw a deer. If a sign was seen, a night hunt would soon follow. Locals totally ignored game laws, nor would judges, all elected, preside over a conviction. In East Texas, they said, "We don't harvest game, we harvest game wardens." And it was so. Today, with education, hunting clubs, deer leases, tough laws, and tougher enforcement, white-tails overstock to overbrowse, to their detriment, these woodlands.

Supervisor's Visit

I was accepted by the Yale Forest School and the FS at about the same time. Explaining the situation in a letter to *Washington: Personnel*, I received word to proceed to Texas, take leave without pay in September, and return to the district the next summer. My job, the letter said, would be protected, and it was. But Supervisor Bosworth didn't like the idea and, in a brief memo, told me so.

I do not feel the end ever justifies the means to that end, but inadvertently that happened on this occasion. I again addressed *Washington: Personnel.* Within a few days the regional forester in Atlanta got word to Mr. Bosworth that the FS would stick by its agreement. I never learned what happened in the RF's office; but P.F.W. Prater, assistant RF for personnel, H.C. Eriksson, regional training officer, and Herbert Stone, the RF, to the present greet me warmly. Only as I write this does it occur to me why they could call me by name 40 years later!

The angry supervisor drove to San Augustine, picked up Ranger Nick, and headed for Bayou Blue where I was deep in the woods marking timber. I suspect they watched a while before HULLOING to let me know of their presence. After the Ranger introduced me to Mr. Bosworth and we exchanged a few pleasantries, they left. Later I learned the supervisor came all the way from Lufkin to severely chew me out for going over his head but, seeing a leg totally shorn from my trousers by the thorny vines, a sleeve in similar disarray, yellow marking paint from baseball cap to boots, and bloody scratches on my hide, he chickened. He'd seen the diligence with which I encountered the briars, brambles, and blackberry vines. Soon Anne and I left by bus and train for New Haven, Connecticut.

Meanwhile, in preparation for returning to graduate school, I spent evenings winnowing words from Bezemer's 1935 French-English *Dictionary of Terms in Agriculture* and Gerschel's, 1911 *Vocabulaire Forestier*. The extensive translation fills a pocket notebook, called Form 289. I still find it helpful in my travels and reading. (The last pages of the 3-1/2 x 6-inch book could be used for map overlays or for personal hygiene in the woods.)

Before I left the Sabine, Ed Ellenberg signed on as a JF. The Sabine soon split into two districts, and Ed and Marge moved to the Yellowpine District, headquartered at Hemphill. Anne and I remained on the newly named Tenaha. As I write, nine professionals and some contractors do the work assigned Ranger Nick, Johnny Johnson, and me in the late forties.

The Sam Houston southern baldcypress on the Sabine National Forest, the largest of the species in Texas. The site would soon be inundated by the Toledo Bend Reservoir.

Caddo State Park, site of author's honeymoon and later subject of published ecological research on a virgin hardwood stand.

8

THE FORESTS OF DEEP EAST TEXAS

A virgin stand of southern pines in Deep East Texas, about 1900 (author's file, source unknown).

Two centuries ago pioneering whites, often accompanied by black families, waded the Sabine River due east of present-day Milam to continue their journeys on *El Camino Real*. They paused at the river's side long enough to be baptized into the Roman Catholic faith, then followed the ruts of the royal road that led them westward through the dark and dank forests of Deep East Texas. A few sojourned for a while in this land rich in lush vegetation. Others continued on to the Spanish missions at San Augustine and Nacogdoches, while those most determined to gain fortunes in land and cattle proceeded beyond the woodlands to the shortgrass prairies of the Spanish colony surrounding San Antonio.

Those hardy souls who remained in these forests of tall pines and deciduous broadleaf trees did little to interrupt ecological processes. With broad axes weighing fourteen pounds, they cut virgin stems of pines from which to hew sills for houses. The knot-free heartwood of the lower boles of the more than two-hundred-year-old loblolly and shortleaf pine trees would last a hundred years, the resin-soaked fiber resisting attack by wood-destroying fungi and insects. Heartwood from buttressed southern baldcypresses in the bayous provided shakes for roofs and exterior walls, while the Spanish moss draping lazily from trees of all kinds served for mattress stuffing and sewing thread.

Occasionally squatters on these red hills, so-named because of the oxidation of the iron that coats the grains of silt and clay in the soil, sectioned a white oak for barrel staves to hold whiskey distilled along the creeks. Other liquids, too, could be contained in these casks, the pores of the white oaks occluded by a gummy substance. Open pores of red oaks, on the other hand, provided material for barrels for storing flour and sugar.

Following Native American culinary practices, persimmons and plums from the woods provided fruit, locust pods served the need for beans, while walnuts and hickory nuts when pounded, cast into boiling water, and strained earned the name "milk of honey." Rich as fresh cream, the oily liquid added sweetness to hominy grits and cornbread. Wild black cherry bark, when boiled, produced delicious tea. The common yaupon, while a popular browse species for deer, made an effective emetic, as its scientific species name indicates: *vomitoria*. Early basket weavers utilized the bark of young shoots of the redbud, the flowers of which—along with dogwood—in the woodlands' understory dazzle the eye with Eastertime's spring ritual.

While some homesteaders passed through the waters at what would shortly become Pendleton's Ferry, a capitalistic venture of long endurance, others chose to enter Spanish, and later Mexican, territory via Blue Elbow, the swamp along the lower Sabine and to the north of today's town of Orange. Dodging cottonmouth moccasins in dugout canoes, called pirogues, whittled and burned from the baldcypress at the river's edge, the rugged pioneers paddled westward to reach solid ground and a woodland of loblolly pines and a variety of hardwood species. Soon they came upon vast expanses of longleaf pines, the string-like needles singing tunes to the cadence of the wind. Though misnamed by taxonomists, the Latin *palustris* means wet site; the tall, straight and cylindrical stems grow on droughty lands. In the geological past, these sandy lands were the sites of coastal beaches.

A little more to the west lay the Big Thicket, a dense jungle of loblolly and shortleaf pines and a myriad of deciduous trees. For men running from the law or a lady, this lair of the black bear and locale for the cavernous dens of rattlesnakes provided an ideal hideaway. According to a folklorist of these haunts, the rattlers grow so old they sprout whiskers.

A few U.S. citizens would follow the route of the early-day botanist Thomas Nuttall (for whom a variety of red oak was named in recent years), entering Texas from the north. Passing through Arkansas territory on a well-traveled Indian trail, they often confused the Red River with others to the south that flow eastward. Squatting on lands they believed to be dedicated for Indian tribal use, the pioneers built villages in what later obviously was

Texas soil.

One of the more southern rivers is Big Cypress Creek, its water and sediment flowing into Caddo Lake, the South's largest natural lake. On earliest maps the swampy Caddo morass, now believed to have originated by an earthquake, bore the name Tsoto in honor of the early explorer DeSoto, whose expedition visited the area in the mid-1500s. As late as 1820 the area remained Caddoan, an abbreviated form of Kadohadocho, meaning "red chief."

Caddo Lake, in Northeast Texas. Indians claimed an earthquake created the lake; Captain Shreve was blamed for blowing up the dam; and Army engineers are credited for restoring the water's level for transportation of supplies by river boat to Jefferson and beyond.

Southern baldcypress and tupelo gum dominated this impenetrable wetland until the U. S. government dynamited a "raft" of logs that dammed the river and constructed ditches for transportation routes. In time water lilies invaded and, with changing water levels, many other hydrophytes migrated to the lake. Leave Caddo, and promptly one encounters pineries and poor lands of scrub oaks and hickories. Branches of old hickory trees appear to weep as they hang from the trunk. Some wag said that on these dry sandy and gravelly soils they have reason to droop.

Logging Begins

A century would pass before the rhythm of the crosscut saw and the whine of the circular saw capture our attention. Loggers in the Lake States had completed cut-out-and-get-out practices there. From the sands of old lake beds and peaty swamps, some lumbermen went west to the great virgin stands of Douglas-fir, redwood, western redcedar, and ponderosa pine. Others came South to begin logging anew, arriving in Texas about the time the calendar turned to 1900. There was nowhere else to go. The white pine and hardwood forests of the Northeast had been depleted before the upper thirds of Michigan, Wisconsin, and Minnesota were entered by the lumberjacks who moved men, mules, and machinery to accomplish in 40 years what it took earlier generations 150 years to do in Penn's Woods and points northward. And during this interim of ten decades, the land we call Texas became a province of Spain, a part of the Republic of Mexico, an independent nation, and the twenty-eighth state.

Harvests of the Texas pineries began in earnest as the century commenced. Still recovering from the Civil War, the nation needed lumber for building homes and schools and factories, crossties on which to rest its rails running north and south and coast to coast, and gum naval stores for paints and ammunition. Cordwood was needed to fuel furnaces, steamboats, and locomotives. And a new paper industry required fiber from the forest. Long days and long weeks the flatheads labored at low wages, without even the glory that accompanied the New England white-water men, Paul Bunyan lumberjacks of the northern lakes, or the tall timbermen of the Pacific slope. Woodsworkers lived with their families in boxcars in company towns into which tram roads were laid for transporting harvested timbers to mill ponds. Others found shelter in tent cities of canvas, the remnant of one called

Ducktown still designated on a national forest map. The more fortunate lived in mill towns, riding the tram rails to the cutting chance in the wee hours and returning home carrying the lantern atop the last load of logs long after dark.

Boat-Builders

The First World War hastened the harvest of the virgin forest. England's prime minister insisted he needed not more of our Army's men to finish off the Kaiser's forces; he wanted only "a bridge of ships" to transport material from American factories to the shores of Britain and France. Wilhelm's submarines ravaged our nation's resources more greatly than did his soldiers' attacks upon the Allied troops in the trenches of Europe. The subs hampered ports of nations friendly to England, sinking vessels attempting to run the German blockades. The general who completed construction of the Panama Canal was enlisted to manage the building of the bridge. When timber cutting went too slowly, he complained to the lumbermen that "birds are still nesting in the trees which must go into the lumber out of which to build ships." The barons, a bit more optimistic, bragged "that in 72 hours they could shoo the birds, cut the trees, mill the lumber, and have it at dockside."

Among the 84 vessels completed by Armistice Day, 1918, was the *War Mystery*. Launched at Orange, "the largest wooden ship that ever took water" required but seven months to build. *The City of Orange* was soon christened, and Mrs. Woodrow Wilson watched *The Nacogdoches* slide into the ship channel at Houston.

At the same time, the military bought 17,000 carloads of southern lumber for cantonment construction. One enthusiast for the task needed but "two weeks (to convert timber) from coastal breezes to the siding of a barracks in Kentucky."

Fifteen years later, the southern forests were no more, the massive boat-building effort hastening the day. Cutover forests of stumps, viewed as far as the eye could see, were now to be sold to the federal government under the Weeks Act of 1911 "to protect the navigability of navigable streams" and its 1924 amendment, The Clarke-

McNary Law, to ensure a future wood supply for the nation.

Mid the Great Depression, in 1935 and 1936, some 600,000 acres of Texas industry lands were purchased for an average of $5.20 an acre. Some lands, with taxes paid in advance, brought as little as $2.00 an acre. These cutover, burned over, over-grazed barrens became the four national forests which some folks today like to pretend or imagine are virgin forests, remnants escaping the pioneers' axmanship.

To understand the change that has occurred over a fifty-year period—a nature drama only the last scene of which is acted out on the stage for us to now view—we need to consider the ecology of these woodlands. How did the forest become what it is?

An Emphraim Shay locomotive, one of many responsible for hauling much of the timber from the South's woodlands (author's file, source unknown).

As the 15 million acres of East Texas' commercial forest lands lay bare, owners happily sold their holdings, some lumbermen utilizing the income to buy into the untapped sugar pine lands of northern California and other species of giant-sized trees of the West Coast and Cascade ranges and the Sierra-Nevada. Millmen assumed these East Texas soils would never again bear merchantable timbers. They seemed not to understand why these forests existed as they did when entered in the early 1900s with oxen-drawn high wheels or mechanical Clyde skidders. These beasts of burden and engines of efficiency dragged the logs to stream sides or to steam-powered McGifford loaders that lifted the heavy timbers onto the beds of log cars pulled by Ephraim Shay's ingenious gear-wheel locomotives.

Eco-Echo

American Indians also played a significant role in fashioning the forests encountered by intruding settlers. The magnificent woodlands that stretched from the Sabine River a hundred miles westward had been home to the Caddoes, the Alabamas, and the Coushatta tribes. A few Cherokees, escaping the infamous Trail of Tears on the march from the Southern Appalachian Mountains to Oklahoma's Indian Territory, attached themselves to local native groups.

These and other American Indians burned the forests with regularity. They needed to expose wildlife upon which they depended for sustenance. They also cleared the underbrush with fire in order to see an enemy. These fires, often becoming conflagrations, consumed the leaf litter covering the soil and opened the canopy of foliage as trees died from the heat. Exposing the mineral soil enabled the winged seeds of the southern pines, when coming to rest great distances from their origin in the tops of trees, to germinate: seeds fluttering to a landing on leaf litter and pine straw lack the wetting and drying and the freezing and thawing necessary for germination. Seeds also require access to inorganic matter, the mineral soil, into the pores of which the new roots grow and attach themselves for nutrients and moisture.

Thinning the canopy by fire enables abundant sunlight to reach the ground, enabling the conifer seedlings to quickly produce carbohydrate through photosynthesis. Light is the catalyst. True, one often sees head-high seedlings under dense overstory trees, but in another year or so these have succumbed as the pines lose the little shade-tolerance they have as seedlings and young saplings. As seeds germinate, a cuplike rosette of a dozen seed leaves appear. As these small leaves fall away, primary needles circling the young tree's stem take over the sugar-producing function. Soon this juvenile foliage too withers, the secondary needles grouped in bundles of 2s and 3s (for southern pines) carrying on the production of carbohydrate. Thus we observe that a source of seed, a mineral seedbed, and full sunlight are necessary for the germination and growth of Texas' pines. Hurricane-force winds and insect epidemics as well as fires provide these conditions. So too do the clearcutting sawyers.

Hence, following the cut-out-and-get-out days of the lumbermen, new forests of pines appeared. The success of the natural regeneration of these pioneer species in ecological succession, however, was often impeded by wildfires that killed seedlings and saplings; by piney-woods' rooters that consumed roots for their nutritious stored starch; and by grazing cattle that trampled seedlings and rode down saplings as the bovine sought a rubbing pole. Where protected from these adversaries, the million seeds that gave birth to ten thousand seedlings on an acre would naturally thin as individual trees expressed dominance. Perhaps a hundred trees per acre reached maturity. In the virgin forest less than a dozen of the Methuselahs, extending into three centuries, would typify the pine stands the lumbermen entered.

What would fill the rest of these old-growth forests? Mostly shade-tolerant trees, like oaks and hickories in the uplands and American beech, river birch, and southern sugar maple along the stream courses. Many other species, climax or subclimax in ecological succession, entered the forest as hitchhiking seeds hanging to the fur of animals, as seeds scarified in the gizzards of birds and subsequently passed in the droppings, as nuts, the shells softened by the acids in animal stomachs, and on the wings of winds. Others of these broadleaf species arise as sprouts from the roots and stumps of previously felled trees. As the pines pass from the stand in a hundred years or so, broadleaf forests of beauty, but of limited utility, capture the site. The hardwoods will remain in control until fire, hurricane winds, insect or disease epidemics, or the loggers' tools prepare the land for the next rotation of pines. Climax forests are simply those stands which if left alone—if mankind's axe and saw are excluded, and there is no catastrophe—will perpetuate themselves *ad infinitum.* In Deep East Texas these are deciduous stems. There is no stopping ecological succession. Neither herbicide chemicals nor the swing of the axe can halt this march: natural processes will continue.

Bears and Birds

These paragraphs must not omit a note about the fauna of the forests of Deep East Texas. Prior to logging, black bear were common; so too, it is believed, was the ivory-billed woodpecker, the largest woodpecker in North America. No silvicultural procedure would have rescued this bird, for its habitat required at least 500,000 acres of undisturbed virgin forest of a mixture of southern pines, baldcypress, and hardwoods.

Ornithologists believe its present range is solely in the high mountains of Cuba. A smaller edition of the ivory-bill is the pileated woodpecker. The 18-inch-high bird is common in pine and hardwood mixtures.

Bringing much forest management to a halt in East Texas is the endangered red-cockaded woodpecker. As trees with red heartrot, a fungus-caused disease, were harvested to avoid spread of the malady and destruction of pine stands, the bird's habitat was inadvertently destroyed. It makes its homes in hollowed-out boles of living, diseased trees. Management practices that leave fungus-infected trees and control dense undergrowth with prescribed fire enhance the woodpeckers' habitat. Birds may excavate a nesting cavity over a several-year period, many generations using the same nesting place until fire, storm, or wind destroys the tree.

Only recently has the American woodcock been considered significant in these forests. The game bird winters in dense brush, the "picket fence" protecting the bird hiding just above the ground and under the brambles from overhead predators.

Bobwhite quail habitat was enhanced by the patchwork quilt of logging, burning, and grazing of the Texas woodlands. Fire in longleaf pine forests, especially, provides for greater populations of this game bird. Alas, imported fire ants infest nests at hatching time, destroying eggs and chicklets. Hawks attack many quail during northward migrations in the spring.

As the cut-out-and-get-out harvest neared completion, habitat for white-tailed deer disappeared. Neither track nor sign could be seen for a couple of decades. Poachers used every means, including tracking hounds, to make a kill if word of a sighting was out. Now, with law enforcement and game management, white-tails are abundant. The absence of predator coyotes and wolves encourages population growth. Harassment by razor-back hogs, along with diseases related to browse shortages, are about the only things keeping deer numbers checked, now that forests of many kinds and ages cover the lands.

Black bear are returning. So, too, is the alligator. Once on the U. S. Fish and Wildlife Service's endangered species' list, the reptile is becoming ubiquitous in the waters of these woods.

So in the magnificent forests of Deep East Texas, lovers of the land and its abundant resources find a panorama graced with the finest species for mankind's use, the finest soils in which to grow them, and the finest climate to enhance their vigor. What a blessing!

* * * * * * * * * * * * * * * * * * *

Excerpts of this chapter provide the Introduction for David Gibson's photo album of the woodlands of the region, *Deep East Texas*, published in 1993 by the Longstreet Press, Atlanta, Georgia.

Mules pulled high-wheels, lifting logs from the ground to landings in the woods of Deep East Texas (Southern Pine Lumber Co. collection).

9

YEAR AT YALE

These loblolly and shortleaf progeny pines were seedlings when the 1994 assembled Yale grads were students at Prof. Meyer's Crossett, Ark., camp in 1946. At age 25, the stand tallied 7,000 bd. ft. per acre.

The family of Gifford Pinchot, the father of forestry in America, founded the School of Forestry at Yale University. Henry Graves, Pinchot's successor as Forest Service chief, served as its dean, and a succession of FS chiefs are among its graduates. Here J.W. Toumey introduced ecology as a course of study; here George Garrett and H.H. Chapman and Ralph Hawley and Walter Meyer and Harold Lutz tutored. These leaders of the profession attracted me to the Yale Forest School. The reader knows by now that the popularly supposed elitism of the second-oldest college in the United States, and among the wealthiest, would not provoke my interest in the institution, even though clergy founded the parent college to train ministers for the Calvinistic Congregational Church of colonial times.

About thirty of us enrolled in the 1948-49 school year in Yale's one-year program for students with an undergraduate degree in forestry. About that many more followed the two-year route for those with bachelor diplomas in fields other than forestry.

"Sage Boy" guarding the entrance to the Yale Forest School. Women students of recent years don't appreciate the icon (photo by Lisa L. Diekmann).

The Faculty

Professor Chapman had recently retired, but maintained an office and remained active in national as well as Yale forestry matters. Herman Haupt Chapman probably has published more words about forestry than any other person, even to this day. Some of his writings were textbooks, some editorials, and some highly opinionated essays. A colleague years later wrote of Chappie as "always busy with his pen and used to buy Waterman fountain-pen ink by the quart. However he used to dilute it figuratively with oil of vitriol...." I suppose it was true.

Legends, possibly apocryphal, tell of Chapman's verbal skirmishes with Louisiana's state forester, V.H. Sonderegger. The professor, leading his students in their summer camp at Urania, Louisiana, observed the height growth of some seedlings in a longleaf pine nursery seedbed. Concluding these elongating seedlings to be the hybrid cross of loblolly and longleaf pines, he named them *P.* X *sondereggeri*, or Sonderegger pine. For years previous to this observation, Prof. Chapman had preached on the need for fire in longleaf pine stands to prepare seedbeds and to control brownspot needle blight. All the while Sonderegger orated on fire prevention, and that included controlled burning. His firefighters spent too much time putting out fires to tolerate the Yankee school-man setting the Pelican State's woods on fire. When the description of the new pine hybrid was published, the appearance of Sonderegger's name surprised those who knew of the professional animosity. When asked why he so honored his long-time adversary, Chappie responded, "A hybrid is a bastard, is it not?"

Another legend tells the story differently, that it was the Louisianan who showed Chapman the hybrid. Naming the tree for Sonderegger earned the Yale professor a line in London's *Index Kewensis*. The botanists' manual abbreviates the authors of Latin descriptions, in this case H. H. Chap.

We used Chapman's loblolly pine bulletin, published by the Yale Forest School, like a bible on the Sabine. I'm confident others across the South held it in similar respect.

Professor Ralph Hawley also had recently joined the

ranks of the unemployed but, we gathered, was so aggravated by Chappie's continuing residency that he seldom came near the campus. It was our loss. David Smith, a young PhD candidate, took over Hawley's silviculture courses.

Prof. Walter Meyer taught management and statistical methods when he wasn't consulting for industry in the West. Yale would not accept for graduate credit the C I got in his "stat" course. I had to add another class to the schedule for the next semester. But, dear reader, Dr. Meyer's course was so thorough that, with timely review, I passed on the first try the doctoral qualifying exam on the subject at Syracuse. Few candidates did.

We were a month into Meyer's biometrics, satisfied that properly conducted and analyzed *sampling* can be depended upon for ascertaining the significance of scientific or managerial data. That was until November, 1948, when Harry Truman beat Tom Dewey for the presidency. The polls showed Republican Dewey so far ahead of the incumbent Democrat that newspaper headlines, already set in type, blared, "Dewey Wins." We then analyzed the analyses in the "stat" class, moving on to the associated *theory of chance.*

Faculty worked closely with students. This is shown by my transcript which lists, of 14 courses, three seminars and two problem courses assigned individually by faculty members. Add to this the thesis that also involved a one-on-one relationship. Also, Dr. Albert Dimond of the Connecticut Agricultural Experiment Station, adjacent to campus, gathered a few of us around a table for a seminar in plant physiology.

David M. Smith kept us on track in the New England woods. We knew every tree in the New Haven watershed, a forested tract surrounded by a 10-foot-high chain-link fence. Managed by the School to assure clear (free of sediment), clean (free of dangerous microbes) water for the municipality, the land was off-limits to the general public. The young instructor tutored us in other Connecticut woodlands, including the locale of his studies in white pine regeneration.

Soon, with a doctorate and as Professor Smith, David became "Mr. Silviculture" to a couple of generations of foresters raised on his text, *The Practice of Silviculture.* In the anti-forestry rhetoric of the 1980s, he starred in a TV silviculture documentary filmed in the woodlands of the region. He also served as the forestry consultant for the unabridged *Merriam-Webster's International Dictionary.* Our warm personal friendship today stirs a memory: Dr. Ralph Steen, the president of the university where I served as dean, had returned from an extended trip, apparently visiting a number of universities that supported forestry programs. The president said something to the effect that wherever he went someone knew me. Indeed forestry is a small, close fraternity.

My major professor, Harold J. Lutz, receiving the Society of American Foresters' research award from SAF President Dewitt Nelson.

Fred Dickinson, wood technologist, moved on to Colorado State University; Gene Zumwalt was soon in the director's office of the Bureau of Land Management; while Dr. Mary Bolton enthusiastically maintained the forestry library in Sage Hall.

George Garratt, everyone's father figure, was dean.

Anne and I entertained him in our home years later in Athens, Georgia. He promised to send Anne a jug of "Vermont moonshine." She, the teetotaler, expecting strong stuff, was quite relieved to open the parcel from George containing the boiled sap of sugar maple trees.

J.W. Toumey's "A Study in Plant Adaptation," published in *Popular Science* in 1902, is an "oldie" I treasure in my library. It deals principally with cholla (*Opunta fulgida*) of the southwestern desert lands, but touches on John Bartram's Southern Appalachian discovery and collection of *Franklinia altamaha* (now extinct in the wild) and the Asiatic *Ginkgo biloba*.

Harold Lutz had us digging pits in the soils of the region. His *Forest Soils*, coauthored with Robert Chandler, was our text. We knew the Cornell professor did little beyond writing a single chapter, the publisher wanting a co-author with the unadulterated credentials of a pure soil scientist. While Lutz' classroom lectures lacked enthusiasm, his field excursions made up for it. The woods and soils came alive. I was soon convinced I wanted him as my major professor and to follow up on the inspiration received from Prof. Richer's soils lectures at Penn State.

Redcedars and Worms

Ralph Read, on leave from the Southern Forest Experiment Station's research unit in the Arkansas Ozarks, was of like mind. It was my good fortune to team up with Ralph, who was older, more experienced in forestry, and with a background in research. A special problem course under Dr. Lutz's direction, which we published in the 1950 *Journal of Forestry*, remains among the most cited publications of my research career. We titled it "Influence of Eastern Redcedar on Soil in Connecticut Pine Plantations." The article noted that physical and chemical properties of surface soil beneath eastern redcedar trees (*Juniperus virginiana*) left to grow in red pine (*Pinus resinosa*) plantations differed from those beneath the adjacent pines These properties were influenced by the chemical nature of the cedar leaf litter, its decomposition products, and subsequent incorporation into the mineral soil by earthworms, the fisherman's *Lumbricus terrebrans*.

The pine plantations have a thick mat of pine needle litter and a thin layer of fermenting or partially decomposed organic matter distinctly delimited from the mineral soil. The litter layer under the redcedars, found randomly among the pines, showed earthworm work so extensive that the needle layer was barely measurable. The lime-loving earthworms formed casts, nibbling a bit of the calcium-ladened redcedar foliage, mixing it with mineral soil, and passing it out as fecal matter. Analyses showed the *J. virginiana* needles contained as much as ten times the amount of calcium as the pine straw. The casts, as shown by statistical analyses, significantly improved infiltration rates, air capacity, moisture-equivalent, pore volume, and volume-weight. All these serve to enhance watersheds, allowing rain water to rapidly infiltrate the soil's surface and subsequently to percolate to lower horizons. There the soil stores the water until it slowly trickles from springs and seeps. Infiltrated water, of course, doesn't run off, carrying silt and clay to streams.

We didn't say so at the time, but since then I've suggested planting this low-value (for timber) tree on watersheds like the Carrizo sand outcroppings in East Texas.

Alas, a Japanese scale insect infestation necessitated the salvage of the red pine plantation, but some cedars remain among the succeeding hardwoods.

Together, too, Ralph and I prepared an unpublished manuscript on reproduction under a mixed-hardwood forest.

Malby Lake Woodlands

The MF thesis dealt with *Some Relationships of Forest Soils and Site, with Special Reference to a New England Mixed Hardwood Stand*. In reviewing the book written in 1949, I find it a fair mixture of field work, laboratory analyses, and bibliographic effort. David Scott, a doctoral candidate who served as Prof. Lutz' laboratory assistant, got me through some tight spots in the use of chemicals and associated equipment in his Bowers Hall lab. David became a notable silviculturist, serving on the faculty of the University of Washington's College of Forestry.

I soon knew real well Permanent Sample Plot 64 in

Compartment 9 of the School's Maltby Lake tract. The soil was virgin, never cleared for cultivation, due, no doubt, to the rocky nature of the Maltby series, the genesis of which is a deep till left after glaciation. Here I studied ecological relationships of an 80-year-old red oak stand under which ash, oak, and maple seedlings appeared between the clumps of shrubby vegetation. Another stand on a soil of good quality included red and white oaks (*Quercus* spp.), hickory (*Carya* spp.), and tuliptree (*Liriodendron tulipifera*), while another plot on a medium-quality upland site consisted of a variety of broadleaf species originating from seed and sprout buds following a shelterwood harvest a few years previously. Thus, the variety of the species within the small area of a single compartment encouraged me to recognize the four factors of site in ecological studies: physiographic, climatic, edaphic, and biotic. The interrelationships are often overlooked. Maltby, once fenced water company land, is now (in 1994) open to the public and provides habitat for deer and turkeys.

Spurred On

For a while Anne and I rented a beach house on Long Island Sound, residing there until Thanksgiving. Stephen Spurr, a neighbor in a nearby waterfront bungalow, invited me to ride with him, as I still did not own a car. Steve, who was paper-grader for Dr. Meyer's management course, passed along some tips, as we rode along, on a variety of subjects, including such mundane matters as the word "data is plural." He, too, had worked on the ecology of earthworms in the sandy soils of the region.

Steve was on leave from Harvard Forest, working on a PhD degree. He later became President of the University of Texas. Our friendship renewed, I carried my management notebook to Austin, barged in on the Prexy, and showed him his hieroglyphics. He still couldn't interpret the red scribbles. Dr. Spurr was among the smartest people I have ever known. Yet it was he who, as earlier noted, while addressing the faculty and forestry students at Stephen F. Austin State University, said, "In my work I meet a lot of people smarter than I, but no one better educated (than a forester)." The caption of this passage also commemorates the greatest punster I have known. An honor came my way when asked to nominate Dr. Spurr for the Society of American Foresters' Sir William Schlich medal. He won the award.

Classmates I've kept up with over the years include Allen Neff (who maintained class correspondence), Bob Webb (Texas industry executive who taught the wood procurement course at SFA), Herb Winer (industrialist and Yale forestry professor), Ken Carvell (silviculture professor at West Virginia University), Ben Bryant (wood technologist), and Frank Armstrong (high-ranking Army officer who established the Foresters in Real Estate organization). Many more I read about.

Yale grads of the around-the-WWII-years met for a reunion in Georgia in 1992. Front row (l-r): Frank McCaney '41, Dick West '42, Wendell Becton '29, Al Nelson '50, Robert Clark '42, Author '49; Middle row (l-r): Silas Little '36, John Gray '42, Dick Williams '50, Zeb White '38, Evert Johnson '47, Gordon Condit '41, Hamlin Williston '42; Back row (l-r): Sid McKnight '42, Lewis Grosenbaugh '36, Paul Y. Burns '46, Don Tufts '48, Ralph Arnold '41, George Hindmarsh '48, John Wishart '46. All worked in the South, for that's where the jobs were in the forties (photo by H. Williston).

No Town for a Bride

New Haven was no place to take a bride of three months. The "nice" apartment with kitchen and bath we rented by mail indeed had a kitchen—a hot plate in a closet and a sink that hung over the bed. The bath down the hall

The 1994 gathering of "fossilized foresters" at Crossett, Ark. Rear (l-r): John Wishart '41, Rev. Lester Nickless '39, Ralph Arnold '41, O.H. (Doojie) Darling '58, Sid McKnight '42, Dick Williams '50, Don Tufts '48, Dick Jorgensen '42, Robert Webb '49, Bob Clark '42, author '49; Front (l-r): Evert Johnson '47, John Gray '42, Jesse Ralston '40, Hamlin Williston '42, Lewis Grosenbaugh '36, Gordon Condit '42; Not shown: Henry Wilson '47.

Regional Forester Jack Alcock and the author at the '92 Yale southern foresters reunion (photo by H. Williston).

was shared with others. That month's rent went down the drain. Hurlon Ray, a fellow student, mentioned he had rented a house on Long Island Sound and needed another couple to share the rent. Would you believe it, we unknowingly contracted with the same landlady who advertised the nice apartment. She got us again. The beach house had no insulation and no way to heat it in winter. The Rays and we endured until Thanksgiving. Frozen out, our plight was desperate until Ralph Read, the fellow forester noted earlier, and his wife, Deenie, and little daughter Ruth, took us in. We shared the expenses, used the kitchen in shifts, and formed lasting friendships. Ralph, a talented pianist, later directed the Rocky Mountain Station's branch at Lincoln, Nebraska, until retirement. He and Deenie finally made it back to their beloved Rocky Mountains, settling at Estes Park.

How glad we were to share the flat near the Winchester Arms plant several miles from campus. Only the bus driver knew how near broke we were at a month's end. I walked Anne to the stop, waited til she counted out pennies for the coin drop, waved "Goodbye," and walked to class. Winter winds blowing from Long Island Sound made me remember I'd frozen my feet before, in France. We'd walk home together from her job as a statistician at Yale's Bureau of Highway Research. Anne is acknowledged in Morton Raff's, *A Volume Warrant for Urban Stop Signs*. If I had my way we'd never live where urban stop signs are warranted!

Digging in the Branford sandy loam soils in the Eli Whitney (inventor of the cotton gin) Forest, I handled a lot of roots, including those of poison ivy. Unfortunately the roots of the vine don't display the trifoliate leaves—nor do the above-ground stems in winter. Myself nearly immune, I carried the oil of *Rhus radicans* (*toxicodendron* if you are an older taxonomist and *Toxicodendron radicans* if you are really up-to-date) home. The light-complected Anne totally lacked resistance to the allergen. She reacted severely. Covered with the itchy hives and none of the prescribed remedies giving relief, innovation became essential; for not even yellow laundry soap was able to cut the plant's oil, the source of the problem. I resorted to the use of ether borrowed from Dave Scott's lab where we used it to cut the lubricant on lids of glassware previously sealed with a grease gasket for air tightness. It worked. But I cringe at the thought of the explosiveness of that hydrocarbon with the aromatic odor.

By graduation time, we had saved enough to buy a used Chevy.

Breaking Out of the Cell

We were like monks in the brownstone Sage and Bowers halls, the buildings together our "monastery" on Prospect Street. Studies restrained us from extra-curricular activities. To meet students enrolled in the several Yale colleges, Anne and I participated in Bible studies sponsored by the University's Inter-Varsity Christian Fellowship. We met some brilliant young men already committed to seminary studies and foreign missions. Years later I remembered these fellows when a young ordinand said, "The best men of my seminary class dedicated their lives to Gospel proclamation in distant lands."

I must not leave Yale without reminding my readers of the beautiful gothic Dwight Chapel on High Street in the old campus. The ornate sanctuary memorializes Timothy Dwight, an early day president (1795-1817) of Yale College and a revivalist preacher and theologian of the American Revolution period. Carved in the wall inside the entrance to the sanctuary are the words he wrote for one of America's grand hymns of praise:

> I love Thy kingdom, Lord,
> The house of thine abode,
> The church our blest Redeemer saved
> With His own precious blood.

Nearby is another marker, this one carved in wood. It commemorates the lives of forestry professor Walter Meyer's brother and his wife. Missionaries to the Philippines, they were killed when the Japanese captured the islands in the early days of the Second World War.

My Yale diploma, in Latin, is worded as were the parchments of Timothy Dwight's era. Translated, it reads, "To all to whom this letter comes greetings in the Lord always..." The certificate, mailed to me in Texas, says I rated second honors.

10

ULCERS AT SYRACUSE

Professors S.O. Heiberg and D.P. White with mean trees of non-treated and potassium-fertilized white pines in the Pack Forest sand plains (source unknown).

Kjeldahl flasks exploded and flew through the air to smash against a wall in the attic soils lab in Marsh Hall. Chemical analysts use the one-liter glass vessels for cooking samples of vegetation and the organic matter component of "the medium in which plants grow." By this method they determine nitrogen percentages in tissues and soils. (Today Kjeldahl lab walls are carpeted with gym mats and workers wear helmets.) Dodging the missiles raised adrenaline levels; it also caused this fellow to ask, "What am I doing here?" I asked the question again about six weeks after arriving at the country's largest forestry school and one of the world's best-known.

Each week I met with the doctoral committee of faculty assigned me by the dean. Each week I came up with a new thesis proposal. Each week it was rejected. In one session it would be a wood technologist who criticized; in another the plant physiologist found fault; and then an entomologist or pathologist. I'd had enough by the sixth such session, informed the gentlemen as tactfully as possible that I could take a hint, a job in San Augustine was waiting my return, and I would bother them no further. The reader needs to understand what I didn't then know. No one yet had received a PhD in silviculture at the College of Forestry, though others had attempted. This candidate was to set an example. Quickly a committee member said, "Come back next week and we'll have some suggestions." I did, and they did. Of the three one-sentence proposals, I chose the one, I learned years later, they considered impossible. That I should tackle the toughest task earned me a feather in my cap in the minds of the faculty. What they didn't know was that I chose the task in the Adirondack glacial outwash sands because it seemed to be the easiest. I thought I could complete it in the two years I still had available on the GI Bill after the PL-16 money ran out upon being awarded the MF at Yale. (Dear Reader, was it welfare? Some neighbors think so. I've paid it back many fold with higher income taxes directly related to earnings because of those educational opportunities.)

I detailed the proposal promptly, got its approval, lost out on a grant to fund it, and began laying in the plots. Working alone on weekends in the Adirondack Mountains and running analyses at night kept me on schedule.

Choosing a School

My reason for returning to graduate studies and for choosing the Syracuse school was not one for which to be proud; but upon completion, I certainly was grateful for the circumstances that led me that way. The motive was the two years of GI Bill still available and which had to be claimed by July 1951. Scot blood in my veins and obviously in my genes (but not in those of my offspring nor in my Scottish wife) prevailed. Neither interested in teaching nor research, and happy working in the woods, I simply didn't want to waste the opportunity. The 1952 College bulletin did its best to discourage non-serious applicants and women with such phrases as, "Men who meet these requirements" and "make demands on a man's time." (They'd change those pronouns shortly!) The Graduate School Bulletin, in referring to PhD theses, notes that

> the student is required to *penetrate the frontier of human knowledge.* In his research and evaluation of literature, in his planning, execution, and evaluation of his own research, and his thesis covering work done, he is required to demonstrate *original scholarship* of high order (italics mine).

I suppose for that original scholarship effort members of the scientific Society of Sigma Xi elected me an associate. Later, in Georgia, I'd receive full membership.

Duke University wanted another year of basic sciences, since my master's degree was an MF and not an MS (Yes, that's what the graduate school wrote.) I didn't know about the Duke DF. Texas A&M would gladly enroll me in the range and forestry program, but the studies would be all in range, requiring much undergraduate work. The department head gave a whole day escorting me around the campus and dining me. Yale, based upon my GRE, would admit me for a DF, not a PhD, with no suggestion of time required for the task. I don't recall the University of

Michigan's response, but Dean Hardy Shirley, then of the graduate school and later the CEO of the New York State College of Forestry, wrote a kindly letter in answer to my inquiry. In effect, he wrote that the catalog says it can be done in two years; however, no one ever has.

Living Conditions

We lived on the first floor of a two-story GI barracks in a community for Syracuse University's married students called Slocum Heights. The university provided paint if the students brushed it on the cardboard interior walls.

Without a washing machine and before the invention of Pampers, diapers were done by hand. No refrigerator, we resorted to the university-furnished icebox. To avoid the twice-daily emptying of the pan beneath, a hole drilled through the floor into which a tube borrowed from the attic lab was inserted served nicely. To save the cost of blocks of ice, the snow bank to the rear of the building came in handy. Packed down tightly, the snow could be cut and hauled to the kitchen. That worked until the fellow across the hall, a wildlife biologist, completed a taxidermy task and tossed the leftover formaldehyde onto the snow bank. There it froze. Until, that is, tamped into a block and carried to our icebox just before bedtime. Tragedy was averted by the tight seal on the box and the tube that drained away the liquid containing the pungent gas. The week's groceries Anne purchased that day were ruined. It was enough to make a grown man cry. Folks, it *ain't* worth it. For those two years Anne and the family, now with Janet Anne and Stephen Laurence in diapers, were neglected.

With two diaper-clothed infants we made our summer home in a log cottage at Pack Forest, in the northern woods of the Adirondack Preserve. Anne washed the diapers in a tub by the spring, hanging them to dry on a line under the crowns of spruce and fir trees. By eventide the cotton cloths were fragrant with the oily droplets of terpenes that saturated the air. No synthetic aerosol dispersant could match the aroma emanating from the small, breathing stoma of the needles. The pleasant smell of the coniferous forest captured by the diapers made the otherwise unpleasant task a joy to remember. Never mind the haze that makes the Virginia mountains blue, the North

Carolina peaks smoky, and the Adirondacks ofttimes gray. One can endure and enjoy that kind of air pollution. (The droplets are "particulates," similar to those emitted from industrial stacks.)

Studies

Apart from Friday night baseball games and church work, San Augustine offered little for advantageously utilizing one's time. TV had not yet arrived, and the AM radio, broadcast from Center, reached the town rather garbled. So, after de-ticking, I spent evenings boning up on my Sorbonne French and Professor Meyer's statistical methods, both to satisfy the College's required qualifying exams. When I arrived in Syracuse, Dean Shirley's secretary wasn't about to let me take the French test without first taking courses in the language. To fail, she said, would embarrass the College. So, dutifully, I enrolled, grateful the course didn't "make." French II, however, did. With my background, the instructor agreed to my auditing the course. I got a 200- or so page book from the Library on *The Forests of Portugal*, written in French. I knew that book like the palm of my hand. And what would you guess? Upon entering the qualifying examination room at Syracuse University and shown my assigned seat, there, with a thick bluebook, was *The Forests of Portugal*. My translation covered the tablet's front and back covers, inside and out. Out of paper, the proctor dismissed me.

Syracuse University, a private institution founded by, and now loosely connected to, the Methodist Church, seemed a strange place to find a *state* college of forestry. The College actually is on state land, sharing a common boundary with the University's property. They share much more than a boundary. I took advanced courses in plant physiology, analytical chemistry, and, since my minor for the doctorate was in management, a course in public relations. Students conveniently walked to the other side of the "big ditch," the stadium, to participate in these classes.

The physiology professor was so impressed with the earthworm-redcedar-red pine *Journal of Forestry* article by Ralph Read and me that I had it "made." The chemistry professor, on the other hand, informed the night class of a dozen graduate students—all the others Syracuse University majors in engineering and sciences who worked for industry by day—that he graded on a curve. We looked at each other. The words stung. We all needed Bs to survive. And to that end we labored industriously, helping each other, yet knowing the one we helped to understand an equation might be the survivor while the helper becomes the loser. The mid-term grades showed us *all* failing. If my reader is a teacher, please never make that threat to your classes. We were all psychologically wrecked; our academic careers ruined. Too late to drop the course. Professor Joseph Neal somehow sensed his error in judgment. An old bachelor, he invited us all to his mother's house near campus where he had set up card tables. With drinks and his mother's baked goodies for us, he went table-to-table, beginning with his first lectures, to patiently show us our errors. I don't know about the others; I got an A on the final and a B for the course.

Harold Ickes, the only Republican to serve in Franklin Roosevelt's cabinet, was Secretary of the Department of the Interior from 1933 until 1946, when a public outburst against Roosevelt's successor, Harry Truman, brought about Ickes' resignation. (The disagreement dealt with the management of federal oil lands, for which the Interior secretary was responsible.) Ickes' public relations chief subsequently became the dean of Syracuse's School of Journalism. Dean Wesley Clark's course was an exciting seminar composed otherwise of professional communicators. The connection between the Forest Service and Interior was not lost on either the instructor or the student. The teacher knew well the old and continuing FS and Interior antagonisms. My term paper was a "Public Relations Program for a U.S. Forest Service Ranger District." That course can be credited for any success I may have had in developing the School of Forestry at Stephen F. Austin State College as its first dean.

The journalism dean told me of a presidential decision not reported elsewhere. FDR had on his desk on the eve of Pearl Harbor the executive order to return the Forest Service to the Department of the Interior. In Interior it had been housed until 1905 when President Theodore Roosevelt, encouraged by Gifford Pinchot, wrested the agency for the U.S. Department of Agriculture. As Pinchot manipulated one Roosevelt, Ickes did likewise another. Only WWII saved the Forest Service for Agriculture.

Studies for the silviculture faculty, directed on the ground by Jack Barrett, required a detailed management plan for the Tully Forest, now Heiberg Forest, a memorial to one of the profession's great foresters. The Courtland County tract contained 102 compartments spread out over 8 soil types. The assortment of northern hardwoods would test any silviculturist's soul. An apple orchard and sugar bush got mixed in with the prevalent pine—hemlock—hardwood type. Jack and I later worked on three editions of *Regional Silviculture in the United States*. We cooked up the idea together; he edited the tome.

C. Eugene Farnsworth was another faculty member whose wise counsel and instruction I valued. Gene began every lecture by asking a question. When no one responded, he would donate the hour to answering his own inquiry. Clifford Foster, director of Pack Forest, and Paul Graves, who opened doors in Washington's bureaucracy for our government administration course, made significant contributions. The imperial Dr. Sergei Wilde of the University of Wisconsin considered me his student, as he did all the students of his academic progeny. Years afterward I was the Russian immigrant's "grandson" because Don White, one of my major professors, was his doctoral student.

For Dean Shirley's course in forest regulation, we wrote essays for his review and class discussion. A draft of one of mine finds me terribly confused in 1952 about the role of the feds in managing lands for which others pay taxes. Even today, I can cite the pros and cons, but am hardpressed to say "Always yes" or "Never, under any circumstances."

Svend O. Heiberg, known throughout the world for his studies in forest soils, brought with him from northern Europe the reason for the chlorosis and necrosis that occurred to conifers planted in the deep sands originating from glacial outwash. For years students in silviculture lost exam points if they attributed the foliar coloration on affected trees to anything but a shortage of nitrogen. But it was he, along with his young protégé, Donald P. White, who correctly diagnosed the ailment as the result of inadequate potassium in the soil. Prof. Earl Stone at Cornell concluded likewise.

So what? To answer that question for both scientific and practical reasons was now my task.

The Thesis

Dr. White emphasized that doctoral research must push back the boundary of knowledge. Rehashing what others had written, as in so many liberal arts dissertations, doesn't pass the test. And, he also insisted, I must know more about the subject at the conclusion of the effort than anyone else. So with maps, field tools, and all the enthusiasm I could muster, work began in the sand plains of glacial outwash origin in the Adirondacks. The glacier receded and the lakes eventually drained into rivers. The deep light-textured acidic sands of weak podzol character now appeared as easily tilled soil. We settled in in the cabin at the 2211-acre collection of abandoned farms south of Lake George, called Pack Forest. Charles Lathrop Pack's gift to the College would be our home away from home. Though the lumberman never saw the tract, a boulder dragged there from a drumlin bears a bronze marker which tells of the dissemination of his ashes hereabouts and on which is inscribed the lines

> Here he lies where he longed to be,
> Home is the sailor, home from the sea
> And the hunter home from the hill.

Level topography favorable for cultivation early in the nineteenth century, before Leibig's Law of the Minimum and cation exchange were discovered, led to removal of the virgin forest. At first the soil produced good crops, fertility maintained by the release of nutrients in the organic matter accumulated over the centuries under the old-growth spruce, pine, hemlock and an assortment of broadleaf species. But with exposure to the atmosphere, the organic matter was oxidized, the soil now releasing potassium cations as rapidly as they became available through decomposition in the beach-like sands. Around 1930, foresters put these lands back to work. As they planted conifers on the old fields, the prehistoric lake beds and terraces again supported trees. But, within a few years, symptoms of a nutrient deficiency appeared on white pines (*Pinus strobus*), red pines (*P. resinosa*), and white spruces (*Picea glauca*) as chlorotic needles, abnormally short needles, abnormally short annual height growth of the main stem, and needles that failed to persist. As noted earlier, Professors Heiberg and White determined the cause to be potassium (K) deficiency.

Gray birch showing potassium deficiency in sand plains of glacial outwash origin in the Adirondack Mountains.

In descriptive studies, abhorred today by numbers-crunching biologists, I soon noticed a retrogression in the quality of vegetation on the abandoned fields: old hay fields→ poverty grass→ *Polytricum* moss→ less demanding mosses and lichens→ erosion and bare land. Adding 200 pounds per acre (that, for modernists, equates to approximately 200 kilograms per hectare) of muriate of potash (60% KCl) to the soil reverses the trend.

For comparisons, I installed plots on a shallow mantle of coarse till, on a brown podzolic mantle of calcareous origin, on a course-textured esker-like ridge in a virgin coniferous forest, and on a fertile shallow field. Fertility plots were laid out in stands comprised of red maple (*Acer rubrum*), paper birch (*Betula papyrifera*), gray birch (*B. populifolia*), American beech (*Fagus grandifolia*), trembling aspen (*Populus tremuloides*), fire cherry (*Prunus pennsylvanica*), wild black cherry (*P. serotina*), and common choke cherry (*P. virginiana*). I sampled leaves, including petioles, taken from east-facing branches (determined by compass) at the crown middle (with a measuring rod).

***Spartium scoparium*, the leguminous Scotch broom, showing potassium deficiency (left) as an indicator plant and fertilizer-corrected seedling (right).**

Shrubs, forbs, grasses, and mosses were also sampled. For good measure, and at Prof. Heiberg's "suggestion," I planted European scotch broom (*Spartium scoparium*) and lupine (*Lupinus perennis*), as these plants in Scandanavian soils served as indicator plants for nitrogen deficiency. The good professor, I think, was determined to justify his earlier insistence that the problem related to inadequate soil ammonia and nitrate.

What a break! Symptoms of K deficiency showed up in several key species. As reported in *Forest Science*, the journal founded by my old friend, Steve Spurr, wild black cherry exhibited striking and unusual discoloration. In August, bright red margins extending almost to the tip of the leaf abruptly appeared. The line of demarcation between the pigmentation and normal coloration was well defined. By late September the anthocyanin margins had widened. Blue and violet hues accompanied the red color that intensified as the season progressed. Autumn coloration differed from this foliar abnormality, although both result from the migration of chlorophyll from leaves, exposing the red and yellow permanent pigments, anthocyanins and carotene, respectively. Potassium is a principle element in the chlorophyll molecule. Orchard-grown cherry foliage, in contrast, exhibits curling of the edges of leaves when K is in short supply. Foresters spotting these symptoms, the French call *le diagnostic foliare*, in regeneration practices can prepare to fertilize the site or, instead, to plant a less K-demanding species like jack pine (*P. banksiana*) or scotch pine (*P. sylvestris*).

Gray birch exhibits similar chlorotic margins, while for red maple the whole leaf is chlorotic throughout the summer. Species of *Rhus* also showed the effects of inadequate K, thus serving as indicators for the silviculturist in regenerating sites. Even Prof. Heiberg's *Spartium* signaled a K-deficient soil.

The studies also confirmed Earl Stone's diagnosis at Cornell that the needle-tip chlorosis on jack pine can be attributed to magnesium shortage. A 25-pound per acre dose of Epsom salts ($MgSO_4$) remedies the malady. Of course, finding any of the earlier mentioned coniferous species with the symptoms alerts the forester to the need for correction in existing stands of seedlings and saplings, whether naturally occurring or planted.

But suppose none of these trees occur in a compartment in the forest designated for regeneration. Then the silviculturist resorts to chemical analyses. As I reported the procedure in the *Soil Science Society of America Proceedings* (now *Journal*), exchangeable potassium in the soil indicates whether the element is adequate to support the more-demanding conifers. (Exchangeable K refers to a measure of the element available to plants for adsorption through roots.) For instance, if exchangeable K in the surface soil is less than 18 ppm, as it is in most of these sands, white pines will suffer. A shallow upland soil derived from granite, in contrast, analyzes 30 ppm. Potassium in foliage of a number of species, both tree and herb, can also indicate the sufficiency of the element in the soil for the growth of commercial tree species. Forest fertilization now became an economic possibility.

Publications reporting these results showed up unabridged in French, German, and Italian journals.

Birch and Pine Potboiler

Some research administrators discourage "potboilers," the diversions that drain resources and time. I don't agree. One such paraexperiment came about from an observation of white birch (*Betula papyrifera*) trees that grow here-and-there on some sand plains. The timberman's weed trees had seeded in randomly. But, beneath their crowns, standing out like the proverbial sore thumb, were dense stands of healthy white pine seedlings and saplings. The small stands of vigorous trees abruptly terminated at the edges of the crowns of the overtopping 50-foot tall birches. Out in the open fields of the cleared and abandoned farmland an occasional white pine seeded in, these always exhibiting the four symptoms of K deficiency: short, chorotic, nonpersistent needles and short terminal stem growth.

Analyses of the nutrient content of the foliage provided answers to the riddle of these crowded stands some distance from white pine parental seed-bearing trees. Pine foliage under the birches tallied almost twice as much potassium as those growing in the open. Leaves of white birch measured about a fourth more than the healthy pines.

Did the birches seed in on brush, ash piles, or fecal matter high in nutrients? I found no evidence of charred wood or darkened organic leachates in the soil. Rather, it seemed to me that the birch trees, by strong nutrient cycling, supplied organic matter which enriched the soil directly beneath their crowns. Deposited leaves decayed, the nutrients in them leaching into the soil. As with Connecticut eastern redcedar found in the Yale forests, the gray birches "foraged" for nutrients far beyond the edges of their crowns. With the element collected from a sizable area and concentrated in the surface soil under the overtopping canopy, the pines grew vigorously.

The American Potash Institute made this narrative the color-photo cover story for a 1953 issue of *Better Crops with Plant Food* magazine. Referees approved the potboiler as a technical paper for the *Journal of Forestry*.

That the potboiler paid off is a continuing reminder of the need for observational curiosity. A medical school professor lectured his class on the fact that one can taste sugar in a diabetic's urine. The doctors-to-be appropriately squirmed. Bringing to the lecture hall two beakers of the micturition, one with and one without sugar, the instructor demonstrated with a finger dipped in each vessel the accuracy of his pronouncement. Now each student was directed, as a learning experience, to file past the beakers and do likewise. When all agreed that urine-filled beaker B contained the sugar, the tricky tutor exclaimed, "You have indeed learned that sugar in the urine can be tasted; but you have failed miserably in a lesson in observation. I put *this* finger (as he raised the digit) in the beakers, but (raising another) tasted *this* one." Long have I forgotten the instructor who told this story, but the tale lives on to remind me of the need for observational skill. The old *Boy Scout Manual* called it tracking, requiring the ability for second class.

Traipsing through the Adirondacks, often relying on my compass for bearings, I chanced upon an old large ramshackle dwelling. What was it doing in this remote and forlorn setting? I never learned, but the resident invited me in for refreshments, showed me rooms full of old silver, and sold me a couple of pieces as souvenirs. Not 'til years later did the possible illegal accession of the bowls and trays occur to me. No, I believe the beautiful pieces, like the elderly gentlemen, belonged to the once-beautiful home. Quite by coincidence, I passed through Oneida on my return to Syracuse, dropping the tea service off for the famous silversmiths to repair and replate. I note now the Rochester factory numbered each item of the van Berch quadruple plate.

Professors Heiberg and White were so pleased with my thesis (the College didn't call doctorals dissertations) that they dispatched a draft to Prof. Earl Stone at Cornell University. The two institutions remained competitive in some areas, even though Cornell lost its forestry program, founded by Bernhard Fernow, in the early 1900s. The State legislature withdrew funds when it learned, through that day's "tree-huggers," that Fernow's students on practicum assignments in the Adirondacks were cutting the broadleaf trees of no economic value and replacing them with conifers for the future needs of mankind. The opponents of those silvicultural practices, procedures which Dr.

Fernow brought from Germany, were folks of immense wealth who owned vast estates in these mountains. They visualized their private gallery of autumn's palate disappearing as the young men, under the tutelage of the former federal forestry bureau chief, cut and sawed their way through the second-growth seedlings, saplings, and poles that hindered regeneration of merchantable conifer species. This led to the drawing on official maps of a blue line around the Adirondacks. Inside of this preserve no timber management is permitted on state-owned land. That's in the State's constitution. Of course privately held tracts within the line may be managed. Lands which appear to belong to the College of Forestry actually belong to Syracuse University and are held in trust for the College so that faculty can do research and students can carry out practicums that require cutting trees. Later, in 1911, the legislature created the College of Forestry, not at Cornell in the Fingerlakes (where the state's agriculture college resides), but in the city of Syracuse.

Back to the thesis, now being reviewed by the friendly professional competitor of Professors Heiberg and White. In time Earl, as I would soon call him, asked for me to come to Cornell to chat about the research. He had many good ideas which I would carry in writing back to the Syracuse campus. The co-chairmen of my committee seemed delighted that I would run all the additional analyses the Cornell world-renowned forest soils specialist suggested, save one. I would not collect data for another year. Eventually Dr. Stone arrived at the same conclusions as I, publishing his work after mine appeared in print. Unfortunately for me, it's always the papers by the esteemed Dr. Stone that get cited in the literature on potassium-deficiency symptoms.

When the committee was satisfied that I was ready to defend the two-year-long research project, the date was set. Dr. John Decker, a brilliant plant physiologist, was on the examining committee. The night before the exam someone called Dr. Decker, telling him I was in the lab going over his text. He came to the attic, peered over my shoulder, took the book in his hands and closed it. "Why are you doing this," he asked. My answer, "Because you blackballed 'Joe Jones' ("Jones" was the only one to precede me to get this far), and I want to be prepared." Prof. Decker: "I voted 'no' because Jones wouldn't admit he could be wrong. What kind of a scientist is it who is that arrogant about his conclusions. I've watched you. You've made a lot of mistakes. And you were always willing to admit error."

The author being congratulated by State University president William Carlson upon receiving the PhD in 1953. Forestry Dean Hardy Shirley and Executive dean Reuben Frodin look on. As the sole doctorate that year, the alumni association was obliged to use this photo on its magazine cover.

Remember the two-year target date! I met it with a month to spare.

With the committee satisfied and the date for completing my teaching tasks still a month away, I headed straight for the University's infirmary. As we had no telephone in the barracks on the glacier-dumped drumlin called Slocum Heights, even telling Anne of the success of the defense must wait. After a rather detailed physical examination, the physician said, "Did you really expect to earn a doctorate here without getting an ulcer? It's this way every graduation."

The Southern Trip

Preceding the fall term each year, the College of Forestry's silviculture department encouraged students to enroll in an elective "southern trip." It fell Don White's lot to lead the tour. He drove a school car. It fell my lot to corral the students. I drove the truck.

The stake body rig behaved like a World War I surplus. Dr. White complained without success to the administration about the condition of the 1-1/2-ton, 6x6 before we set out. Wooden benches lined each side and a canvas tarp protected the students from inclement weather.

The tour led through the Southern Appalachians *via* Skyline Drive in the Shenandoah National park in Virginia. The rain began; the tarp went up. About sundown we arrived at the campground alongside the winding scenic road. When Don Viele, who helped with the driving, and I untied the tarp and lifted out the tailgates, one-by-one the men fell from the truck and into our arms. They were being asphyxiated by carbon monoxide. Would they have survived another ten minutes? We don't know. We found the problem, a leak in the pipe leading to the muffler, and had the offending part replaced the next day.

Almost every day on the three-week trip, the junker was in the shop for repairs. On one occasion, in Arkansas, we had to rouse a shade-tree mechanic from his bed, then we sat around all night while he located and replaced a timing chain. The worst, however, occurred on the two-lane 23-mile causeway that crosses Lake Ponchartrain in Louisiana. The truck caught fire. We couldn't stop on the long bridge. Fortunately the rig carried two fire extinguishers. Viele passed one out the window to the crew as we drove on while, with the other, Don doused the burning oak floorboards in the cab. The wooden floor of the cargo bed and the stakes were only slightly charred. Dr. White had the floors patched in New Orleans, where we spent the night in a stadium and where we provided tasty blood for mosquitoes the size of the Pelican State's prized bird.

The stake sides of the truck were cabled together, lest the weight of so many heavy backs leaning against the slats would break the wooden pins that held them to the truck body. It happened anyway, on St. Charles Avenue in New Orleans. No one fell out, but without the rope some would have.

We spent a part of each day visiting a forest. Among the stops was the Crossett Experimental Forest where we met Russ Reynolds and Sam Guttenberg who showed us work with loblolly and shortleaf pines. Sid McNight gave us an in-depth tour of bottomland hardwoods and the Mississippi bluff hills. Other state, federal, and industry foresters guided us through their operations. At the conclusion of this trip, all 18 students were well aware that the opportunities to grow trees and to make a contribution to society lay in the South.

Did I say 18 students? Yes, I did. For three weeks I'd been counting to 18 several times a day to make sure all were aboard. Yet, it didn't register when, at a campground in northeastern Ohio, on our last night together, one asked if I'd ever played "18 pile up." I'd heard of 52 pickup, but not 18 pile up. "Yes, I'm willing to learn." That quickly I was under a stack of 18 stout fellows.

These seniors were anxious to reach Syracuse. Some had brides there, as I did. They prodded the driver with chant and song in the predawn hours. (Dr. White always bunked in a hotel or motel; he'd catch up with the truck.) Now through this town and that we raced, but not through Jamestown, N.Y. The city policeman had a list of misdemeanors that nearly filled his pad. All of us had to wait for the crippled city judge to be aroused from his bed to hear the case. He was grumpy. While I was pleading my defense, the students were changing green money to pennies. When they heard the judge say "$10," they rushed in and dumped a thousand pennies on his bench. Not 'til later did Prof. White find us.

A Second Heiberg

Prof. Svend Heiberg, the tall mustachioed dapper Dane, had invited "my cousin, H.H.H. Heiberg, chief forester of Finland" to lecture to forestry groups in the Northeast. The

Syracuse professor had never met the visiting Scandinavian who, upon arrival and thereafter, always was attired in rumpled tweeds that bore the marks of wear. His white shirt seemed never to have been subjected to a thimble full of clorox. The two Heibergs were a study in contrasts: the tall slender immaculate Dane and the very short shabbily dressed Finn (who, by the way, brought along his young wife, the stereotype of the beautiful statuesque well-garbed Swede). Of course Svend was obligated to introduce the traveling orator, now called a "friend." On one occasion he went through the usual introductory academic listing of the speaker's pedigree, and then with a burst of enthusiasm announced in a strong Danish accent, "H.H.H. Heiberg, NO RELATION."

Back to Pack Forest

I was the Great Dane's assistant at the Pack Forest senior internship required of all forest management majors. Somehow I got locked into the assignment, preferring of course to be with Anne and the kids in the Lone Star State. Prof. Heiberg was the director of the program in the woods south of Lake George, other faculty coming and going to participate in the concentrated refresher course. No Syracuse forester would ever embarrass the institution for lack of knowledge and field experience: "Silviculture I will master, I can grow trees even faster,..."

Mr. Heiberg embarrassed me by constantly referring to me as "doctor." I think he was proud of his product. Later, he, with a Master's diploma, was appointed dean of the graduate school.

The professor rose every morning at about 5 o'clock to light the fire and to turn on the phonograph in the faculty cabin. The music, always the same, was a half-dozen long-playing platters of popular Scandinavian-style songs sung in English. Over and over we heard, "Wonderful, Wonderful Copenhagen."

I erred grievously. I figured it was my turn to do these predawn chores. The professor was in a bad mood all day. Another staff member explained that this ritual of fire and song by the director, established years before, was not to be violated.

The director had a garage for his big Roadmaster Buick.

He drove like he lived, always in a hurry. Once he failed to stop before knocking out the wall of the garage. The heavy tank bore nary a scratch. A work crew replaced the boards while the students self-dispatched to appropriate a highway stop sign to post on the back wall of the garage.

On another occasion, the seniors jacked up the Buick's rear wheels an indiscernible fraction of an inch. Behind the steering wheel, Professor H zoom-zoomed as was his usual way of gunning and taking off. Three times he got out of the car to find the problem. Pranksters gathered to observe at a safe distance.

Svend Heiberg had established experimental plots throughout the forest in stands of white pine, planted Scotch pine, and other conifers. One series of sample plots, for which he personally supervised the cutting of every tree, involved a shelterwood harvest of *P. strobus*. Especially on west- and southwest-facing slopes he showed that this regeneration system works, younger trees with thin bark otherwise suffering from heat injury. Blisters appear on the bark of the main stems when exposed to the sun's rays.

IVCF

The Inter-Varsity Christian Fellowship at Syracuse University couldn't meet in Hendricks Chapel without a faculty advisor. Yes, I'd serve in order to provide the opportunity for the evangelical group to gather for prayer and praise. Some fifty or so assembled weekly. It was an opportunity for me to meet and even now to correspond with these articulate students from every school and college on campus pursuing careers in a variety of disciplines.

Crossing the Stage

Time came to parade across the stage with the graduates of Syracuse University and its stepchild, the College of Forestry (later to be named the College of Environmental Science and Forestry, a unit of the State University of New York, and with a president at the helm) to receive the coveted degree. Dean Shirley was now in charge of the whole college, replacing the retired Joseph Illick, and no longer the graduate school dean. To Hardy Shirley, who I respected so much, and who turned over his teaching load to me in my last semester, I made my plea: I wished to be excused from graduation exercises. (I lost my junior high school diploma on the way to Chris' house, remember; I avoided the high school ceremony for other reasons-they didn't wear robes in those days; I ducked out of Penn State's processional to attend Anne's at Wheaton College; and got excused from Yale's ritual in order to get back on the Forest Service payroll at San Augustine.) Dean Shirley's response to my entreaty: "Laurence, if you are not here for graduation, whose picture will we put on the cover of the alumni magazine? You are our only PhD." So I put Anne and the children on the train for her Texas home and used the month prior to commencement teaching the course in silviculture for seniors at Pack Forest. We finished the race set before us, owing no one. And my likeness was on the cover of the magazine. I attended graduation in June 1953.

The College's Alumni Association honored me with its Distinguished Achievement Award on the 40th anniversary of receiving the PhD. I appreciate the State University's motto, "Let each become all he is capable of being."

And now back to the South.

11

RESEARCH ON THE ESCAMBIA

Natural stand of second-growth longleaf pine established as a result of fencing against cattle-grazing and hog-rooting and prescribe-burned for brown-spot needle disease control.

"**Brewton** is really a nice town. Don't let the main street fool you." That was my introduction to the South Alabama mill town in a letter from my new boss, Walt Hopkins. I visualized arriving on a Saturday afternoon, the downtown crowded with men in bib overalls and women in flower sack dresses, both with tobacco slurry drooling down their chests. I arrived on a Saturday afternoon in June 1953. People in bib overalls and flower sack dresses lined the storefronts. The chew or the dip was evident.

Brewton really is a nice town. We settled into a newly converted barn, every room richly paneled in figured heartwood of redgum trees harvested from nearby bottomlands. Walt, the research center leader, officed in Marianna, Florida, left day-to-day management of the Escambia Experimental Forest to Thomas C. Croker.

No one could have had a better supervisor than Tom, the son of an old-time Presbyterian cleric. By old-time, I mean he didn't believe in insurance: The Lord will provide. Tom's novel, *Land of the Free* (Vantage, 1988), is, I think, his autobiography. The book is a fine place to learn about longleaf pine (*Pinus palustris*) silviculture, for T.C. Croker, even in retirement, remains *the* authority. 'T'was my good fortune to be his subordinate.

Tom turned over to me studies that he initiated. When the data were analyzed and papers ready for publication in the *Journal of Forestry* and the Southern Forest Experiment Station's *Southern Forestry Notes*, he insisted on my being the sole author. One of those trials showed how walls of timber surrounding clearcut openings and seed trees impede the growth of seedlings for a distance of at least 55 feet. Another, published in the *Journal*, shows the need to control scrub oaks (*Quercus incana*, *Q. laevis*, *Q. marilandica*, and *Q. stellata*) before grass-stage seedlings reach 4 years of age if you want them to make height growth. That account was popularized for the *Southern*

Lumberman. A "potboiler" involving decomposition of sawdust received attention too. Unable to acquire thermocouples that could withstand the heat of bacteria-inoculated sawdust, the project was abandoned.

At the Loxley (Baldwin County) thinning plots, released by the Tennessee Coal, Iron, and Railroad Company for our study, I learned how severe a burn one may get handling creosote. Working alone, using an angle prism to line up plot corner posts, face and hands suffered. FS safety instructions failed to mention this. These natural longleaf pine plots, established in 1934 and thinned at that time to 200, 300, and 400 trees per acre (the check plot had 1724 stems per acre), showed no apparent response to the treatments in 1954.

Walt's instructions were not to tell anyone that I had a doctorate. “They'll think you won't know anything," he said. It was the local pharmacist who learned of the degree because I had to utilize his instruments and chemicals for some simple experiments. Writing "Dr." on a prescription blew my cover, the delivery boy thoroughly confused when he dropped off the medicine for a doctor living in a converted barn.

From then on, I was on everyone's free-speech list. (Yes, I believe in free speech.) Even the Baptists welcomed me. So too, the Presbyterian Church in the lumbering town of Jackson, Alabama. We drove the 150 miles each Sunday to serve as a lay worker behind the sacred desk in the beautifully ornate wood structure.

The Sheriff and the Hogs

T.R. Miller Mill Company in Brewton and the Alger-Sullivan Lumber Company in Century, Florida, and five other smaller companies paid my salary. They wanted some longleaf pine regeneration answers which the existing staff, consisting of Tom and two subprofessionals, Virgil Davis and Al Thomas, couldn't begin to provide. The industries contributed our office supplies, labor, and other necessities, the Station budget amounting to only $15,000 for the fiscal year. The Miller mill folks made sure we had a hog-proof fence around 3,000 acres so that we could rely on the data collected. One pineywoods rooter, dear reader, can pull from the ground as many as 800 *P. palustris* grass-stage seedlings in a 10-hour day. They thrive on the starch in the long, carrot-like roots. The T.R. Miller Company, concerned over the failed natural regeneration of this prized pine species, leased the acreage to the Forest Service for 99 years for research. Techniques for utilizing a modified shelterwood system resulted from this endeavor, enabling this "finest forest" to reclaim some of its original range.

Thomas C. Croker in a natural stand of longleaf pine seedlings ready for a prescribed fire. Needles will be consumed, but the unscorched bud will burst forth with new, vigorous needles.

Before my arrival, Tom initiated a trapping program within the fenced tract that required a visit to the Escambia County Courthouse to identify the hog owners, according to the ear nicks that serve as brands. The shape and numbers of these notches are dutifully recorded in the clerk's office.

Kindly delivering the razorbacks and their kin to rightful owners, Tom soon found himself recapturing the same animals. He made another visit to the Courthouse, this time calling on the sheriff, and at the same time made a deal with the weekly newspaper editor. If the sheriff, "Doc" Fountain, would deputize the biggest hog thief in Escambia County to round up hogs on the Experimental Forest, the editor would publish this as a front-page story under a headline in second-coming type, the big block letters saved for the return of the Lord and other noteworthy events. With the word out, the farmers got their pigs, and research could again commence. Croker's official account, published by the Regional Forester in 1987, notes that the deputy, Albert Harris, "built a dozen wooden traps..... Owners, alarmed by the loss of their hogs that always 'died' in the pound, hastily built fences and penned their animals. A few, angered by the trapping program, stole the traps and set the woods on fire."

Public Relations Pays

Walter Hopkins, as research center leader, took me with him on occasion to raise money for research. He was an expert promoter and insisted on our using nicknames on publications. Tom for T.C. Croker, Larry for me, and Walt for himself. A more personal approach, he thought, would facilitate fund-raising. As readers will know, I resisted.

PR's importance was also demonstrated annually as the public gathered to see the logs harvested from the "Farm 40." The locals heard inspiring talks on forest management. That some heeded what they heard is evidenced by the abundant supply of wood today in South Alabama.

Departing Longleaf Country

Seven of us PhDs in the Southern and the Southeastern forest experiment stations were stuck at GS-7 grades. That was my salary as assistant ranger on the Sabine. So I had some troubling emotional pains when I was asked to interview for a research job at the University of Georgia. The Forest Service had been good to me; twice the organization allowed me leave without pay with the guarantee of a job when I returned from school. It is not my nature to ask favors; I didn't ask for a promotion. I simply took the Georgia job, as it would pay a third more than I was making. My letter of resignation went through channels to the Station director. Phil Briegleb, a gentleman's gentleman, quickly responded, "Why was I leaving?" My answer, "the kids are hungry," or words to that effect. After we settled in at Athens, Director Briegleb sent an assistant director, Lew Grosenbaugh, to encourage me to return, offering a GS-9 rating. I would have stayed on the Escambia for that. Squeaky wheels do get the grease. I think all but one of the GS-7 PhDs left within a few months. The forestry schools were expanding, and needed instructors with doctorates. Two years later Tom Croker was dispatched to lure me back. He found me, making use of accumulated leave, at the Forest Service's Coweta Hydrological Laboratory in North Carolina. The trip was in vain; I'd found a home, as we'll note in the next chapter. The near-decade in Athens would be a challenge.

12

GEORGIA CHALLENGE

The nation's oldest plantation, apart from a small stand established by Russians in Alaska, along Tobacco Road in South Carolina. The account of these trees, lifted from nearby woods prior to the Civil War, was published in the *Journal of Forestry*.

High o'er the hills of Habersham
Veiling the valleys of Hall,
The hickory told me manifold
Fair tales of shade, the poplar tall
Wrought me her shadowy self to hold,
The chestnut, the oak, the walnut, the pine,
Overleaning, with flickering meaning and sign,
Said, *Pass not, so cold, these manifold*
Deep shades of the hills of Habersham,
These glades in the valleys of Hall.

How better to introduce one to Georgia than through its poet laureate of the Civil War period. The hunt for ten-tine buck in "The Revenge of Hamish" reads as a Scot ballad, and with reason. Sidney Lanier, for whom the north Georgia lake is named, was a child of the mountains settled by the Scotch-Irish peasantry looking for a home. A patient reader of "The Marshes of Glen" hears a lyrical description of the wetlands of the coastal counties down by the sea. Every Georgia schoolboy learns the "Song of the Chattahoochee," the third verse of which appears above. Should not a young forestry professor also grasp, through these toneless melodies, the glory of the Peach State?

Anne continued to recover at her parents' home in Texas following the miscarriage of twins while I, with her consent, packed for the move to Athens. The van's (pre-U-Haul days!) delay gave me time to rent a house on a little residential street, Sylvia Circle. I wrote Anne daily and called Tyler frequently. She wrote me often, but I never got her letters. Finally, the University's postman happened to mention another older faculty member with my name. He lived two houses away on Sylvia Circle. Receiving Anne's messages, *that* Dr. Laurence Walker panicked and promptly burned them.

William Andrew Campbell, the forester and philosopher who played a major role in my joining and leaving the Georgia faculty. "Andy," a USFS pathologist, received honors for his work with littleleaf disease, attributed in part to the *Phytophthora* fungus, that attacks shortleaf pine (Source unknown).

Abundant Resources

Donald Weddell was dean of the George Foster Peabody School of Forestry founded by the philanthropist several decades past. (The School now bears another's name; universities "auction" these monuments to highest bidders.) The dean, in poor health, seemed to appreciate the leadership of the research center leader of the Southeastern Forest Experiment Station located in the forestry building. William A. Campbell located me as a potential candidate for the new research position created by the lobbying of school graduates who belonged to Governor Herman Talmadge's fraternity. The legislature only infrequently passed a budget. Continuing resolutions kept the money flowing, and in those growth years, any receipts to the state's treasury that exceeded budget requirements went into the governor's discretionary fund. The school even had money to buy trivia, like curtains for its auditorium. And I, with Mervin Reines, a plant geneticist, had new pickups, money to establish and equip laboratories, and funds for graduate fellowships.

"Andy" Campbell, the dynamo who recruited me on behalf of Dean Weddell, rates these lines for being the most energetic man I'd ever known. Ideas flowed from his mind like flood waters over a dike. The Forest Service researchers, the school, the Georgia Forestry Commission, the Southeastern Section of the Society of American Foresters, and the Georgia Forest Research Council all benefited from the proposals he pushed. A leaf from his book explains, I think, my managerial methods as dean at Stephen F. Austin, though not nearly so effective.

The Georgia Forest Research Council, an independent unit of state government promoted by Andy and others, received largess from Talmadge's kitty. These moneys were distributed to the organizations named above for research. I got my share. The Council, with one professional employee, Ed Ruark, promptly published research findings. Landowners didn't need to wait for referees to approve journal articles to tell them how to manage their timberlands. The Council continued to function until Ed's retirement and then was absorbed by the Georgia Forestry Commission. In my day, the Commission highly admired spark innovation. Its director, Guyton DeLoach, like Ruark, was an old buddy of Talmadge. They reported to the governor, although working through gubernatorially appointed boards.

Talmadge's father, Gene, was not so enthused with foresters and forestry. His appointment of a state forester who was a logger and who lacked professional credentials, as required by law, was challenged. The senior Talmadge's answer to the complaint was, "He's a woods-ranger, right?" Weddell replaced the woods-ranger, shortly thereafter moving to Athens to head the oldest forestry school in the South.

Prescribed fire continued to provoke discussion and rancor. Ernst Brender and I had successfully worked out the prescription for regenerating shortleaf pine (*Pinus enchinata*) and loblolly pine (*P. taeda*) stands on the Hitchiti Experimental Forest located along the junction of the Piedmont province and the Coastal Plain in central

Georgia. Convinced that the procedure improved regeneration, diminished damage from wildfire, and enhanced wildlife habitat, the governor (now Herman) directed the Commission to use his own forests near Lovejoy to show the world: if *he's* willing to chance a holocaust, anyone else would do so, too. That's an example of the public following the two Talmadges had from Brasstown Bald to Savannah's marshes. They could put people in their pockets. It was a political cult. The whole fleet of GFC trucks and a host of foresters were on hand for the Talmadge burn. After the fire, the stands looked clean and green.

The Talmadge machine worked so effectively that once, when called upon to speak at a Lions Club meeting at Royston, I mentioned the Senator's interest in forestry. Three days later my mail contained a letter from him: "I am deeply grateful for what you had to say in my behalf."

Coshocton wheel at the base of simulated burned-over watersheds measured soil and water movement from the site.

Multiple Opportunities

Hired as a researcher, I equally enjoyed other tasks. When sickness struck someone of the small (eight of us) faculty, I'd happily respond to the call. Silviculture, fire control, public relations, and a soils seminar in the Agronomy Department come to mind. When a guided tour was needed, I was ready. On one such occasion I escorted in

Ernst Brender, USFS researcher, worked with the author in determining a prescription for a Piedmont province prescribed burn. Note the portable instrument shelter.

a roundabout way a bus load of forestry deans from the Atlanta airport to the campus. Geology, soils, vegetation, climate, and the kudzu explosion kept the conversation going. The handout for the visitors in 1958 follows:

ATLANTA TO ATHENS
A Tour Guide (sort of) for Forestry School Executives

The road from Atlanta to Athens passes through typical mid-Piedmont countryside of Georgia. This province, the name translated "Foot of the Mountains," joins the Southern Appalachians to the north and west, while the Coastal Plain lies southeasterly. The Coastal Plain boundary, marked by the Fall Line, is readily recognized by an axis joining many major cities of the East: Trenton, Philadelphia, Wilmington, Baltimore, Washington, Richmond, Raleigh, Columbia, Macon, and Columbus, Georgia. Here the falls (1) prevented boat travel further inland, (2) provided clean water for industry, and (3) were a source of power. The rolling peneplain—not truly a plateau, though often referred to as such—has a less well-defined junction with the Highlands to the north. Piedmont elevations range from about 500 to 1800 feet: Atlanta is 1,050 feet and Athens is 700 feet.

Igneous and metamorphic rocks of the Piedmont are among the oldest known geologic formations. Soils derived from the acidic granites and gneisses are usually relatively course-textured, predominantly sandy loams. However, virtually *all* of the Piedmont has suffered from severe erosion—leaving the elevation about a foot less than it was a century and a half ago—and, thus, heavy subsoils of fine texture are exposed over wide expanses. Waters in creeks and rivers are relatively clear in contrast to a generation ago.

On much of the route traveled, one sees the transition from row cropping to forest lands. The youthfulness of the stands along the expressway—cut cross-country through field and forest about 10 years ago—attests to this.

Hardwood invasion is pronounced in older pine stands (usually loblolly and shortleaf mixtures here). These encroachers are usually inferior species, although white and red oaks also enter. Oak-hickory types form the ecologic climax for most of the Piedmont coniferous forests. Indeed, Bartram reported in his *Travels* of 1775 seeing, between Athens and Augusta, black oaks 12 feet in diameter 5 feet above the ground. Site index for these lands is now about 70 for loblolly pine and 60 for shortleaf pine. Ninety-five and 80 are maxima, respectively.

Lawrenceville, the first town after leaving the expressway, is a typical Georgia textile town. Winder boasts two furniture factories. At the eastern edge of Winder is the village of Russell, where bachelor Senator Richard E. Russell's family home may be seen to the right behind a screen of large hardwoods. Lesser towns along the way struggle for existence as agricultural acreage diminishes and the labor force moves or commutes to the cities.

And then Athens, "City of Learning" and named for its old-world counterpart, was founded in 1801 when the University moved to its modern log structure in the midst of Indian territory. In that year, "Old College," a brick structure which still stands, was also built. Indian hostility toward the intellectual whites was so great that for years the sheriff provided protection for the graduation exercises. Even today, in pioneer regalia and saber in hand, the sheriff precedes the Chancellor of the University System in the academic procession.

Syracuse Seniors

Each year, I met the school bus load of Syracuse seniors, leading them from the Appalachian Highlands, through the Piedmont, and south to the Coastal Plain. These were presegregation times and tensions between the races ran high. A black student insisted on making the journey in spite of an admonition by The New York college's Dean Hardy Shirley to refrain from the journey. Ernst Brender was showing the students experimental plots on the Hitchiti when a cloud dropped its stored-up supply of gully-washing water. Prepared for such an event, Ernst, a German immigrant, took us to a one-room clapboard, bare-walled community house, once a rural school for white kids. Lights out, a slide program commenced with Ernst lecturing. His crew chief, a local subprofessional named Goolsby, sat next to me in the back of the room. Suddenly there was a pounding on the door; it was opened slightly. The intruder, seeing nothing in the dark, retreated. Goolsby, recognizing the visitor, went outside to chat, and later returned to his seat. I asked, "What was that all about?" The response, "They wanted to know if they is any n....s in here.' "And what did you tell them?" "They ain't none!" And then loudly, "Hey, they ain't, is ther?" Silently I edged my way to where the lad was sitting, told him I'd give him a signal before the lights came on, and he was to huddle with me in my raincoat as we made a dash for the bus. Safely inside, he hid himself low in the seat; we got out of there.

Natural Areas Study

The Southeastern Section of the SAF wanted a study of

all the natural forests remaining in the three states covered by the Section: Alabama, Florida, and Georgia. I located some 19 on which I tallied the species, mensurational data, soil, site condition, and human encroachment history. The *Journal of Forestry* published the article in 1963. To see those woodlands now, to know if they have survived, or, if not, how they met their fate would make a fascinating venture for a young ecologist. Almost all those stands dated to the 1830s. Discovered records, though few, suggested a relationship to the mass migration of Indians to Oklahoma Territory during the heart-rending Trail of Tears. Could the militia have torched the woods to route the natives? Or could the Indians have ignited the forest to confound the army? The coniferous forests all came into being following the clearing of the site that exposed the mineral soil. Fire would have been the cause; but why else should the woods all burn simultaneously?

The article in the *Journal* and some local publicity through the SAF's Section *Newsletter* led to an invitation to look at a prospective candidate for inclusion in the list of natural areas. (Natural areas need not be old-growth; the ones I studied were.) I was to be the guest of a Boston banker and, while visiting his vast game preserve in longleaf pine country, to go duck hunting. I was unprepared. Jim Jenkins loaned a 20-gauge shotgun. At the small-town Thomasville airport I was met by a uniformed chauffeur from the Jeptha Wade Plantation and greeted thusly: "Dr. Walker, I presume; Mr. C. is waiting at the lodge." Arriving at the lodge of the villa, some 30 miles distant, a fellow dressed like an Austrian hunt master promptly took the shotgun from the station wagon to lock it up in the gun room while a butler escorted me to Room 10 of the guest wing and announced the hour for dinner. That meal, and all the others, consisted of wild game, some not harvested locally. Lawyer clients had flown in for the special occasion. A maid turned down the bed and drew the water for the morning bath, the first time that had happened since Momma did it when I was three.

Guns were cleaned and oiled, and ammunition provided by the host only after being posted to your blind. I brought down a few birds, as did these moneyed market men. I froze in the blind. Ferried by boat before daylight, the guide promised to return by noon. I fared better in the afternoon quail hunt, sitting high in elevated benches in a rubber-tired horse-drawn wagon. Expertly trained dogs, caged until the moment needed for the point, promptly retrieved the quarry, and then were returned by the nimrods' vassals to their kennels. Prescribed fire, food plots, and appropriate thinning prepared the forest for quality hunting that included turkeys.

As for the longleaf pine stand, it didn't qualify as a natural area. A few residual high stumps, infiltrated with protective resin, and some crude gutters indicated earlier use for gum naval stores production.

Other Old Forests

Because of my interest in the Marshall Forest, near Rome, Georgia, the Nature Conservancy requested my help in planning for the "passive, non-consumptive recreational use and research" for the 200-acre preserve, a gift of Maclean Marshall near the Shorter College campus. Dr. Lewis Lipps, of the College and earlier a student of the University of Tennessee's Appalachians ecologist Royal Shanks, chaired the committee assigned the task. Lewis, visiting in Athens, liked our kids. Steve treasures a beautiful chess set she gave him during a trip to the University.

Lewis later became a Distinguished Professor, winning for the Marshall Forest in 1986 the Nature Conservancy's Stewardship Award. The tract, bordering on the City of Rome, provides habitat for a rare plant of the genus *Scutellaria*. To protect it, the committee nixed a Floyd County road through the woodlands. A "salvage" garden of northwest Georgia plants covers the ground.

A face-off with an Atlanta college professor came about over the virginity of a stand of trees along Cooper Creek in the Chattahoochee National Forest. The biologist, noting trees marked for harvest by the Forest Service, brought a newspaper reporter to photograph the stand of beautiful big white pines (*Pinus strobus*) and eastern hemlocks (*Tsuga canadensis*). I was asked to review the situation. Detailed in *Forests* (John Wiley & Sons, 1990), we'll note here that the white pines seeded in on soil newly abandoned from cropping, due to the agricultural depression of 1895, and the more shade-tolerant hemlock stems seeded in under the pines. The plow sole and the overburden of rich silt that washed from high elevations into the cove were readily identified. One-inch annual radial growth occurred on the boles of both species. I took both sides: the stand should be preserved for its beauty and further study. It was not a virgin forest, nor a remnant.

I often visited Fernbank Forest in the city of Atlanta. It is a near-virgin Piedmont stand. By now intrusion by man probably has interrupted natural succession.

The Okefenokee Swamp fire of 1955 was reason to visit the area cutover in the late 1800s and later designated a federal wildlife refuge and a state park. The fire, burning for weeks or longer on an island, had been carefully watched. The swamps of organic soil dried suddenly, like a sponge. A change in wind direction sent burning embers across U.S. 1 (before Interstate Highways) to encompass a vast area and to encourage the Congressionally approved Interstate Fire Compact. Now personnel and equipment, unlike highway patrolmen, move freely across state lines to control wildfires.

Soon after the fire, a pretty and petite congresswoman, Iris Blitch, introduced legislation to provide funds for building a road around the swamp. Swayed by her femininity, the old men approved the money. It was spent, but the road was never completed. The best of engineers can't build roads on peat and muck, although one can yet see the elevated tram log roads laid out in the vast alligator-infested wilderness.

My visit to Ossabaw Island, accessible only by boat, involved the mining of ore. There I was spooked by the largest rattlesnake I ever expect to see. The lanky fellow was sunning across an 8-foot wide graded lane. Neither head nor tail were visible; they were hidden in the "rough" on opposite sides of the road. Its girth was that of a man's thigh. I almost stepped on it, while examining the crowns of the giant loblollies and live oaks under which lay the low-grade bauxite.

A small grant from the Highlands Biological station in the mountains of North Carolina provided a nice family vacation in the scenic Southern Appalachians. The station, on Duke Power Company lands, needed a survey of humus types in these forested watersheds. The study appeared in the *Elijah Mitchell Society Bulletin* of the North Carolina Academy of Sciences.

Hard Labor Creek State Park was available for prescribed fire trials; for in Georgia the state parks' people wisely let the Forestry Commision manage their timber. We had good success regenerating loblolly and shortleaf pines where brush had captured the site.

At nearby Whitehall School Forest we tried "shock polarization," a sort of artificial lightning strike to kill weed trees. The wheel-mounted power plant was too unwieldy.

Perhaps the 1959 publication entitled *Forest Soils and Silviculture in Georgia*, coauthored with Jack Perkins, was my most notable effort in the Peach State. Overlays of the state map showed the geology, soils, and forest cover types. It was well received from Oregon to the South. Al Folweiler, director of the Texas Forest Service, put it "in the hands of each of our district foresters."

Student Relations

While I only occasionally taught undergraduate courses, graduate students kept me busy and happy. We spent a lot of time in the woods together, imitating the old adage about good teachers being those who sit on one end of a log while their students sit on the other. I think the saying had something to do with the "log colleges" that sprouted west of the Appalachians as the nation grew rapidly toward the Pacific shore.

A personal encounter with a graduate student while sitting on a log may have saved the young man's career. His peers disliked him. He said he didn't care, that he could buy them all out. He could indeed. A long discourse followed: "If a man would have friends, he must show himself friendly," a Solomon proverb. What an attitude change! The student reached the highest rung of a career ladder; writing later to say he "enjoyed the talks we had, not just about forestry but about life and people."

On one jaunt along the Savannah River, northwest of Augusta, the students discovered a partially submerged petrified palm log. Did *Palmaceae* once grow here? We hauled the 12-inch diameter, 6-foot-long silicaceous "timber" out of the swamp, got a tombstone cutter in Elberton to saw a one-inch cross-section for us on the condition that he could have one, too. A single cut dulled the black diamond teeth of the saw so badly he decided against securing a sample for himself. The rest of the log is in the Children's Museum in Savannah.

On Warp and Woof

Georgia's forestry faculty was torn asunder. The ill dean was unable or unwilling to give direction. With enrollment growth and new faculty added, turf became sacred. For example, G. Norman Bishop taught at least five courses. He didn't want to give up dendrology because he wrote the book on Georgia trees. Nor surrender to another faculty member the course in public relations because he had been assistant state forester for public relations in South Carolina. Nor let someone else teach fire because he originated certain control techniques. A good man, an enthusiastic instructor, I think he worked himself to death.

Archie Patterson, management professor, was largely responsible for the SAF's original Code of Ethics, but we couldn't make the code work among ourselves. (Archie was repeatedly elected a county commissioner.)

The death of Donald Weddell brought the faculty together, but only for a brief respite, at the dean's home to console the widow and to help out as needed.

Bishop Grant, a lovable senior (in age) wood utilization instructor, was appointed acting dean until Al Herrick arrived from Purdue University as the new dean. Bish promoted me to associate professor.

By his own admission, Dean Herrick had little interest in forestry and probably knew little silviculture. He went to the Syracuse College of Forestry because the city was his home and tuition at the state school minimal. He talked unceasingly about management by objectives and other IBM-style business nomenclature with little meaning to me.

Al Herrick tried to bring the faculty together. If he didn't plagiarize from a personnel management manual the "Warp and Woof" lecture that he delivered to a sober-

Forestry dean, A.M. Herrick, on the dedication of the G. Norman Bishop (with boutonniere) Arboretum. John Cooper (left), one-time supervisor of the Texas National Forests and UGA Alumni Association president; O.C. Aderhold, (center) University president; and Professor Archie Patterson (right) look on.

minded faculty, the treatise should be included in such a text. It was a classic. The collective faculty was the whole cloth, each member being either a thread that ran horizontally (the warp) or a thread that ran vertically (the woof) to hold together the cords of warp. Some left the seminar room quietly despondent; others thought it a great pep talk describing the strengths of each faculty member that, when woven together, made strong cloth for the sails of the academic ship.

After eight years at UGA, I was offered a full expense-paid year as the first Bullard Fellow at Harvard Forest, the esteemed University's research station in Massachusetts'

Plateau Region. Permission denied. Had I known that within a year we'd be on our way elsewhere, I would have resigned in order to continue tree nutrition studies at Petersham.

When the Texas offer came, Dean Herrick said I was his highest paid associate professor, but should not expect to go further. He said he was saving full professorships for "hard to get people." In those nine years, over 20 papers appeared in refereed journals, often by-lined with the names of graduate students and other associates.

Faculty Characters

The faculty consisted of a bunch of characters. Reid Parker and Bill Moss, the latter earning his master's degree working with me, enjoyed playing tricks on each other and on others. Reid, a top-drawer teacher, couldn't get enthused about a PhD. Bish Grant, his mentor, had me clandestinely try to encourage him through the ordeal. Never mind, Reid had a timberland services company on the side that others envied. And in time he replaced his mentor on the University's athletic committee, eventually becoming athletic director with Coach Vince Dooley. Bill Moss went on for a PhD in psychology. He's remembered for slipping the unwrapped Baby Ruth candy bar into the punch bowl.

Prof. J. Reid Parker, who became the University's Athletic Director.

Bill Moss, graduate student and later a PhD in psychology.

Bill taught psychology on closed-circuit TV. Reid snitched President Aderhold's stationery, addressing a letter to Bill objecting to something the psych instructor said. Moss self-dispatched to visit his department head. Trouble was narrowly averted; the boss had a sense of humor, too.

Jim Jenkins, pharmaceutical chemist-turned-wildlife biologist

Prof. Mervin Reines, tree geneticist.

Jim Jenkins, educated in pharmaceutical chemistry, preferred life in the Okefenokee Swamp. A doctorate in biology won him the task to organize a wildlife program within the School of Forestry. He did more. The much-used drug-loaded dart gun for immobilizing wild animals for study or transplanting came from his fertile mind. Veterinary medicine and pharmacy school scientists also contributed to the project. Jim showed a photo of himself sitting on a log in Africa, a huge tranquilized adult female gorilla (the first ever captured by this means) hugging him. Then came Snake Stop, a powdered substance that, when spread across the threshold of a tent, repelled creeping reptiles. To demonstrate its effectiveness, he caged two students at one end of a 4x12-foot box. In the other end were one each of the South's four poisonous snakes. A stripe of Snake Stop across the middle of the cage kept the dare-devils and the reptiles safely separated. The snakes, when venturing from their "lair,' would approach the chemical and soon retreat. I recall the concoction was the odor of musk. Audubon folk forced it off the market, claiming injury to birds. Jim also developed Dog Stop, simply household ammonia carried in a spray can by postmen to discourage vicious dogs. The pets don't like it in their eyes.

Mervin Reines, the big but gentle tree geneticist, arrived in Athens the week I did. He, too, was funded by the Georgia Forest Research Council. An extremely patient experimenter, he finally got pine tree needles to take root in a nutrient gel in petri dishes. Thus, from one genetically superior tree, thousands of clones could be produced, each with the specific high-quality characteristics for which the parents were chosen.

The Trusting Turk

Among the graduate students was a post-doctorate on leave from a forestry school faculty in Turkey. Dr. Ibriham Atay always wanted to travel with me, and I enjoyed his company. On one such occasion, we were passing near Tallulah Gorge in North Georgia. I thought he should see this, for at the north end, northern forest species, such as white pine (*P. strobus*) intrude into the valley, while southern forest species, such as shortleaf pine, grow at the southern terminus. We began the descent down the canyon wall, discussing soil-vegetation-elevation relationships along the way. Suddenly Atay asked, "Are we coming back this way?" "Yes," I assured him. Another fifty-foot drop and he asked the same question. And he received the same answer. Then Ibriham said, "Give me the car keys and I'll wait for you here." To avoid embarrassment I hiked to the bottom alone, for I'd heard that, if Near Easterners don't trust you, they hang on to something precious to you. Like those keys.

At Christmas break I had an assignment for Union Bag Corporation in the flatwoods of South Georgia. My task on the cold, rainy, miserable day was to map the soil. The Turkish professor insisted on going along. At noon Ibriham was exhausted. So I said, "I'll go to the car for our lunches. You wait here and, no matter what, don't leave this hill." In that part of the world, a hill is a site a foot above the surrounding 10,000 acres. Unfolding the map, I laid it on the ground, placed the compass on the map, and took a reading. Rising, I closed the compass. Atay grabbed the map, folded it, and put it in his pocket. As we've noted, if one from that realm doesn't trust you, they keep something dear to you. And that map was soon to be precious. So as not to offend, I said nothing about the map, took a bearing, sighted the compass, and was off to get our lunches.

Two swamps waded, I came to the end of the tract. It was not the right boundary. I had gone off on the wrong end of the needle. (Never would I dare tell this to a class of students.) I aimed back to the "hill," called Atay but, due to the wind, couldn't hear his response, and proceeded on for the lunches. Offsetting ten chains, I headed back. There, late in the day, stood the shivering Turk. Running to me, he threw his arms around my shoulders, loudly exclaiming, "Now I know what they mean in America when they say, 'Glad to see you.'"

My fear just before reaching Atay: a headline in the next morning's *Atlanta Constitution*, "Forestry Professors Lost in Woods."

Western Wandering

Never budgeted to teach, I enjoyed the task. I had two sections of the same silviculture course. Having given the observation lecture you've previously read in these pages, I had the students quickly jot down within a certain time all they noticed. The next day the second section arrived at the same edge of a soggy swale. But now, 20 feet from dry ground was a post with a sign that could be read only from the water. No student volunteered. I had to wade out to see for myself or eat my words. The sign read, "Instructor's observation test." It was erected in the hours between the meetings of the two sections.

If I'm going to teach silviculture, I'd better learn the subject, so I set out for a six-week tour of the West in 1960. I met up with a second cousin, Fred Johnson, a 1909 graduate of Yale. An aunt "connected" us a few weeks prior to the trip. Together with a Mr. Phillips, then 84 years of age, we retraced the steps these two old-timers took in the early days of forestry in the Rampart Range of the Colorado Rockies. Fred, arriving in Colorado by stage coach, finished his career as assistant regional forester; Phillips, who joined the FS in 1898, retired in 1940 as supervisor of the Pike National Forest. We bathed in "Virgin's Bath," a natural pool cut into granite rock and discovered and named by Fred in 1915. We looked at plantations planted by Phillips' crews early in this century. And I learned of Bill Kreutzer, their friend to whom the badge for Ranger No. 1 was issued in 1898. Kreutzer's life is chronicled in a book by Len Shoemaker.

Stands thinned 51 years prior to our visit were readily identified by the dated galvanized nails driven into the tops of tree stumps. Before the days of marking axes, timber-markers drove the double-headed nails into the bases of trees to be harvested. When the trees were felled, the loggers removed the nails and hammered them all the way into the top of the stumps. These nails, still with little rust, were dated 1909 and 1910. Slash in this high, cold, dry climate lay on the ground as it had fallen from the woodmen's axes.

This was my first visit to the Fraser Experimental Forest on the Arapahoe National Forest where Fred was once supervisor. The lodge was being readied for President Eisenhower's visit. The cook said I was sitting in Ike's chair. It felt like any other to an exhausted low-altitude fellow catching a breath at 11,000 feet.

(Fred, I learned later, spent three months at Trinity, Texas, in 1911 with the Yale foresters. The school held summer camp studies on industry lands there.)

My notes say I went from Laramie, after studying a timber sale at 7500 feet, to Yellowstone National Park by bus, the only way to traverse the 400-mile distance. It was a 12-hour ride in a station wagon.

The bus driver, an effeminate lawyer from New York, and I rode the front seat; three old-maid school teachers from Houston, the center; luggage filled the back. When the pinnacles of the Grand Tetons hove into view, I mused aloud: "Grand Tetons, how'd they get that name?" Whereupon the big burly bus driver, with a booming bass voice, loudly responded, "That's French for big tits." Then the lawyer took his turn, "Perhaps we should refer to them as large bosoms," and the school teachers remained so mute I thought they had jumped from the wagon. How come I never heard that expression in France?

Copious notes and faded Kodachromes for experiential lectures at Georgia say that Montana's Yellowstone area of black obsidian soils lacks natural reproduction, even of lodgepole pine (*P. contorta);* that Douglas-fir (*Pseudotsuga menziesii*) in northern Montana might as well be clearcut as there is no response to release and no reproduction; and that the principal alternate host to *Cronartium comander* on lodgepole pine is bastard toad flax, a range herb. In the Bitterroot National Forest, foresters were already spending $25 an acre to remove inferior grand fir in sanitation cuts in order to produce 100 superior crop trees per acre, and Douglas-fir was pruned to 17 feet before reaching 9 inches in diameter. Acti-dione, a brewery by-product, seemed to dry up white pine blister rust. On Idaho's Coeur d'Alene National Forest, the four factors of site—edaphic, biotic, physiographic, and climatic—became obvious for the variety of conifers found there. One could yet see evidence of the Great Idaho Fire of 1890.

Dr. Chester Youngberg, forest soils professor at Oregon State University, met me at the Pringle Falls Experimental Forest and, as we traveled, lectured through the western edge of the Northern Rockies to LaGrand with its basaltic lava flows of pumice silt poured from volcanic craters of western Oregon's Cascade range. The infrequency of ponderosa pine (*P. ponderosa*) seed release, the 18-inch deep ruts of D-6 tractors logging pumice soils, and the early use of prescribed fire that increased forbs and grass by releasing nitrogen were helpful observations. Chet, who later served as Monsanto's chief forester and as president of Inter-Varsity Christian Fellowship, escorted me through the Cascades and to the spectacular big, deep blue hole more pleasingly known as Crater Lake.

When Chet resigned from Monsanto, he recommended me for the task. I visited with Marion Barnes, their chief scientist, in St. Louis, but about that time all the major fertilizer manufacturers pooled resources to set up the National Plant Food Institute. I hired on for a year as "chief forester" for the industry's educational arm. In those months, on leave from the University, I covered the South.

Fertilizing Southern Forests

Setting up the southwide forest fertilization project for the National Plant Food Institute in 1958, I was obliged to find my replacement for the University. James Pallas, a new PhD from California, carried on the work, developing a technique to freeze-dry plant tissues. Cryosurgery, now a method for halting tumor growth, bears a striking resemblance.

Graduate students improvized a crude growth chamber in which we were able to show that the ability for slash pine (*P. elliottii*) to absorb nutrients varies by the species' provenance.

Vigor of slash pine seedlings fertilized and growing in an improvised growth chamber depended upon provenance.

We called a MS thesis on the Economics of Forest Fertilization, "Moss' Manuring Manual." The National Plant Food Institute published it in *Plant Food Review* under the title, "Will Fertilizing Forests be Good Business?"

Herbicide Research

Hired to do research in hardwood control in pine stands, my interest lay elsewhere. However, the publication record shows the School of Forestry got its money's worth in recording the effectiveness of the many new herbicide chemicals coming out of industry laboratories since 2,4-D was discovered during the Second World War.

2,4-dichlorophenoxyacetic acid was supposed to kill rice when sprayed on Japanese paddies. The Nippons would surrender or starve. But, when tested in the fields of southern Louisiana, the rice survived; the chemical killed the broadleaf weeds competing with the rice. An improvement over 2,4-D for killing shrubs and trees was 2,4,5-T, a third chlorine added to the phenoxy ring to form 2,4,5-trichlorophenoxyacetic acid. 2,4,5-T, of course, was made famous as Agent Orange dumped on the jungles of Vietnam to kill trees in order to expose enemy soldiers. Of the dozens of salts tested, we found the triethylamine salt of 2,4,5-T the most effective for controlling weed trees, the kind of "weeds" now chipped for export.

Our research showed gibberellic acid stimulated growth of longleaf pine (*Pinus palustris*) in the grass stage, that mineral spirits is an effective herbicide, that complete girdling was essential for killing broadleaf trees, etc.

To keep control on the kinds of chemicals, methods and times of treatment, and the multitude of species subjected to herbicides, we utilized IBM punch cards. That was innovative in those days!

Nuclear Research

To learn more about the translocation of these and other organic herbicides (for a while called silvicides and dendrocides, neither term enduring in the foresters' vocabulary), I enrolled in a summer course at the Oak Ridge Institute of Nuclear Studies. Located on Tennessee's Clinch River, the site gained fame for the Manhattan atomic energy project of the World War II effort that resulted in the disastrous bombs dropped on Hiroshima and Nagasaki. The famous William Pollard of nuclear energy fame lectured daily at ORINS, often in his priestly attire, and then shifted gears for classes at the Episcopal Seminary at the University of the South at Suwanee, Tennessee. Some folks suggested it was guilt over the holocausts in which he, as a physicist, participated that led him into the ministry. The 30 or so students, half of whom were MDs, often attended worship at the Oak Ridge

Episcopal church where the Rev. Pollard sermonized on the Sabbath.

Our "diplomas" showed 32 of us graduating in 1956 from Old Bremsstrahlung University with the degree of Dabbler in Radio Isotope Procedure (DRIP). Some of the studies were too deep for us nonphysicists: we called it the Institute for *Unclear* Studies. Here's the first verse of the *Alma Mater* for "The Clinch College of Nuclear Knowledge":

Far above the old Clinch River
There's a sacred Pile;
Neath it stands our Alma Mater
Made of yellow tile.

The *Pile* referred to the original radiated material for nuclear reaction; *Bremsstrahlung*, German for the deceleration of electromagnetic radiation.

We learned that the then-known smallest particle, the anti-nutrino, was described as being like an invisible man looking at himself in the mirror. And physicists vote on their data while biologists measure theirs. Periodically nuclear scientists meet to vote on whose half-life data to accept. In the 1950s, C_{14}'s half-life was 5670 years. It may be different now, depending upon the ballot.

The license I received from the Atomic Energy Commission in 1956, the first at the University of Georgia and the first issued to a forester, enabled me to utilize radioactive carbon-14 for tracing the movement of weed-killing chemicals. The unrestricted license also made possible the use of phosphorus-32 in fertility studies and iodine-131 for tracing feeder roots.

The AEC was pretty loose in those days; the lab was never inspected and the dump for waste tissues containing radioactive contaminants amounted to a foxhole surrounded by barbed wire in Whitehall Forest, a half-dozen miles from campus. The dump was eventually checked and approved by a Washington official, provided I add a second fence and post the orange signs just then becoming a standard practice.

Research involving the use of these Beta-emitting elements appeared in *Forest Science* and the *Journal of the Agronomy Society*. The latter work involved tracing phosphorus through the runners of coastal bermuda grass, then the best forage for cattle. Drs. "Jack" Perkins and Joel Giddens of the Agronomy Department were the lead researchers in that study.

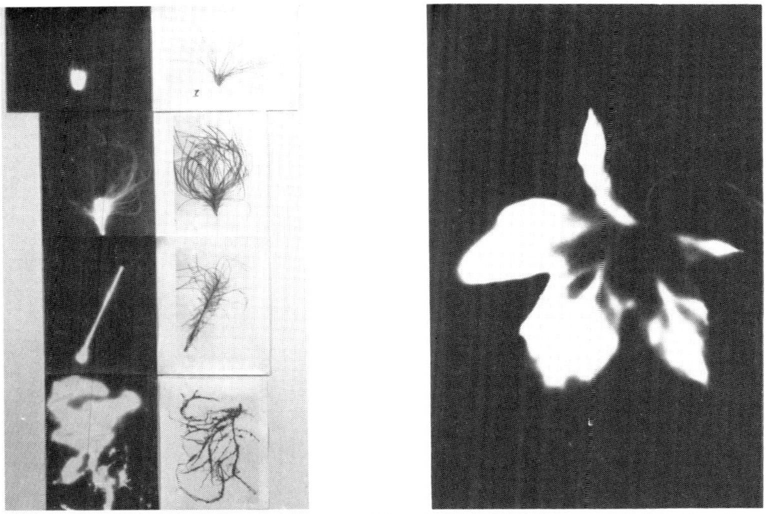

Studies utilizing phosphorus 32 showed the nutrient element concentrated in buds, stems, and roots. Radioactive carbon-14 served to trace the movement of herbicides like 2, 4, 5-T in broadleaf weed trees.

Spreading Out

A "perk" for being a researcher in Athens was the opportunity to establish plots in places as diverse as the Appalachian Mountains and the Atlantic Coast. Experiments in which I teamed up with an agricultural engineer at the USDA Tidewater Experiment Station near Brunswick were particularly rewarding. Dr. R. Lamar Green and his laborers laid out with precise engineering procedures the ditches, dikes, and water controls (from an artesian source) that enabled us to determine survival and growth rates for loblolly and slash pines where water levels were maintained at ground level, 4 inches below, 8 inches below, 4 inches above, and 8 inches above ground level. Four inches below won out, but trees survived as long as the terminal bud remained above water. Then we superimposed fertilizer treatments on the water level plots. Suffice to say here that the fertilized seedlings in the "-4" plots soon became saplings. From Aukland, New Zealand, I received a request for more information about this study. It was addressed simply, "L.C. Walker, U. Georgia, Athens, *Greece*."

Growth of these loblolly pine seedlings related directly to ground water depth, according to a study in the lower Coastal Plain.

Coweta Sojourn

With leave accumulated, I took off to moonlight one summer at the Coweta Hydrologic Laboratory, just north of the Georgia boundary in North Carolina. The U.S. Forest Service facility provided sites for studying the production of clear, clean water in monitored watersheds managed in various ways. Here we found 12-12-12 commercial fertilizer added to lime defoliated white pine seedlings. Ammonia gas was apparently released when the nitrogen reacted with $CaCO_3$.

Here, too, I mapped "unit-area control" silvicultural treatments for two watersheds. Reasons for treating small stands were outlined: Ericaceous vegetation removal to increase water yields by reducing transpiration and interception while preparing the site for reproduction and to enhance regeneration already present, and thinning yellow-poplar for water-yield improvement and to enhance growth of residuals were examples.

The family joined me in a cabin provided by the USFS. From there we could see the landscape that rose from perhaps 1500 feet to more than 4000 feet. Not too far away, within the Nantahala National Forest, grew a stand of American chestnut (*Castanea dentata)* trees not yet inoculated by *Endothia parasitica.* I wonder if the spores of the pathogen ever reached the disjunct stand at an elevation of 5000 feet.

Working the woods of Coweta, I chanced upon a split-rail chestnut fence. Trees growing amongst the 10-foot-long rails and the sturdy posts exceeded 50 years in age. The now-dense forest had been a pioneer's farm until, in all likelihood, erosion washed away the top soil and some subsoil, too. The fence had marked the settler's boundary or retained his livestock. I borrowed a truck, salvaged the rails and posts, and erected the fence around our Athens home. To some neighbors, the Lincolnesque fence was ugly.

Low-grade oaks to be removed and replaced with valuable conifers under unit-area management to increase water availability for plants and man while raising the quality of wood produced in the future, as suggested in a study at the USFS Coweta Hydrological Laboratory in North Carolina.

Just south of the North Carolina-Georgia line, at Rabun Gap, was a mountain mission school founded in the early days of settlement by Scotch-Irish Presbyterians. It was the only high school in the county. Kids came from afar to live, study, and work there. Every student worked. Anne and I became friends of Dr. and Mrs. Carl Anderson, the president and his wife. Until we left Georgia I managed the Rabun Gap-Nacoochee School's vast timber holdings in exchange for the use of the guest house that looked out upon one of the Creator's masterpieces.

Fort Valley

"Pack two lunches and eat in your lab." That's how experiment station director George King suggested I handle the meeting with Dr. L.C. Ellison of the Valdosta State College. Apparently George thought it was time for the black agricultural college at Fort Valley to begin a research program. In those pre-integration days of the 1950s, for Dr. Ellison and me to eat together publicly would have aroused some of the citizenry of Athens. On my trips to Peach County to assist in research development, the segregation system was equally strange. I'd get a free meal in the student dining hall but was obliged to eat alone at a table set just for me. A thousand eyes stared at me, like Jack Horner, sitting in a corner. A decade had passed since I, a lowly enlisted man, was segregated from society's "gentlemen," the officers of the U.S. Army.

Peach County in the upper Coastal Plain, as its name suggests, was great for growing peaches. But the white personnel of the several units of the experiment station dotting the state adequately handled problems with fruit trees. The director's oral instructions were to come up with something that could utilize the few resources, other than labor, which Valdosta State College possessed.

The college owned slash pine plantations of various ages planted by students. So, together, Dr. Ellison and I designed fertility experiments. The fertilizer industries provided the materials and the students pitched in. They laid out plots, set in corner posts, measured tree diameters and heights, and carefully distributed the salts. Our cordial relationship faded when I left the Peach State.

Airline Thrills

All three of my commercial airline scares occurred during the Georgia days. Newspaper accounts said the Capital Airlines Viscount "came within 20 feet of striking an Air Force B-47 jet bomber 19,000 feet in the air..." and "dove 600 feet to avoid collision." The second occasion was on a flight from New York to Atlanta that took us out over the Atlantic. Quickly the pilot ruddered the ship for a return to Newark and announced the rapid loss of fuel; the ground crew had left the cap off the tank, the draft sucking out the gasoline. Later, Tom Nelson, deputy chief of the Forest Service, and I left Washington for Atlanta via Charlotte. Unable to lock the wheels, the crew prepared us for a skidding landing. The wheels locked. After assurances that the problem was corrected, we took off, only to return from Raleigh to Charlotte to repeat the preparations to crash. Tom and I finally made it to Georgia. Seated with us was Mrs. Ed White, widow of the recently killed astronaut (a fellow graduate of Western High School). She called her father-in-law, the general, to come get her; she wasn't getting back on that plane after the second preparation for a crash.

Miscellany

I was serving on the University Council when *The Saturday Evening Post* accused Coach Wally Butts of "throwing" a football game to the University of Alabama's famed Bear Bryant. Some attributed the magazine's collapse to the litigation and remuneration that followed. President Aderhold told the Council he drove the coach to the medical school at Augusta to try to get the psychiatrists there to say Coach Wally had lost his marbles and should therefore be excused from the misdemeanor. The excuse didn't fly. Some wag said they should rename the journal *The Bear Butts Post*!

When Al Worrell left Georgia for Yale (I had introduced him to Dean Garrett), he left behind a fence-post economics bulletin. E. Broadus Browne, director of the Agriculture Experiment Station, had me edit it for publication.

Dr. Browne wanted monthly reports from all of us on research budgets. While we were confident he never read them, they helped us keep tabs on our own work. Merv Reines always put humor in his "renditions."

About this time, Dr. Aderhold had me appointed to the Governor's Scientific Manpower Conference. I took on these chores; inevitably they led back to forestry.

Although initiated into the agriculture honorary Gamma Sigma Delta at Penn State, I became a charter member of the Georgia Chapter "for services to agricultural science" in 1961.

Anne often entertained foreign visitors: Western Europeans, Asians, Yugoslavians, and—notably—a group of Pakistanis. After dinner, they belched, each succeeding burp with greater enthusiasm. After five minutes—and heard next door in those pre-air conditioning days—they gradually lowered the decibels. We didn't know that this is their custom to express appreciation for good cooking. They would've stopped had Anne said "Thanks."

I soon was assured of an international reputation! This time the letter from New Delhi, India, was simply addressed to "Prof. Dr. Laurence C. Walker, University of Georgia, Athens, *Greece*."

Georgia was the first state to license and register foresters. My certificate shows me R.F. 347.

Exodus

A year before our Georgia sojourn would end, Wendy Elizabeth made her debut. We called her "postscript." Before we loaded the van, another "do-it-yourself grandchild" was kicking around inside Anne.

That same Andy Campbell who encouraged me to come to Athens also played a role in our leaving the "oldest college in the South." This you will read about in our journey to the "oldest town in Texas." I'd almost left two years earlier when George Stanley offered me the chief forester job for the Kirby Corporation, of which he was vice president. A few days visiting the East Texas operations convinced me that I was not the one to oversee these half-million acres.

We began this chapter with lines from the devout Sidney Lanier. I think it appropriate to end this narrative with verses from his pen I call the "Foresters' Hymn," although the sage called it "A Ballad of Trees and the

Master." Christians will recognize in the allegory Christ's stroll to the Garden of Gethsemane and then the gruesome trudge to Skull Hill.

Into the woods my master went,
  Clean forspent, forspent;
Into the woods my master came,
  Forspent with love and shame.
But the olives they were not blind to Him,
  The little gray leaves were kind to Him,
The thorn tree had a mind to Him,
  When into the woods He came.

Out of the woods my Master went,
  And He was well content;
Out of the woods my Master came,
  Content with death and shame.
When death and shame would woo him last,
  From under the trees they drew him last,
'Twas on a tree they slew Him last
  When out of the woods He came.

If the decade in Georgia was a challenge, which it was, it couldn't begin to compare with the demands that must be met and opportunities seized in the years now before us. In my letter of resignation, I told Dean Herrick, "... the high respect with which I regard your leadership is well known. Many of your administrative techniques will be adopted for Texas conditions."

Stands like this 23-year old South Georgia naval stores slash pine plantation chipped for gum responded favorably to nitrogen fertilization.

Lamar Green, agricultural engineer, at a diked and/or ditched water-level plot in the Tidewater zone of Georgia. Seedlings grew best when water levels were maintained 4 inches below ground level.

13

TO THE "OLDEST TOWN IN TEXAS"

Stephen F. Austin State University campus, the shortleaf and loblolly pines seeding in when the land was abandoned from cotton production by Gov.Thomas J. Rusk during the agricultural depression of 1895.

I put out the fleece. Whether to leave a home we all enjoyed and personally rewarding work as a researcher at the University of Georgia to move to Deep East Texas was the second-most difficult decision of my life. Anne, born and reared but 75 miles from Nacogdoches, refused to even participate in the debate with which I wrestled, on one side and then the other. Considering the situation since receiving President Ralph Steen's invitational letter and Dean Joe Gerber's telephone calls, it would be a tough call. So, prior to my visit to look over Stephen F. Austin State College, I'd determined I'd not consider the assignment unless it paid at least one-third more than I received at Athens. The fleece was dry, though the morning grass was wet with dew. The agreed salary was to the penny one-third more than the Georgia paycheck.

I talked the matter over with a lot of folks, many of whom volunteered opinions that I was ready for the loony bin to think that a forestry program at a teacher's college (until 1949) could be anything but frustration. And Anne, from her high school days in East Texas, knew well the college's reputation as a place to play. SFA lacked academic respectability.

I accepted the task with a great measure of reluctance. One of the reasons for hesitation was our home in Athens, built to a large degree with our own hands. Each of the older kids had a big upstairs room in the story-and-a-half structure. Wendy, now 16 months old, soon would have the guest room. And Jean or Jim, on the way, would share with Janet or Stephen, depending upon the Providential toss of the coin.

Here's where William Andrew Campbell again stepped into my life. Andy, heading the Forest Service research center on the Athens campus, realized my predicament. He dropped in to ask a theological question. "Understand you can't make a decision?" "That's so." "I thought you were a Presbyterian." "That's so, but what's that got to do with this matter?" "Well," Andy said, "I thought Presbyterians didn't put such value on material possessions." That's all the pathologist-turned-philosopher said, but it was enough. I wired Nacogdoches, "Offer accepted" in July 1963.

The city's birth, and thus its claim to be the oldest town in Texas, dates—without proof but by the narrative among several that I like most—to the directive a Caddo Indian chief, encamped along the Sabine River, gave his son to run swift with the wind to the setting sun for two days and there make camp. At the confluence of Lanana and Banita creeks, the brave named the place Nacogdoches (Spanish spelling) in honor of his father. The Caddoan leader then sent another lad the same distance to the rising sun. He settled on the edge of Louisiana's Cane River—probably earlier the bed of the nearby Red River—and, as his brother, called the stamping ground Natchitoches (French spelling).

The college had 2700 students, 90 in forestry. Bachelor of Forestry (BF) and Bachelor of Science (BS) degrees were offered, the former for professional foresters and the latter for those going into the teaching profession but, having a certain number of hours in forestry, "do not wish to specialize in the field of forestry."

Robert Owens, organizer of the original Department of Forestry at SFA in 1946-7, and Professor David Smith, forest engineer, at the 1970 dedication of the Forestry Building.

Robert Owens and Arne Kemp, my predecessors, had assembled an excellent faculty of 6, all but one with a doctoral degree: Victor Bilan, Robert Baker, Nelson Samson, Gerhardt Schneider, Ellis Hunt, plus Arne, the department head. A NIBCO draftsman, Leo Effenberger, taught engineering drawing part-time. A librarian, two technicians, and two secretaries were also on board. The budget was about $100,000. By that time 170 foresters had been graduated. (The College's bulletin for July, 1946, shows Owens as an assistant professor and lists three forestry courses along with a promise that more would be added. The 1947 catalog made good the promise.)

The faculty of forestry in 1963. Front row l-r: Leo Effenberger, Robert Baker, Lacy Hunt (lumberman philanthropist), Leonard Burkart; rear row: Ralph Steen (university president), author, Victor Bilan, Ellis Hunt, Gerhardt Schneider, Joe Gerber (dean of the College).

As Arne was a wood technologist, now departed to be assistant director of the Lake States Forest Experiment Station, my first task was to locate and employ, *in absentia* while still in Georgia, his replacement. Leonard Burkart, with a new PhD from the University of Minnesota, joined us sight unseen.

I cannot say for sure, but I couldn't help but feel that SFA was a holding tank for faculty until greener pastures would open up. No one owned a home nor seemed interested in such. I think that our buying Arne Kemp's lot and immediately beginning to build a house may have had an effect; within a year four faculty members had houses under construction. We were here to stay, I thought.

I could not tell the faculty that, shortly after my arrival, the president called me to his office to discuss several matters. One was that if we did not succeed in getting accredited the next time we tried for the coveted Southern Association/Society of American Foresters status, the department and all its faculty would transfer to Texas A&M University. The last failure was the previous year, in 1962, possibly a factor in Arne's departure. Dr. Steen, a historian from the West Texas plains who literally hated trees, had made a deal with TAMU's president, General Earl Rudder: A&M would pick up the tab for faculty relocation if we were not successful in the effort. The general, a WWII hero in the Normandy invasion and one-time Texas Land Commissioner, wanted a forestry program. This would be an easy way to get one. The message to me was a heartbreaker; there would be more agonizing moments ahead.

At the same session, the president pulled out a set of blueprints, prepared by architects Kent and Marsellos of Lufkin, of the "new" forestry building. It called for a $175,000 renovation of the university's warehouse. That, Dr. Steen said, would satisfy our requirements *ad infinitum.* Mustering the courage to say, "This won't do," I asked for 24 hours to redraw an overlay of the architects' plans. The faculty agreed that this should be a laboratory building for instruction and research in silviculture, soils, and wood technology. We included in the drawing a shower room, a balance room, and a darkroom. The faculty also agreed that if we could have quality labs, we could endure the old brick-veneered army barracks (probably moved from Camp Fannin, near Tyler) for offices, library, and classrooms. Dr. Steen accepted the proposal, provided we'd not ever ask for more. (Latecomers to the scene: the Forestry Building, erected in 1970, sits far back from Raguet Street so that we could continue to use the two barracks while the brick and mortar facility was under construction.) A research kiln and nursery were already in use. We paid Texas A&M $100 for a pair of old greenhouses at its abandoned experiment station on North Street.

A refurbished warehouse became the Forestry Laboratory.

One of the two original forestry buildings, brick-veneered army barracks. The baldcypress sign of heartwood, from the last harvest from pre-inundated Lake Sam Rayburn, was provided by W.T. Carter and Son. The sign marks the dedication of a square block for the School of Forestry and associated facilities.

The greenhouses we bought from Texas A&M for $100.

Within a few months a third matter came up. Now, I questioned the reliability of the fleece. Dean Gerber called me to his office to say I was away from campus too much. I explained that building a program required knowing the forestry clientele. After a long silence, Joe said, "You better go across the hall." To the president I went. That was that—for a while.

Defending the Turf

One of my first chores was to get the department out of trouble with industry, occasioned by truthful public statements of a professor about the effects of the forthcoming abundance of timber on stumpage prices. Industry, of course desiring to keep prices low, wanted the professor muffled. The first lumberman with whom I conferred seemed satisfied; for by the time I arrived at the next industrialist's name on my list, the first had called the second to tell him to give me a hearing.

Defending faculty was not uncommon. Once I had to defend myself. It happened this way. The father of one of our students was a Dallas lawyer who lined me up to speak to the Audubon Society there. Its members and friends packed the Science Auditorium at the Fairground. After the introduction and as I arose to speak, the moderator suggested that I "put the notes away and let us just ask you questions." I happily agreed. The discussion went on during refreshments and, when the custodian locked the building, under the street lamps. Driving back to Nacogdoches in the wee hours, I thought I'd really made points for SFA's forestry school. Again, that's what I thought.

The faculty and staff in 1966; rear (l-r) K. Watterston, V. Bilan, Nell Crews, Jerry Ann Davis, Walter Dickey; front (l-r) author, N. Samson, E. Hunt, R. Baker, A. Verrall, H. Wiant, L. Burkart; not shown: W. Wilford, J. Singer.

A couple days later I received a well-written letter from a Dallas woman who had attended the gathering. She wrote that I had sold out the school to the wood-using industry. As I didn't recognize the name, I folded the letter and left it on the desk until I could learn something about my aggravated correspondent. The letter remained unanswered when I received a call from one of the state's industrial forestry chiefs. He and two others of similar stature wanted to visit me. Ah, I wondered, would they provide a generous grant for our struggling program?

After a few minutes of introductory niceties, their leader came to the point: "You're selling the school out to the radical environmentalists." Thank you, lady, for your letter. I unfolded it and handed it to the trio's headman to read to his associates. Read and folded, he passed the missive back and said, "Let's all go to lunch."

I did learn the woman's identity. Her husband's picture was on the front page of the Dallas *Morning News*, a conservative paper, shortly thereafter. He, the chaplain at Southern Methodist University, was photographed cavorting in his clerical robes down the aisle of the University's ornate chapel with a mini-skirted coed during a Holy Communion worship service. Subsequently the minister's wife and I communicated cordially. But I learned that deans should learn to walk a tightrope: it's a prerequisite for the job.

Who'd want the job? That's what Robert Burns (b. 1759) wondered in the 1780 ballad about *The Dean of Faculty*. Yet folks will compete for it:

> But Scot to Scot ne'er met so hot.
> Or were more in fury seen, Sir,
> Than 'twixt Hal and Bob for the famous job,
> Who should be the Faculty's Dean, Sir.

Bob got the job by vote of the Faculty of Burns' *alma mater*, much to the chagrin of the poet:

> With your honours, as with a certain king,
> In your servants this is striking,
> The more incapacity they bring,
> The more they're to your liking.

Accreditation

Society of American Foresters (SAF) accreditation would be won this time or never. For 20 years the program had endured the stress and struggle to satisfy the professional organization. Heartened by the fact that I knew all the members of the Visiting Committee and knew them to be fair, the faculty, students, and alumni prepared for the "trial." That the department was elevated to School status in 1965 was an asset. (Despite being the only undepartmentalized school in the Texas higher education system, the Coordinating Board nevertheless gave its blessing. Not until 1969 did the legislature agree to elevate the college to a university.)

Just before arrival of the review committee, however, a monkey wrench fell into the machinery. One of the committee members, unable to attend, on his own volition selected his replacement. The substitute, head of a neighboring school, previously had told me in a friendly chat that he couldn't see SFA ever being accredited. Disappointment was evident when I learned this good friend (who remains a dear friend) would join the team. I complained of his prejudgment. The SAF subsequently wrote into its rules that personnel from neighboring schools are ineligible to serve on accreditation visiting committees.

Between the time of the visit and the meeting of the SAF Council, which reviews, approves, or rejects the report of the visiting committee, the committee chairman was elected to the SAF Council. I was told the committee could not agree, pitching the decision to the Council. The Council, too, was unable to decide until the councilman who had chaired the committee and who was head of a central state university's department of forestry, spoke up. The SAF president, Bill Hagenstein, related to me that this member told the Council, "I'd swap them (SFA) professor for professor." The council's attitude was that, according to Bill, if the faculty is *that* good, the school should be accredited. And so, in 1966, it was.

Only as I write this have I learned that the team member I thought had misjudged us was our strongest supporter; that he was one of the 2 votes (on the committee of 5) that favored our accreditation; and that he, in the presence of a dissenter, argued vigorously before the Council on the school's behalf.

Hagenstein wired President Steen of the decision even before he found me in the Seattle hotel lobby. At about 6'4", Bill leaned way down with his arm over my shoulder as though to give kindly support to a discouraged soul and, in a very serious tone, began to speak. I thought the worst. Then, he brightened with the good news. SFA was finally accredited, and for the full ten years. I'm told Dr. Steen encountered Nelson Samson on the campus later that day. One asked the other, "Do you think Walker will have a beer to celebrate?" "No, but he might put cream in his coffee," was the response.

Recruiting

To increase student enrollment in order to justify a larger faculty, recruiting was essential. We needed specialists beyond the six on the staff who were covering the vast field of forestry. Among the opportunities for attracting students were the technician schools in Florida, Pennsylvania, New York, and Kentucky. We let the faculty and students there know of the credits we would accept. We also provided advanced placement exams for some courses. The success of that effort got me called on the SAF Council's carpet. When I said we simply followed the leadership of a Lake States' college that was pursuing Mont Alto students, the answer was, "Larry, they are old and established; they can get away with it. You can't."

As Oklahoma State University and Arkansas A&M (now U. Ark. at Monticello) were not accredited, we happily responded to inquiries from those states. After all, many Texas residents for whom the state paid tuition to accredited schools had enrolled in programs beyond the Lone Star's boundaries. We also spread a brochure throughout New Mexico, a state without a forestry educational program. Out-of-state tuition was so low that students could come from afar more cheaply than they could go to a forestry school in their home state. A program originally established to allow African-American students in the South to go for medical training to those southern states with schools for blacks in the days of segregation aided our effort. The Southern Regional Educational Board, funded by member states, provided the money. Because Kentucky had no forestry school, the SREB paid for Kentucky collegians to attend SFA. Patrick Angel, recruited from the Blue Grass state, later noted that "we had to look toward (SFA) for a way to break into the forestry profession."

By the time of my leaving administrative duties, total enrollment tallied 990, ten percent of whom were graduate students, and 10 percent of these were doctoral candidates. The faculty numbered 36. About 10 percent of the University's enrollment were foresters. Many we recruited remained at SFA in other fields of study.

Victor Fain, editor and publisher of the Nacogdoches *Daily* and *Sunday Sentinel*, frequently focused on the School of Forestry with special promotional issues. Sam Malone, writer and photographer for the *Sentinel*, served as the School's self-appointed publicist. The publisher paid Sam by the column-inch and pictures printed. We fed him plenty of material. Malone and his cat migrated to San Augustine, my "hometown," where he wrote, edited, and published *The Rambler*. Ramblin' Sam's rag wasn't competition for the *Saturday Review of Literature*, but the fishermen on East Texas' lakes loved his reporting on where the bass are biting.

A 1965 *Sentinel* supplement covered 16 pages about the school. The paper made a big deal out of our having 13 out-of-state students.

Three writers for the Dallas *Morning News* gave us a lot of lines. We'd find stories for Frank X. Tolbert's travelogue called "Tolbert's Texas." His accounts of the champion farkleberry tree (*Vaccinium arboreum*), found nearby, brightened a reader's day. I floated remote Caddo jungle streams with him in Ray Butler's boats and he reported on our work with cottonwood research in US Plywood-Champion Paper's Brazos and Trinity river bottoms, calling the school a "brain factory." Frank named a chasm in the Pecos Mountains Pass Pass in honor of Fred Pass whose *Morning News* beat was East Texas. Pass carried our stories to Dallas, too. Donnis Baggett, who reported on the School as an SFA journalism student, continued covering us when he moved to Big D.

A brochure, *The Professional Forester*, for which our high-school-age son Stephen—later the professional artist—did the layout, was purchased in bulk by such institutions as the State University of New York College of Environmental Science and Forestry.

Having success with our Honduran program, in which 19 students enrolled the first year, I continued correspondence with Costa Rica's president (see these accounts later) to send his countrymen to us for training. He knew he needed some good foresters for the immense undeveloped and underutilized timberlands of the Rich Coast.

Learning President Gerald Ford's son, Jack, was graduating from a forestry school in 1975, while his dad served in the White House, we attempted to get his interest in graduate studies at SFA. The effort brought a courteous letter from the presidential offspring.

A letter received in 1970 from a Michigan man included a card I had given him in 1965 while in Sawmill Canyon along a mountain trail I hiked with Boy Scouts in New Mexico. "The card has been carried through Georgia and Vietnam...." Paul King was "looking for...opportunities (to) study at Stephen F. Austin State College...."

Essays written for the *Presbyterian Journal* helped pull in good students, too. A Dallas father, an attorney according to the letterhead, after reading one article, wrote me "... my third son plans to begin a course of study in forestry this fall. He has been admitted to Texas A&M, which I consider to be a very fine University. However, ... this editorial cannot but lead me to suggest to Daniel that he consider Stephen F. Austin State University."

The faculty spent a lot of time with students, potential students, and parents. A 1968 letter from a Sam Johnson expresses appreciation for the effort in these excerpts:

> ...generous and unselfish gift of time must be, I'm sure, unique in the annals of education...obvious interest in the students and their future...impressed with the policy of assigning new students to a counselor who will be with them...throughout their college career...obvious interest in our son and his future.

Programs

The forestry degree was changed in 1969 to the BSF simply to be consistent with all other US schools. One other college had a BF on the books, but was not awarding it.

The Texas Coordinating Board for Higher Education approved Master of Forestry (MF) and Master of Science in Forestry (MSF) degrees in 1966. The faculty designed the former, a two-year program, for students with bachelor's degrees in fields other than forestry but who wished to be professional foresters. The MSF was developed for those with specialized undergraduate degrees who wished advanced work in that specialty. Wildlife and outdoor recreation are examples. Foresters with BS and BSF diplomas could apply for either degree. All master's work required 6-credit theses, this dean insisting that original research and writing are essential to successful forestry careers.

We wanted permission to offer the PhD. Texas A&M successfully blocked it before the Coordinating Board. In an attempt to be fair, however, the CB made possible the PhD through a cooperative arrangement. Candidates would take 30 hours (1 year) of graduate courses at SFA, then transfer to A&M for a second year of course work, and finally return to SFA to complete the dissertation. A problem became evident: we recruited the doctoral students; then when they had finished the year at A&M, the candidates saw no reason to come back to Nacogdoches. Only two did. The CB staff and the Board itself, recognizing the problem, agreed to our request to offer the Doctor of Forestry (DF) degree. We modeled it after the professional terminal degrees awarded by Yale and Duke universities.

The DF was not to mimic a PhD. It was a professional, not a research, degree. Every candidate was to take ten 3-credit advanced courses; i.e., advanced soils, advanced pathology, advanced entomology, etc. They would then research and write a 30-hour thesis in the subject of their choice. Qualifying exams were to be in the writing of English and in verbal and written Spanish, the latter because we visualized graduates going south of the Rio Grande for employment. Forestry schools, industry, and the FS experiment stations thought highly of this arrangement. At a meeting of Southern Association officials to go over the plan, the dean at N.C. State boldly stated, "With this plan, Larry, you'll get all the students."

So many forestry grads were being employed as environmental managers that the faculty sought, and received from the CB, approval to establish a curriculum. This environmental management degree would be distinct from environmental science, offered at many institutions, and environmental engineering, a program at others.

Enrollment increased to the point that we could support curricula in forest game management (cf. wildlife biology), forest recreation management, and forest range management. The latter was a cooperative effort with Sul Ross State University in Far West Texas where students would spend the junior year taking courses in range management, range plants, range soils, range animals, and range ecology. With this degree the graduates had job opportunities in the West as well as in the South and East. Under Everett E. (Double E) Turner's leadership, these animal science people conspired in 1974 to nominate me for the institution's presidency. They felt the college had not had leadership that envisioned opportunities to enroll out-of-state students in natural resources—water (it's scarce, therefore very valuable, there), range, wildlife (refuges surround the place), recreation (two national and several state parks nearby), and minerals (the mountains abound with many kinds).

Harry Wiant, our assistant dean, and I made a trip to Sul Ross University at Alpine to better understand range management. As we wanted to spend some time on a working ranch, Double E directed us to a spread belonging to a graduate of the Davis Mountains school. Stopping to confirm our directions at a store near what would become Guadalupe National Park, a lassie of 10 years or so directed us "back the way you came until the first cattle guard on the right, go straight on that road, passing a road to the left, and you will come to Mr. F's place." "About how far?," we inquired. "Twenty-eight miles," she answered. We did as directed, though night had come upon us. Suddenly, high beams faced us on the narrow rutted lane bounded by boulders and cacti. We had to stop. Then a gruff voice from the darkness growled, "Get out. Get in front of your lights. Who are you?" When we explained our mission and that Prof. Turner sent us, the speaker's voice mellowed. He came into the light, hands were shaken, and he then cupped his hands to bellow in all directions, "Amigos, Amigos, Amigos." From behind the rocks the wetbacks emerged to return to their shelter.

The rancher was courteous. He put us up in his nice home (equipped with an electric organ), fed, toured the vast acreage, and discussed with us his problems. Opening the ranch ledgers, he showed the necessity for utilizing Mexican laborers if he were to break even grazing cattle where the carrying capacity is 20 or so head to a section of land.

At the time of my duty transfer, we were working with Lamar Tech at Beaumont for a cooperative program in forest engineering that would provide expertise in logging, land surveying, and road layout. Industry and government yet needs people with these abilities. When Senator Max Sherman, with whom I had worked on natural resource matters left the capitol to become president of

West Texas State University, he wrote me in anticipation of joint programs.

We all agreed the School should be a single entity, not divided into departments. Keeping other schools on our campus from nipping at our heels was more than enough to keep us busy holding on to turf. We didn't need competition for territory among ourselves.

Feud with the Aggies

Persuading Governor Preston Smith to order one-half of the federal McIntire-Stennis research money to be allocated to SFA angered the Aggies. Gaining graduate programs through the Coordinating Board for Higher Education further fueled the flames. Eliminating the out-of-state tuition payments by Texas taxpayers for Texas A&M students who had completed the preforestry program at College Station didn't encourage peace. And a director of the Texas Forest Service, a unit of Texas A&M, who refused to let us have a copy of the TFS directory didn't help matters.

And, indeed, we felt severely threatened when the Aggies proposed adopting a four-year baccalaureate program. Dick Townsend, Yale class of '39 and highly respected forestry consultant, came to our rescue. A letter on his stationery and over his signature was sent to every legislator, CB member, and others who might want to save the state some money. In the end, the Aggies got a forest science department (the CB voted 17-1 in their favor, only a San Antonio Hispanic seeing no need for two schools), Dick's son enrolled in forestry there, and a change in Forest Service directors brought harmony to the State's forestry educational enterprises. The CB ordered the two groups to meet together annually in order to share common concerns. As I write, I cannot say the fear was unjustified, for the faculty at College Station is productive in research and the "products" are superb. Jim Hull, TFS associate director and an SFA graduate, and Director Bruce Miles do a remarkable job maintaining harmonious relations. My absence from the scene and Scott Beasley's "deaning" no doubt contribute to the institutional accord.

The Commissioner of Higher Education, Jack Kenny Williams, directed us in 1967 to prepare a long-range plan for forestry education in Texas through 1987. The faculty, working as a Committee of the Whole noted that

> a review of trends and philosophies in conservation and renewable natural resource education substantiates the inter-disciplinary approach advocated to develop forestry education within the context of the total environment. Thus resource managers and scientists knowledgable of man's increasing relationships to the forest ecosystem and capable of responsible planning will be available for the state and nation.

The report suggested renaming the school "The Texas State School of Forestry at SFASC.'" Over 100 correspondents and two consultants contributed to the effort of the 179-page document.

Forest Service Laboratory

Time had arrived for the U.S. Forest Service to have a decent facility on campus. Until now, these researchers had been housed in various facilities, the latest in one of the brick-veneered army barracks. Be it known the Washington office of the FS didn't want this unit at SFA. It had been established by an Act of Congress in 1944 as a result of lobbying by Lacy Hunt, a local lumberman; Dr. Paul Boynton, the College president; Congressman Nat Patton; and Senator Tom Connally. (Ergo, only an act of Congress can eliminate the program.) The unit was thus the "only one in the nation specifically instructed to work with a named institution."

Walt Myers of the Forest Farmers Association in Atlanta helped us win the building. He shared with me some of his allotted time for testifying before the House and Senate appropriations committees in each of four years. Senator Ralph Yarborough was a major mover and shaker, while Representative John Dowdy gave support when requested.

Senator Yarborough and I had a close relationship, not just on this but on other conservation issues. He wanted to keep one natural resource matter on the front burner. Padre Island National Seashore, the Big Thicket Natural Area (he wished it to be a national park), the Guadalupe National Park, and the Forest Service research facility were special concerns.

The Senator gave me his private office phone number in order to bypass receptionists when need be. I only used it once, about 9 pm, and, sure enough, he was in his office. When Lowell Halls, the FS Research Center leader, and I met with the senator at his Austin office, we accompanied him to make a payment on his Chevrolet automobile!

The public testifies before Congress on a certain day; members of Congress usually make their presentations on the same matter the following day. On one occasion I hung around to see how the senator would handle his testimony before the House Appropriations Subcommittee for The Department of the Interior and Related Agencies. This Committee has responsibility for FS budgets. Twice the good senator got off the subject of the FS building to brag on the School of Forestry. Twice the chairman called him back to the task at hand, saying, "Yes, we heard from the college's dean yesterday." When his testimony was over, he rushed from the room. I think he didn't know I was in the back of the hall.

The first year's testimony got us architect's planning money. I thought, hey, this is easy. But I had to face these committees three more times before construction could begin. Ottis Lock, former state senator, former member of the SFA Board of Regents, vice president of Southland Paper Corporation, and at that time Nixon's finance campaign manager for Texas, came to our rescue. Congress' allocation for the laboratory had been frozen by the administration. Mr. Lock called the White House. The ice suddenly thawed. Construction got underway.

One fellow taught me how to get attention for your project from a congressional chairwoman of a committee during a hearing. He sent to her home early in the morning a bouquet of flowers, clipping one for his own lapel. When it was his turn to speak, she immediately recognized the boutonniere and thanked him.

Meanwhile, "back at the farm," President Steen was not willing to lease six acres to the Forest Service, the minimum to which the Washington Office under Deputy Chief for research M.B. Dickerman would agree. Dr. Steen finally would go along with six acres, but only at the extreme northeast corner of university property in what was then a roadless wooded area. (The Homer Bryce Stadium now stands there.) I couldn't go along with that, for I wanted close association and collaboration between the School and the Forest Service. We needed more federal researchers as part-time instructors in their specialties, just as Lowell Halls and John Stransky had been teaching wildlife management courses (which, incidentally, gave them Social Security and Medicare benefits not available at the time to federal employees). We also wanted FS financial assistance for graduate student research. Eventually, the Forest Service agreed to accept a site in the square block dedicated as the SFA Forestry Center.

An additional 1-1/2 acres was set aside for a similar Department of the Interior facility. At that time, a USDI researcher, Phil Goodrum, labored with the FS, and the Denver office wanted to expand Phil's program. Also to utilize a new Fish and Wildlife Service laboratory would be Dan Lay of the Texas Parks and Wildlife Department. Both Goodrum and Lay worked on wildlife problems of the woodlands of the region. These two were the first *bona fide* professional wildlife biologists in the state, receiving the first two master's degrees in the subject awarded by Texas A&M University.

Just as the USFS never included the Nacogdoches unit in its 1940s plans, so the Service didn't include the laboratory building in its asking budgets. We lobbied it through without the support of the Executive Branch. And also, these lines must record that the Southern Station's Silviculture and Wildlife Laboratory was, and is, to be an integral component of the School of Forestry. The unit's publications would clearly state this, thus adding to the prestige of the School.

Regrettably, the Nixon administration sent Senator John Tower to make the dedication address for the FS building. Tower played no role in obtaining the funds: Yarborough alone of the elected politicians provided the congressional persuasion.

Senator Yarborough's Visit

The first Earth Day, in 1973, on campus was principally the work of the students of the School of Forestry. David Hyink, now a Weyerhaeuser research manager, worked with other students to line up Senator Yarborough to keynote the affair. He was happy to oblige. Upon his return to Washington, the SFA News Bureau sent him photographs of the event. On one of these he wrote, "To Dr. Laurence C. Walker: with my lasting appreciation for your conservation leadership. Your friend, Ralph W. Yarborough, Nacogdoches 1973." Hyink's fellow workers also got commitments from three other high-ranking public officials. Dr. Steen wasn't happy to learn from the student newspaper, *The Pine Log*, that they'd all be here the same day. Fortunately for me Senator Tower, soon-to-be-candidate Lloyd Bentsen, and then State Senator Charles Wilson didn't show.

In the frontispiece of the Senator's biography, *Yarborough of Texas*, by William Phillips, the politician wrote "To Dr.—Dean Laurence C. Walker—with gratitude for your dedication to the improvement of this Spaceship Earth, and particularly in our native East Texas, and for your work to preserve the Habitat of Man, and with best wishes for your continued success."

Senator Yarborough wrote in the last letter I have from him, dated October 12, 1989, following a bout with a virulent virus that puzzled neurologists and neurosurgeons: "... two of the jobs I did in the Senate of which I am most proud was that US Forest Service Laboratory at Nacogdoches, and the Guadalupe National Park in the West because the Forest Service opposed the Research Laboratory, and the Interior Department opposed my national park. It shows that if you are on the right side, and stay with it, you can accomplish something." The Park surrounds West Texas' 8,751-foot mountain.

Disagreeing with him on some issues, I trust, is why he valued my comments and sought my counsel. I thought highly of him as a statesman of highest integrity.

Alumni Relations

In my earliest days on board, the Forestry Alumni Association was especially helpful. Marvin Davenport (president), Jack Dillon, Jerry Lumpkin, H.L. "Rip" Byrd, Winston Jones, Stan Lindsey, and Jack Irwin were leaders in the movement which focused its attention on the needs of the department. An executive committee served as a steering group for the faculty, publishing a *Newsletter* at Christmastime throughout my tenure at the helm. The Alumni Association's interest waned with accreditation; the members felt their job was done. A student annual promoted by alumni, the *Texas Tally*, was notable among forestry yearbooks. Founded in 1946, it continued annual publication throughout my deanship.

SFA Experimental Forest

The Stephen F. Austin Experimental Forest at the south boundary of Nacogdoches County on the Angelina River once was a federal Land Utilization project. The Farm

Entrance to the Stephen F. Austin Experimental Forest. The plantations at the entrance memorialize a lumberman who encouraged research.

Author with students on field trip to the Experimental Forest.

Security Administration bought LU lands during the Great Depression and turned these rundown farms with worn-out soils over to the Soil Conservation Service in the USDA (formerly the Soil Erosion Service of the USDI) for rehabilitation. The government moved the farmers to the Trinity River flood plain west of Crockett where a levee protected the cleared land. Riding west on State Hwy. 7, travelers can see the barrier betwixt the river and the new ground. Transfer of this LU acreage to the Forest Service as a unit of the Angelina National Forest for road maintenance occurred in 1944-1945, timed with the creation of the research center on campus. State Senator Ben Ramsey, State Representative Jewell Helpinstill, and Lacy Hunt participated in the politics of the transfer. From the beginning, the Forest has been available for faculty research. Victor Bilan's plantings are among the original experimental plots.

The Forestry Building

We had a little less than a million dollars with which to erect a principal facility in 1969. Tired of being frowned upon as non-academics, we planned the Library as the centerpiece of the new structure, readily recognized by its glass walls as a place for serious business.

Forestry Building at time of 1970 dedication.

Chairman R.E. McGee of the Board of Regents at the dedication of the Forestry Building, 1970.

Losing the feud with the architects, who wanted wide halls and non-usable space for appearances' sake (I could not convince them we were not building a high school where hundreds of kids change classrooms every hour), we determined to make the best of the waste of valuable area. John Stransky suggested big game mounts high on the walls of a stairwell. Several museums contributed to the collection of the trophies. Someone added a pair of white-tail deer heads, antlers locked in their last desperate attempt to conquer or survive. Another brought elephant tusks from a war-time Indian expedition.

Portion of big game collection, a great attraction for school children.

An alcove was set aside for a museum of artifacts. Naval stores tree butts and clay drip pans for collecting resinous sap; a one-man hand-operated saw dating to the 1800s, for cutting boards from logs; and fire posters tacked to trees in 1925 by the Texas Department of Forestry (then so-named) and 1927 by the Texas Forest Service were contributed. Students, alumni, and friends added early-day planting dibbles, a 1940s Panama injector for treating low-grade hardwoods with herbicide, antique bark spuds for debarking pulpwood sticks in the woods, and a double-torch or signal lamp (with wicks of rope and fueled with signal oil) carried on tramroad log cars. Robert Hatcher, a graduate student during my Georgia days, brought a 1880s two-man crosscut saw. The summer silviculture class studying the watersheds at Wagon Wheel Gap, Colorado, thought the broken stadia measuring rod found nearby the caved-in cabin on the mountain should be displayed here.

Later the wall of species, domestic and exotic, would be labeled.

We dedicated another stairwell to the Southern Pine Lumber Company's collection of enlarged photographs of industry's operations at century's turn. Joe Denman, Temple president, having just discovered them in a storage shed being demolished, personally delivered the treasures. These pictures later inspired the writing of *Axes, Oxen, and Men*, which the Temple company published.

Mrs. R.R. Childers was a special friend of the School. Her late husband, a veterinarian in Jasper, owned considerable acreage of timberland, giving her a vested interest in forest management. She was also an early-day graduate of SFA. To the School she gave the 14-foot pirogue displayed in the entrance foyer. The southern baldcypress (*Taxodium distichum*) log, cut from a tree on the bank of the Neches River in Southeast Texas, was whittled and burned to form the dugout in 1895. Dr. Childers and his Boy Scouts restored the old vessel. In fashioning pirogues, artisans drilled holes through the log, thereby able to determine the thickness of the hull during carving. Later they plugged the finger-width holes with wood dowels.

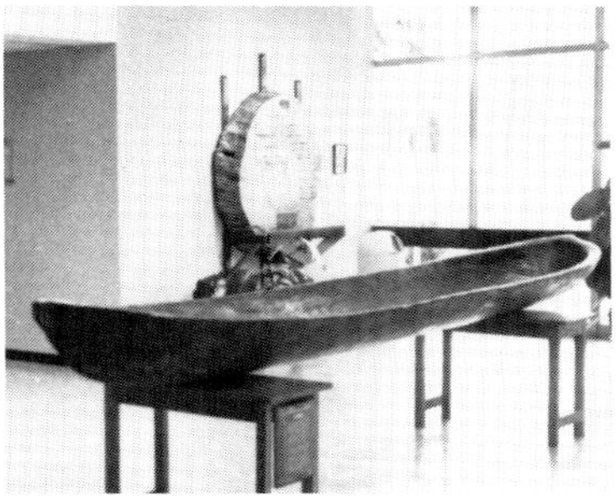

Dr. Childers' pirogue.

Probably an invention of swamp-dwelling Indians, pirogues were an important mode of transportation for the Acadians traversing the bayous of southern Louisiana and for later Anglos and Blacks in similar haunts in Deep East Texas. Pirogues aided rounding up cattle during river bottom overflows, the stranded herds otherwise left to starve on small islands of high ground.

Our association with Mrs. Childers led to her furnishing the Seminar Room of the Forestry Building, on the wall of which hangs her motto, "'Give and it shall be given unto you' Luke 6:38." Charlotte Griffith, executive secretary of the Deep East Texas Development Association, told of a substantial contribution Mrs. Childers gave to DETDA in my honor.

The great tree stump cross-section removed by students from a virgin tract in the north end of Nacogdoches County shows its birth in 1825 when Austin was colonizing Texas and the Fredonia Rebellion was taking place in the area. The 5-foot-plus dbh tree was a young sapling when Texans built their first steam mill in 1832, a mature sawlog when, in 1907, the Lone Star State was the nation's third largest lumber producer (2.3 million board feet that year), and was ripe for harvest when paper makers ground their first pulp, at Orange, in 1911. The old loblolly pine, a remnant of the virgin forest, finally died at age 138, in 1963. That's about the age you read about earlier of the oldest pine stands in the southeastern US.

Prof. Eugene Hastings (l) with student at the chronology log, dating the history of Texas from Colonial times until the tree's death in 1963.

Across the foyer hangs a large plaque of walnut wood on which is inscribed the words of President Theodore Roosevelt during his address to the Society of American Foresters in 1903:

> We have reached a point where American foresters trained in American forest schools are attacking American forest problems with success. ... You must instill your own ideals into the mass of your fellow-men and at the same time show

> your ability to work with them in practical and business fashion. This is the condition precedent to your being of use to the body politic.

I had asked FS Chief Ed Cliff for another quotation, the one by Gifford Pinchot about managing forests "for the greatest good of the greatest number in the long run." Ed said he had better lines, more fitting for a school. Indeed he did. The FS sign shop prepared the marker. While President Theodore Roosevelt delivered the speech to the fledgling forestry profession, I sense some Pinchotian rhetoric in the remarks.

Forest Service chief Ed Cliff visiting with Alyse Moore, SFA student.

We named the reading room, formerly the Forestry Library, for W. Goodrich Jones, the Temple banker who is called the "father of forestry in Texas." Jones' daughter, Doris, and son were benefactors of the School of Forestry and provided my life membership in the Texas Forestry Museum Society.

W. Goodrich Jones, for whom the Forestry Library was named.

The Lacy Hunt Seminar Room was named for another benefactor who was a lumberman. The room's large tropical mahogany tear-drop table was designed by Paul Boynton, a psychologist who served as SFA's president. One of four huge tables on campus, each with a different shape (U, O, or ❑), the design is said to have some psychological significance. At final exam time, I think I know why one is in the form of a tear.

Hall of foresters and lumbermen. Gifford Pinchot heads the line-up.

An art department graduate student gave the School a strange collection of walnut branches for display in the Seminar Room. Faculty and students wondered, "What is this?" Not willing to turn down a gift horse, much less look in its mouth, we entitled the wall hanging The Forestry Gate. (I've been encumbered by much greater entanglements in the southern woods.) Rebecca Townsend commissioned a painting of the Southern Pine Lumber Company mill as it appeared about 1900, also for the seminar room.

Some 17 industries contributed paneling for offices and halls and parquet flooring. So, too, the collection of boards from North American and exotic trees were scrounged from here and there. The sassafras (*Sassafras albidum*) lumber, tongue and grooved on site, for the dean's office came from the Mississippi bluffs. Sid McKnight and his colleagues at the Stoneville Research Center of the Southern Forest Experiment Station recommended the heartwood of this species, as it glows more golden with age. Then they helped me locate the mill managers willing to contribute the paneling. While our aim was to erect a monument to wood, we settled on 24 species of interior paneling. Fred Gragg, International Paper Company vice president, personally hung a Company picture.

The fire simulator was modeled after a West Point battle simulator and a USFS fire simulator, combining what the architect's engineers considered most appropriate.

Thousands of feet of wire connect the student stations to the control room for the 3-dimensional display of photographs and for 8-track stereophonic sound at the call by radio from a student "on the fireline." Five coordinated focal-length projectors focus on a wide screen for studies of the physics of fire behavior and wildfire control. The system contains a student decision-making voting system.

Library

Before my arrival, Prof. Baker and others had begun assembling a good forestry library. I got my major professor at Yale, Harold Lutz, to give us his collection upon retirement. Don Canterbury, retired forester for Champion Paper's Texas operation, entrusted us with his; so did others. Walter Garstka, an internationally recognized hydrologist from Colorado, added to the collection. It included Pinchot's 1899 and 1905 manuals of forestry. When visiting Washington, I'd go by the Library of Congress with a pad of Congressman Dowdy's franked mailing labels. The folks there took me to shelves of discards and I'd select boxes full and ship them to Nacogdoches. In the basement of a used book store in New York City I discovered a copy of John Evelyn's *Silva* in mint condition. Published in 1729, the 329-page 8x12-inch tome with marginal drawings in color, is bound in leather. The University paid $40 for the volume. From studying auction records, I'm confident its value is well above several thousand dollars. The partial title in Old English lettering:

SILVA:
OR, A
DISCOURSE
OF
FOREST-TREES,
And the Propagation of Timber in His
Majesty's Dominions:
As it was delivered in the ROYAL SOCIETY the 15th of October, 1662,
upon Occasion of certain Queries propounded to that Illustrious Assembly,
by the Honourable and Principal Officers and Commissioners of the Navy.

Segment of the W. Goodrich Jones Library, centerpiece of the Forestry Building.

Before my time, too, the Forest History Foundation (now Society) designated the College as the official archives of material on forestry and its industries in this State.

Jerry Ann Davis served as librarian; then in 1966 Mary Jo Linthicum signed on. She maintained the School scrapbooks, now housed in the University Archives. Laura Norman, and Susan Hearnsberger ably assisted her. Together they could lay a hand on every needed reference. Mary Joe served from 1966 until her death in 1972. Alumni, students, staff, faculty, and friends placed a Regulator Clock and bronze plaque in the library as a permanent memorial to her always-cheerful helpfulness.

I fought the loss of the Library. A rental truck was used unceremoniously to move the 80 thousand volumes to the new main library, a memorial to President Steen.

Budgeting Research

When I arrived at SFA, the McIntire-Stennis federal research funds were wholly utilized by the Texas Forest Service. I called attention to this violation of Congress' rules for spending these monies. They were earmarked for educational institutions. Finally, with the help of Governor Preston Smith, the Coordinating Board took control of the money, designating that, beginning in 1965, it would be divided evenly between SFA and the Forest and Range Department at TAMU.

Mary Jo Linthicum, principal forestry librarian, and students.

In 1963, the research budget amounted to $10,000, all of it assigned to Victor Bilan for salary and research. (It approached $300,000 at the time of my reassignment.) Thus, it angered Victor when, upon entering on duty, I asked him to help with silviculture instruction. As department head, I wanted a relationship with students that came with teaching the introductory course. I also wanted the initial instruction to be field-oriented. And as the author of "The Southern Pine Region" in *Regional Silviculture of the United States*, I enjoyed teaching the senior course and, thereby, more fully knowing the students as they left these academic halls. So I needed help with the sophomore silvics class. Victor insisted that he was hired to do research, *period*. He was greatly disturbed. But I wanted his vast store of knowledge shared with students in the silvics course. Language problems probably attributed to his reluctance. Nonsense, his accent and humor made him a good lecturer. Within a year or two, he thanked me for forcing him to teach. In time, he won many teaching awards and served as major professor to a great company of graduate students. Readers will find a eulogy to Victor, spoken before an annual district conference of Kiwanians, in the Appendix.

To stimulate outside funding, the School circulated a flyer entitled, "An Invitation to Cooperate in Forestry Research." Our annual report laid on the desk of every state legislator in the House and Senate chambers seemed to produce results; each year we received more money for research.

"A Ten-Year Plan for Forest Research at SFASU," provided for the Coordinating Board's Research Advisory Committee in 1966, spelled out why the School was in a strategic position to gain national recognition in research: faculty, locale, alumni, and the economic position of the timber-based industries in Texas.

National Science Foundation lecturers included world-renowned plant physiologist Theodore Kozlowski (who once tried to hire me for the University of Massachusetts). Francois Mergen, forestry dean at Yale; and Andrew McComb, hydrologist at the University of Arizona. NSF research grants were readily obtained in those days.

SFA school deans with the president, Ralph Steen, front, (l), Joe Gerber, graduate school; rear l-r Robert McKibben, education; William Turner, fine arts; Theodore Kallsen, liberal arts; Edwin Miller, science; and the author, forestry.

The Insistent Mother

Years before, as a layman and moderator of the Presbytery of Northeast Georgia, I sat in on a lecture by the president of Columbia Theological Seminary to the presbytery's ministers. One of his points related to a certain problem for pastors. It was good advice for deans too. "Never," he said, "let your office door be closed with only you and a woman inside, regardless of the confidentiality of the conversation." I passed these instructions to Nancy King, our administrative secretary, as I had earlier told Nell Crews, Nancy's predecessor. Fortunately so.

A certain student had been expelled from the College. His well-dressed and attractive mother of my age came to press for his reinstatement. Upon entering my office, she closed the door, sat down, and hiked her skirt above her knees as she crossed her legs. Nancy played her role well. Calling on the phone, she asked a non-question to which I answered something like, "Oh, yes, I need to sign that now." I signed the fake papers, while Nancy noted that I was in bad trouble. She left the door open as she returned to her office. The mother promptly rose, shut the door, and went through the same routine. Nancy found another excuse to enter, again leaving the door ajar. I think the mother finally caught on, as I routed her to the vice president, a fellow who admitted to his vanity. Later that day, I chanced upon the mother on campus. On this occasion she grabbed me by both lapels to tell me how nice I looked in my new (really old) suit. She was a determined woman.

Persistent Mother

A three-week silviculture field trip I taught carried students *via* a rickety school bus to the West. This particular summer the turnaround was Yellowstone National Park in Wyoming. One coed, among the first females in forestry, registered for the course, every night of which would be spent on the ground during the 3,000-mile sojourn. Told that another female must accompany her, she was a bit incensed. A sister, friend, even her mother would satisfy the requirement. Unable to locate another female, she sent her mother to plead her case.

"Tell me," the mother asked, "Why must there be two girls?" "To protect the professor," was my reply. "And who is this professor you cannot trust?" she asked. "It is I," I said.

I needn't have worried. Another girl enrolled for the course. At each day's end, the gang of 30-plus would split, one girl going with one group to unroll a sleeping bag; the other with the other men. Both were well protected.

On Writing

To call attention to the School and to the forestry profession, I tried writing. Among the efforts was *Ecology and Our Forests*, published by A.S. Barnes in 1972. Some members of the profession didn't like it, though it was designed to tell laymen what we do and why we do it. When the editor of the *Journal of Forestry* published two negatively critical reviews, one as a lengthy letter to the editor, I wondered if it was because colleagues from this region of the country had nominated me, still an outsider, for the SAF vice presidency. This treatment, in turn, brought other letters, laudatory in tone, and one suggesting the membership was an "association of perfected saints." W.A. Wertz wrote a letter to the *Journal* editor: "Dean Walker has sparked a dialog seldom heard in our circle of professional foresters—a dialog which centers on writing for the public." To all such critics, H.M. Silver has "One last memo... . If in the printed article you find that despite every precaution an error has been printed, do not despair of humanity ... . No book is completed until Error has crept in & affixed his sly Imprimatur."

When Captain J.A. Moir chanced upon our *Silviculture of Longleaf Pine* bulletin in a London used book store, an interesting correspondence followed. The seafarer had planted all sorts of exotics from around the world at "56 North." The latitude marked Glen Eden Cove, his home near Dunbartonshire, Scotland.

More-or-less regularly published articles over some years in the *Farmer-Stockman* instructed farmers on forestry and let folks know we were here to educate their offspring. Miller-Freeman Publishers collected these squibs into a book, with 46 short chapters, called *Farming The Small Forest*, published in 1988.

Faculty Additions

Harry Wiant had been one of my graduate students at the University of Georgia. Later he earned the PhD at Yale. Now on the faculty at Humboldt State College in California, he was the most likely candidate for the assistant dean position in the School of Forestry. This was before the days of search committees; an administrator simply selected subordinates with whom he (there were few shes in administration) thought he could work. We made a good team, principally because, unknown to the faculty, he leveled with me. Had he not responded to the call, "Come Home to West Virginia," the turmoil that would later wrack if not wreck the School (discussed in a later chapter) could likely have been avoided. Harry continually pleaded with me, "Don't try to change them," but I didn't listen. I wanted every faculty member to be a producer and to keep their laboratories tidy. "Don't worry the faculty," Harry would say.

Faculty joining the staff during my stint as dean, in addition to the aforementioned professors, included Mingteh (Mike) Chang, Jack Coster, Eugene Hastings, Ray Hicks, James Howard, James Kroll, Michael Legg, J. David Lenhart, Gerald Lowry, W. Thomas McGrath, Edwin Miller, Hershel Reeves, J. Robert Singer, David V. Smith, Seymour Somberg, Arthur Verrall, Kenneth Watterston, Charles Webb, George Rink, Sherwood McIntyre, Robert Gara, Carlton Yee, and Bill Wilford. R. Montague Whiting, Malcolm MacPeak, Bruce Cutter, G.H. Weaver, James Martin, H.J. McCarty, and Robert Wendling were recruited as part-time faculty while working on doctorates. Billy Early, Bob Webb, Robert Bloom, Finis Prendergast, Ernest Golden, Lowell Halls, John Stransky, and Henry Short joined the team as lecturers in their specialties. Kenneth Mace of the Biology department taught pathology in an interim. George Stephenson, retired Forest Service researcher (as were Verall and Wilford), signed on as part-time editor of the School's series of bulletins and research papers. Alan Britton and Kent Smith served briefly. Robert Maxwell and Ben Dickerson provided instruction in forest history and rural sociology, respectively. K. Kadambi, India's chief conservator of forests, joined us with a limited visa to provide some additional international flavor. SFA was among the first forestry schools to teach courses in law, wood procurement, and air quality.

Robert E. Tucker, a Louisiana forestry graduate, rose to become Southland Paper's chief forester. Then something happened. Mr. Kurth, the company's president, called to ask if we would like Bob on our faculty. Southland would pick up the salary. Of course we were delighted. Tucker had a hidden talent. Upon departing SFA, he plunged into my hand a book. Titled *Louisiana Birds*, the large volume was wholly illustrated by Bob. He earned his way through LSU as the illustrator for ornithology professor George Lowery, Jr.

Secretarial Nuggets

One nugget followed another as administrative secretary for the School of Forestry. Nell Crews, a recent bachelor degree recipient, held the post for two years—until a new forestry graduate, Phil Higgins, asked for her hand. What good fortune for a less-than-easy-to-please dean than for Nancy King to apply as Nell's replacement.

Nancy King Pase in the sassafras paneled office.

A nervous and excited young lady, also with the required Bachelor in Business Administration (BA) diploma, took over the tasks after Labor Day in 1966. As I was away at the SAF meeting in Seattle when she entered on duty, it was to Nancy I called to pass the message to the faculty about finally winning accreditation. Nancy later wrote that she had little idea of the significance of that announcement. But she soon realized she was an important member of the team and continues a strong sense of loyalty to the College of Forestry.

Hear Nancy on a particular occasion.

My job as your secretary was sometimes frantic and tiring, but was never boring. In the ten years that I worked there, my respect for you grew more and more each year. As I observed your life from the perspective of my secretarial desk, I learned that there were three things of utmost importance to you—your God, your family, and your profession.

Even though I knew you as an administrator at an institution of higher learning, I began to see that you never stopped learning yourself. You had an insatiable desire to see as many types of forests in this world as you could. So you traveled, participating in meetings in many locations from Alaska to Australia. I was impressed that at any meeting you attended you always visited the forests of that area (taking myriads of photographs!). And on many occasions when the burdens of your responsibilities were weighing heavily on you within the School of Forestry, you would come to work dressed in field clothes and spend a good part of the day in the forests of East Texas. You always returned seemingly refreshed and ready to tackle the task at hand. Now that I am married to a forester, I have a greater understanding of what the land and the trees mean to those of you whose desire in a career is to manage the forests for the "greatest good of the greatest number in the long run."

Another thing I quickly observed about you was that any time you traveled, you took what I came to call your propaganda package—a folder of material about SFA's School of Forestry—which you shared with any prospective student you met upon your travels. For example, once when you were driving in Colorado, you stopped to help a young couple change a flat tire. The young man left that meeting with a stack of material from you and later enrolled in the graduate program in the School of Forestry.

(Well, almost. The young man, a hiker, helped me replace a water hose at timberline on the Frazer Experimental Forest in Colorado. He went on to earn a DF at SFA.)

Nancy continues. (I'm enjoying this too much to scrub it.)

> In addition to observing your devotion to the School and the profession of forestry, I was also in a position to see your love for your family. You were never too busy to talk to Mrs. Walker or one of the children when they called or came by the office, and the tenderness in your voice when you spoke to them or about them was obvious. In the early years, I saw you pull Wendy and Jean to the office in a little red wagon many Saturdays when you came to check your mail and keep any appointments you had. In later years, THEY drove YOU there in a more sophisticated vehicle! Your children's devotion to you was a joy to see; and although by your own admission you lacked patience, I saw nothing but patience and love as you spent time with them. Mrs. Walker, of course, is a precious lady, and the perfect balance for you. Her soft voice and gentle ways have always impressed me.
>
> As your secretary, naturally, I found out early that you were a prolific writer. I doubt if you napped on any plane trip you took, because you always came back with a yellow legal pad filled with a manuscript of some kind. Your distinctive handwriting is quite familiar to me! You filled up the dictaphone, too, with many of your manuscripts, as well as correspondence of all types. I suppose you answered every letter you received. I typed your signature at the closing of so many letters over the years that even today, when I type the word Sincerely at the end of a letter, I have to be careful not to go down four spaces and type Laurence C. Walker, Dean! Word processors would have made the manuscripts and correspondence much easier!

During these years, a fellow appearing as though he'd just emerged from "Tobacco Road" asked Dr. Kenneth Watterston, on the forestry building parking lot, where "I

The author's family in the early Texas years. They rate cheers for the frequent absence of a husband and father, much too wrapped up in his work. Janet, a graduate of Wheaton College, became a surgical nurse; Stephen, a product of the Los Angeles College of Design, is a creative director for an advertising firm; Wendy, inspired to be a deaf education major at SFA because of two deaf nieces, teaches school; while Jean, a Texas A&M Aggie, teaches science. Anne, in her wisdom, promised, "We pay for no wedding until you have a college degree." The strategy worked.

could be found?" Ken, figuring the fellow was my kin from Harlan County in Kentucky's Appalachians, directed him to the building, also informing the man I was out of town. The ruffian responded, "He better not be," and raised his clenched fist which was in a cast. Nancy gave the man the same answer, whereupon anger swelled, as he shook the fist and said that I was to take off the cast. Nancy: "He didn't put it on, did he?" "Yep, riot chere in this hospital." A few phone calls and Nancy discovered that she needn't be an attending nurse after all. There was a new young surgeon, Dr. Larry Walker, in town. Apparently to the injured logger the Forestry Building appeared like a hospital. Nancy and the real Dr. Walker still chide me about moonlighting at the hospital, and without a license, too.

One young man, apparently fresh from the Navy, thought Nancy's name plate "Adm. Secretary" meant Admiral's Secretary. She carried the title with dignity.

Nancy, indeed, was "the gal who runs things around here," as she reminds how I introduced her to visitors. It was so.

She left the job when I was reassigned. It came about this way: Joe Pase, a homesick-for-Texas forestry student in Florida, wrote for information about our programs. He got the packet and, with it as always, a hand-signed letter. Transferring to SFA, he earned the bachelor's degree, served in the Vietnam War, mounted a fantastic collection of Southeast Asian insects, and returned to Nacogdoches for a master's degree in forest entomology. As with Nell, Nancy married the forester, and they rode off into the sunset together.

Other Benefactors of the School

Florence Frazier, of Marshall, had expressed a desire to help the School of Forestry. She was the sole heir of property received in a land grant in the Conroe area dating to the establishment of the Republic of Texas. Named for her forebears, Florence called it the Ballenger-Long Endowment. We were to divide the proceeds of her estate with the Baptist Seminary in Fort Worth, her late husband having been the minister of Marshall's First Baptist Church. Mrs. Frazier's interest in forestry was occasioned by her father's employment with the Texas Forest Service as a fire patrolman and rural educator, carrying portable "lantern" slide show equipment to communities to discourage woods-burning. Subsequently she provided scholarships annually to forestry students. We lined up Charles Stealey, Jr., Yale forester of the class of '50, to manage the land for the Widow Frazier.

Texas Farm Products and Dow Chemical Co., sometime

about 1970, nominated the School for the "Be a Hero" award. I never learned why. It didn't matter. The award was a much-needed new pick-up truck. Tom Wright, TFP vice president, made the presentation.

The Yellowpine fire tower during a dry run to determine if an Army helicopter could hoist it for relocation to campus. Major Brian McWilliams of the ROTC staff led the effort.

When Don Canterbury moved to Massachusetts to be near his daughter, a long letter from his wife told of his wishes. "Don has always said...he would like his office furniture to go to you for Stephen F. Austin." We subsequently took the old stake-body truck to his home near Rice University and moved desk, chair, map table, voluminous maps in containers, files and cabinets, and books. The School could use these materials from one of the nation's leading foresters, a Yale graduate of the class of '22.

E.O. Siecke, the State's second state forester, also provided for the School in his will. The gift of another notable beneficiary didn't materialize till 20 years after University Regent Harold Bates and I spent a week in Houston. Harold, a lawyer from a family of attorneys and jurists, knew all the buttons to push as we met appointments with managers of foundations headquartered in the Bayou City. Mrs. Henry M. Rockwell, the sole remaining recipient of a trust, depended upon its income. At her death the 7-figure proceeds were to go to the School of Forestry. This they did.

Two supporters of the School of Forestry; Don Canterbury (l) and E. O. Siecke, retired Champion chief forester and director, Texas Forest Service, respectively.

Texas Friends of Conservation and Houston's Bank of the Southwest also provided scholarships for forestry students.

Lacy Hunt Visiting Lecturers

We utilized money from the Lacy Hunt endowment to bring in practicing foresters for a week to visit with students. The sojourners spoke in formal and informal lectures, while an office along the main flow of student traffic was set aside for one-on-one conversations.

Stephen Spurr, (r) president of the University of Texas, autographing his text for a student while making his "inaugural" address in Texas at SFA.

Several visitors deserve note: Charles Connaughton had served as director of the Southern Forest Experiment station and Regional forester for the Southern Region, extending from the Carolinas to Texas. Charlie closed out his career as regional forester for the Pacific Northwest, land of the big trees and vast federal estates in Washington and Oregon. Edward Cliff, while still chief of the US Forest Service, also served in this capacity. Then, Arthur Greeley, formerly Ed's associate chief, added significant input to the visiting lectureship. Cliff, the jack-leg Mormon, by his own admission, and Art, the second-career Methodist minister, were studies in contrast. For Greeley's first lecture, faculty and visiting foresters filled the 500-seat Kennedy Science Auditorium. Art, following the customary introduction, began his lecture with a prayer for guidance for future managers of the Creator's benevolence. The forester who entered seminary following early retirement and subsequently ordained a clergyman spoke and later wrote to say that "faith is the basis of our continuing to find purpose in life."

Charles Connaughton (r) with student while the retired regional forester served as visiting professor.

Industry executives also occupied the visiting professor's office. Paul Dunn, one time Oregon State University's forestry dean and now vice president of the giant St. Regis Corporation, was among these helpful people. The program folded when a state audit ruled it improper to use gift money for first-class accommodations in a local inn. The visitors had volunteered their time!

Paul Dunn, vice president of St. Regis and former dean of Oregon's School of Forestry, makes the 1970 Forestry Building dedicatory speech.

Smokey Bear Creator

A long generation of foresters in the South witnessed the continuing 35-year weekly saga of Smokey Bear cartoons by Harry Rossoll. Harry, co-creator of the anthropoid bear, also illustrated uncounted bulletins for federal, state, school, and industry publications. Old-timers recall the folksy, comical, fat bruin with a pointed snout wearing baggy dungarees and a World War I campaign hat with chin strap. That was in 1943, when the FS began free distribution of the public service cartoons to 3000 newspapers. Harry's prominence as a forestry illustrator and his warmth as a friend encouraged me to ask the regional forester to allow Harry to leave his easel in Atlanta and to paint a mural for our new building. Roy Bond willingly would send a lesser-known artist. No! It was Harry we wanted, as much for Harry as for his artistic ability. Roy relented. Harry brought his paints, set up shop, and spent as much time kibitzing with the students about forestry and life (he'd had some tough blows) as he did spreading oil on canvas. When the task was completed, students honored Rossoll with a dinner. But we didn't know Harry. He, in turn, hosted a party. And what a party!

Harry Rossoll at work on the mural.

The Rossoll mural depicts in vivid colors the broad

scope and various aspects of professional forestry employment. We asked him to include a female, as coeds were beginning to show an interest in our programs. And we suggested he reduce the size of the instrument shed at a dam wier, lest some wise guy paint a half-moon on the door. The box does appear like an outhouse; and I've seen such "comfort" structures alongside streams in many woodlands of the world.

Harry Rossoll's mural.

I treasure a print of a 1973 Rossoll portrait, entitled "ch. Orion Flush," the famous bird dog on the point for bobwhite quail in a sparse longleaf pine stand. The large framed scene, commissioned by a wealthy Louisianan, greets a visitor entering my study. Twenty feet away, at the entrance, Joe Richardson, trainer of both hounds and athletes, promptly identified the dog as the champ blood line pointer, sire of other champions of the breed.

The Prince of the Pines

John Henry Kirby, Plowboy of Peachtree Village in Tyler County, Texas, became the Prince of the Pines, whose dominion covered extensive oil and timberlands. The lumber baron to some was an agent of the Lord and to others the Devil's representative in southeastern Texas and southwestern Louisiana. SFA's late Robert Maxwell and Robert Baker record his story in *Sawdust Empire*. But our concern here is the massive hand-crafted office furniture assigned to the School of Forestry during Thomas Orth's presidency of the Kirby Corporation in the 1970s. Kirby's daughter, so the story goes, had the figured redgum (*Liquidambar styraciflua*) carefully selected from the river bottoms and second bottoms of the lumber pioneer's vast holdings and the cants from these trees shipped to craftsmen in the Adirondack Mountains of New York. There, at Saranac Lake, the huge desk, chairs, and other pieces were fashioned from original designs. Each company president, through Orth, utilized the furniture, the School finally receiving it when Louisiana-Pacific bought the Kirby Company in 1984. Kirby ruled from behind the ornate *secrétaire* until his death in 1941. Orth and James Webster also presented the Kirby records to the SFA Archives.

2,4,5-T Study

Texas Tech's School of Medicine had a contract from the federal government to study the effects of the herbicide 2,4,5-T on humans. While this was an early-day Environmental Protection Agency effort, the continuing saga of the alleged damage to Vietnam soldiers no doubt was involved. The Lubbock school needed someone with a knowledge of the use of the chemical. I was enlisted. Then, rather suddenly with the election of President Carter and his appointment of Barbara Blum as EPA boss, Tech's contract was cancelled. The study never got started.

And the EPA outlawed the use of 2,4,5-trichlorophenoxyacetic acid. I wondered then, and since, if cancelling its use in forest management could have been related to Blum's ineffectiveness in controlling the wood-using industry while she served as Governor Carter's natural resources point-person in Georgia. The industry there managed to protect the Georgia Forestry Commission from entanglement with her new broad agency of state government. Farmers legally can still use 2,4,5-T in rice fields and pastures!

Presidential Candidates

Two men whom I knew well campaigned for the presidency of Stephen F. Austin State University when the

Board sought Dr. Steen's replacement. I was asked by the chairman, and spoke candidly of what I knew of each. Reagan Brown headed up the Agricultural Extension Service, always getting applause for his inspirational talks. I had been associated with Dr. Anson Bertrand when he managed the USDA Agricultural Research Service program in Georgia. Either prexy would have been ideal as a follow-up in an institution too long steeped in educationism and liberal arts. Sciences, many thought, had been neglected. Now dean of Agriculture at Texas Tech, Anson was ready for the task. When passed over by the Board, he promptly proceeded to Washington as the assistant secretary for the Department of Agriculture. Reagan soon became the State Commissioner of Agriculture.

The Shay Locomotive

The old No. 6 locomotive lay rusting in the graveyard at the W.T. Carter and Bros. Lumber Company mill at Camden. Tommy Carter, by 1970, the principal with whom we dealt, was willing to let the school have the engine with some provisos. It would be a loan; we must move, restore, and maintain the antique logging engine.

Having shaken hands on the agreement, the faculty, Sylvans Club leaders, and I pondered how best to get the job done. Colonel James Calahan, commandant of the ROTC on campus, interested his students in assisting. Railroaders of the Southland Paper Company, owner of the Angelina and Neches River Railroad, took control; while Lufkin Foundries loaned two of its largest truck-tractors and a lowboy trailer for dragging and hauling the heavy engine to the campus. Cross-ties were laid on which rails were attached to form a ramp from the track on the ground to the track on the trailer. With the two giant tractors harnessed with stout cables to the Shay, the rusty engine wouldn't budge. Five gallons of diesel oil sprayed into each wheel hub did no good. Then an old-timer came ambling along, saw the predicament, and informed the crew that just before the engine was retired, it had been refitted. Maybe the pistons were stuck? That was it! Unbolting the pistons from the cylinders, the engine was slowly inched

Views of the Shay locomotive and log car. This locomotive last moved logs for the W.T. Carter and Son Lumber Company in Polk County. Carter flatheads were the last to use oxen in the Texas woods, about 1950. Lousiana-Pacific provided the log car once used by the Angelina County Lumber Company.

by the tractors onto the lowboy. News people picked up the

scene as the old Shay lumbered down the main streets of Lufkin and Nacogdoches en route to the schoolyard. When personnel of Lufkin's Louisiana-Pacific's plywood plant learned of the event, they called to ask if we wanted a log car, too. Indeed we did. The A&NRR men, under Superintendent Jack McMullen, Jr.'s, direction, were willing to take on the added task, requiring another trip with the tractors and trailer and laying more rails. The Sylvans Club went to work on refurbishing the engine.

Nancy King, administrative secretary for 10 years, driving the golden spike for the Shay locomotive track.

They added to the locomotive a bell, source unknown,

that had been used for years at the Summer Camp to call students to meals. Rummaging in the attic of a farmer's barn on the Upper Melrose Road, I came upon a heavy iron piece threaded at one end, hooded at the other, and open on two sides. Molded at the factory on the "thing" was the word BASS. My initial reaction was that its pronunciation would be that of the fish, not that of the musical term. With further reflection and discussion with its owners, the iron gadget indeed was the whistle to a locomotive. And Mary Ann Bussa's grandfather, Clyde Woodward, had been involved in logging. He likely engineered this very Shay for W.T. Carter. The whistle's threads fit perfectly the Shay's air outlet pipe. Ah, now to see how it sounds. It took only about 60 seconds of the deepest bull bass you can imagine for the entire compressed-air tank to empty. Folks heard the sound two and three blocks away.

Sylvans Club members replaced the wooden floor of both the cab and tender, persuaded a college cabinet maker to refit the door that opens out to the boiler's exterior, poured gallons of fiberglass into the rusted-out holes, painted the engine and log car, and redid the company's name in old-timey letters.

This 36-ton Shay, Carter's No. 2, welded together in 1907, first labored, until 1915, building the Panama Canal. In the tropical climate, its wood cab rotted out. Upon arrival in the South's pineries, a blacksmith replaced the wood with sheet metal. (Another account has the manufacturer's No. 2005 purchased by Carter from the Behring Company soon after its arrival at the latter's Texas mill.)

Emphraim Shay was a blacksmith for a sawmill in the Michigan woods. He figured out a way to use geared wheels, like those in a clock, and to offset the boiler to balance the weight of the wheels. The great Lima, Ohio, C/H locomotive manufacturing company bought his idea and made these "beasts" for world markets. The Shay could negotiate grades up to 7-1/2 percent and curves as sharp as 60 degrees. Readily "walked" back to the rails when it jumped the tracks, and fueled with pine knots, these locomotives facilitated logging the virgin forests. It is said they could pull ten times the weight of ordinary locomotives of the same class, make ten times as much noise, and travel at one-tenth the speed.

Summer Camp

Paul Wilson, state representative from Sabine and adjacent counties, selected the site for the Summer Camp, an "educational experience enjoyed" by every forestry graduate. Paul chose the location simply because it was close to his homeplace. A knowledgeable forester could not have done better. Nine forest-cover types lie within short distances of the facility near the Red Hill Lake Recreation Area in the Sabine National Forest: pine—hardwood, upland hardwood, loblolly pine, loblolly pine—shortleaf pine, bottomland hardwoods, southern baldcypress, American beech, longleaf pine, and slash pine (plantation).

The peckerwood mill at summer camp.

Paul was also chief of facilities maintenance for SFA. With his own crews in the mid-1950s, he built the dormitory and classroom-dining room. Wilson's Waldorf would be long remembered by students who received instruction there over a 30-year period. Land surveying, logging and milling, mensuration, and applications of silviculture classes would be endured along with chiggers, ticks, mosquitoes, and *reptilia* of all sorts.

Martin Luther Mason cooked for the students at the Summer Camp. "Cookie" was a favorite until the enrollment changed: foresters began coming from the cities, brought up on McDonald's, whereas their forerunners were sons of loggers and lumbermen, accustomed to eating food cooked in deep fat.

Querying Cookie on the complaints, I learned he saved the grease from one meal for the next, storing it in a number 10 lard can behind the stove. He didn't know it was rancid, for all his own meals were bowls of Post Toasties. He said of his own cooking, "My stomach can't handle it." Suggesting to Cookie that he not save the grease, my next visit with the students found the chef with a gallon jug of Mazola in the elbow crook of his arm. T-bone steaks were still cooked in hot grease. He couldn't help himself. After all, he fed thousands over the years in his "Greasy Spoon" astride the railroad tracks in Nacogdoches. The cafe died with the demise of passenger service.

When Mr. Mason, with the beautiful tenor voice heard among the First Baptist Church choristers, retired, Orie Wilson took over the task. For two decades the widow loved those students and they loved her. Orie commanded the kitchen when the first young ladies arrived at the Milam installation. Alyse Moore was the first to room with Orie, while others stayed in the homes of local folks. When student numbers increased further, we housed the women in a trailer and Orie happily shared the chef's private bathroom. With enrollment topping a hundred, Mrs. Wilson's daughter, Doris Burrow, hired on to help mash the potatoes and scramble the eggs. Orie seldom brought her pet chimpanzee to camp.

In 1994, Monty Whiting, long-time Field Station director, published the students' favorite recipes of these two lady cooks. From 1968 to 1988, they probably "fed more professional foresters than anyone else in history." Tom Boggus, class of 1978, described their breads:

> SPLASH! Another survivor of surveying has made it to the refuge of Red Hills Lake. . . all pledge never to leave this haven for briar-scarred field station foresters. But wait, what's that smell coming from the camp kitchen? Mrs. Wilson and Doris have dealt this young conspiracy a fatal blow, for the smell of those homemade yeast rolls is unmistakable. Serving these golden brown dinner rolls with any meal, made (with) your own TLC is sure to squelch your rebellions too!

Continued growth in student numbers required refitting Wilson's Waldorf. Carpenters and plumbers converted wash rooms to dorm rooms and erected a new large shower room. Empty containers left lying around were not required to convince the observer that neither plumb bob, spirit level, nor carpenter's square were utilized for the new construction. The whole building tilted. I wasn't alone in needing to stand on a box to reach the shower faucets and for other necessary functions.

Student grumbling to President Steen about the summer camp requirement called me to the prexy's office. Some claimed they'd had enough Army; others thought it would be like Boy Scouts. When we discovered the complainers had not been to the Sabine National Forest installation but had been to "Nam," and my recalling Blue Jay on the Allegheny (Chapter 6), I realized they had a point. Changing the name to Field Station solved the problem.

When the worn-out installation had seen better days, closing in 1986, Kyra Sanders memorialized its demise: "Contented in old age like grandparents (and) happy to know the strength it has passed on to its children and grandchildren was not in vain. Rest well, friend."

Big Thicket

The Big Thicket problem would not go away. Big industry and small private landowners near the proposed setaside fought against the federal effort; while Dempsey Henley, mayor of Liberty and real estate magnate, encouraged it. For a while Senator Yarborough was president of the Big Thicket Association. Often public attention focused on Lance Rosier, an elderly self-taught botanist living in Saratoga who knew every wild, and especially the rare, plant in the area.

I tried to spell out the pros and cons of the Thicket purchase in an *American Forests* article, and accompanied Supreme Court Justice William O. Douglas on a tour in preparation for his book, *Farewell to Texas*. Dempsey lined up a large oil company helicopter for our visits to various sites, the judge carefully avoiding the industry's logo when the press took photographs. Some environmentalist hangers-on accompanying the judge as we rode from place to place in a fleet of chartered, air-conditioned Cadillacs refused to believe the rings in my increment borings, accusing "assassins" (it's in the book) of "poisoning a thousand-year-old magnolia" so the "forest would be less attractive for a park." I couldn't find the five holes into which "arsenate of lead" was allegedly poured. The tree was too far gone for sampling with the Swedish borer. So, a stranger to Texas with a prominent name tours the whole state in a few weeks, writes a popular book, and proclaims it "a vanishing wilderness." All the while he insisted the US Forest Service clear a view of trees for his vacation home adjacent to a national forest in his home state of Washington. I offered to review his manuscript, in which he refers to my helpfulness in the field, for technical correctness. He wrote from his Supreme Court chambers, in 1966, "... I appreciate your offer of assistance."

Supreme Court Justice William O. Douglas and Mayor Demsey Henley in the Big Thicket.

About this time, I met Campbell Loughmiller. He and his wife Lynn had two dissimilar, yet similar, interests. For 20 years they operated Camp Woodland Springs, a place for treating emotionally disturbed boys. They also knew natural history and loved to roam the Big Thicket. Although professional sociologists, the Loughmillers were equally enthusiastic about wilderness adventures, taking these kids on raft trips, canoe treks, extended mountain or Thicket hikes, and anything else that would enable the youngsters to become self-sufficient, no longer predators on society. Even in their base camp in the East Texas woods, the boys cooked their own food or would die of starvation. They became young men as they learned to conduct their own worship services. Campbell writes of this effort, funded by the well-heeled Salesmanship Club of Dallas, in *Wilderness Road*. The Loughmillers and I endeavored to find ways our School's students and graduates could assist in his programs, particularly work in the Big Thicket. We were only marginally successful.

Letters to the editor in *Texas Parade* appeared in the summer of 1976 between Dudley Lynch, author of a "Business Memo" column, and me over the Big Thicket National Preserve. Setting aside the Thicket was no longer a matter of "preservation of biomes," I wrote. It was now a social-political issue. "Students and faculty have every right to take positions on such issues; but for the school dean to do so, in my opinion, would draw the school into a controversy which would affect its working relationship with the industries, the government, and the preservation-type groups which employ its graduates." I had previously noted in those pages that "all of the Big Thicket has been harvested once, most of it twice and much of it three times." Now I noted, "That the Big Thicket has been harvested can be used, of course, as a rationale for both setting it aside and for continuing to use it for the production of fiber. Again, I saw my task as spelling out pros and cons. After the Congressional Act to establish the first natural preserve was signed into law, the National Park Service retained me to make an ecological study. Feeling among Big Thicket citizens ran so high and hot that the NPS instructions were not to put a foot on the land to be purchased and to do the study along public roads.

In 1975, State Senator Don Kennard had worked up legislation to provide funding for the University of Texas to make a study of a half-dozen natural areas. The University obliged by hiring Don as the project director. Not being a naturalist, but a knowledgeable outdoorsman, he contracted with several of us at SFA to do the field work. The story of Blue Elbow, the swamp astride the Sabine River at the northern boundary of the city of Orange, was published in *Texas Parks and Wildlife*, the Houston *Chronicle's Texas* magazine, and in a UT report. Several faculty members joined me for similar studies of International Paper Company's Woodland Cathedral in Panola County and of Harrison Bayou, a virgin tract on the shore of Caddo Lake. These stories also appeared in *Texas Parks and Wildlife*.

Bicentennial Dedication

Sabine County, much of which lies within the Sabine National Forest, arranged for the dedication of a monument on the courthouse square commemorating the nation's 200th anniversary. I reminded the large gathering that history is "His story." "Our times are in His hands" in spite of the fact that the preacher who first represented the Baptists in this area, in 1822, was arrested during an exhortation. I carried them through the changes in the forest, from virgin stands to devastated lands to restored forest. Here Juan Ignacio Pifermo settled in 1796 in return for 11 leagues of land, his houseplace somewhere east of Palo Gacho (poly gotch) Creek. Not long till James Gaines built his store and opened a ferry boat operation across the Sabine River. Nearby he laid out the town of Pendleton, a family name, on *El Camino Real*, the royal road for which Chevrolet pick-ups would gain a name. As Gaines had two wives simultaneously, a Rev. McMahon became the subject of Gaines' wrath. A major in the local militia, McMahon only a year earlier had become a Christian while engaged in a secret prayer meeting on the bank of Ayish Bayou in the then Mexican territory and not open to non-Roman Catholic services of worship. Soon the militiaman was ordained. George Crockett tells about it in *Two Centuries in East Texas*, noting McMahon's willingness "to take a whipping as well as any man on this ground." McMahon lies beneath the pulpit of the remote woodland chapel that bears his name. If present orations from behind that sacred desk are not faithful, rumor has it, the ghost of the late cleric shakes the marble slab beneath the preacher's feet.

Chamber of Commerce types reported that east of Milam flowed "pure fountains of excellent spring water." There, with fervor, folks "sang the songs of Zion: 'Jehova rules the tide/and the waters he'll divide.'" This was a fair warning to the likes of the Gaines clan and possibly for the sheriff, appropriately named *Lynch*. So a Bishop Morris in 1841 announced, "There are in this neighborhood some Methodists, Baptists, and Presbyterians—and plenty of sinners."

Though Sabine County dates to 1837, minerals and oil never brought wealth to it. The 1910 census records 1551 goats and 1488 sheep. At Sabinetown on the River, Civil War Yankees built a breastworks. Across the Divide, the soldier's hats were inscribed "Texas or Hell." Mrs. Lillian Cordray in a 1955 manuscript noted that "the invaders didn't get to Texas, so it was supposed they took the other route."

About that time, too, the gazebo on the courthouse square in San Augustine required dedication. I was happy to oblige the good folks in my "hometown."

Parent loblolly pine tree of the stand that seeded-in in 1895 on land that became the campus of Stephen F. Austin State University. The tree died in 1978 at 138 years of age, close to the established physiological maturity of this species.

14

LITIGATION AND TERMINATION

The environment is dynamic. Ecological relationships change with time. Here American chestnut in the Appalachians succumbs to the blight caused by *Endothia parasitica*. Sprouts originating from the chestnut stumps and roots and seedlings of other species, like yellow-poplar, take over the site. And so it is also with human ecology (USFS photo by R.M. Charttan, 1925).

"**When** disappointment happens
And fear your heart annoys,
Be brave like Captain Lawrence
And don't give up, my boys."

My mother quoted these lines so often I thought they were written especially for me. Apparently as a child, disappointment often got me down. Now as the episode outlined in this chapter unfolds, I needed to be reminded of the rhyme. Mom, once a schoolmarm, always had a Ben Franklin quote or a Bible verse for appropriate occasions. Not until a literature class in junior high school did I realize the lines refer to a US naval officer whose dying command, "Don't give up the ship," became the watchword of the Navy. James Lawrence, commanding the *Chesapeake* in a battle against the British *Shannon* on the Great Lakes during the War of 1812, received a mortal wound. Of the occasion, the poet, whose name I cannot find, wrote,

"A grapeshot struck the captain,
They laid him down to die.
And in the noise of battle,
I heard the captain cry,
'Don't give up the ship.'
So, when disappointment happens..."

I report with reluctance the events noted in this chapter. Honesty compels me to state the case as I remember it. Yet, ethics suggests I hesitate as two of the major participants are dead.

Dr. Seymour I. Somberg was highly recommended by his colleagues at Auburn University. However, he specifically asked that I not contact Professor Wilbur Devall, head of Auburn's forestry department. I agreed, for I knew some folks hold it against subordinates who look for work elsewhere. Such folks consider it an insult to their leadership ability. Yet, I had known Wilbur many years;

and from my Alabama days knew him to be totally honorable. I used to lead his students on field trips on the Escambia, and we talked on one occasion about my joining his faculty.

Si, as we called Dr. Somberg, was an excellent economist, had done well in the consulting business, was an officer of the Consulting Foresters Association, and spoke fluent Spanish. We could use all these attributes when he hired on as assistant dean for research.

The faculty endured for a year or so and then, one by one, told me, "He's destroying us; he's got to go." I waited for the annual evaluation session and with Dr. Lenhart, assistant dean for academics in the School of Forestry, present suggested to Si that he take the balance of this year and all of next to try to find employment where he would be happier. With that he went out, promptly telling all the faculty that I fired him. As dean, I chose not to inform the faculty of the facts nor to remind the professors of their earlier assessment of the situation. Apparently, Si spent his time complaining, playing havoc with faculty morale.

I thought it would all blow over when I got a call from the president's secretary that "someone" was on his way to see me so I shouldn't leave. The visitor was the federal marshal from Tyler, a servant of the notorious anti-government court of William Wayne Justice. Quickly, the officer thrust the summons in my hand and left. I wondered if Dr. Steen used the back door of his office to escape the judge's orders. Lawyers in East Texas knew well that if they have a case against government, any government, get it under Willie Wayne's jurisdiction. At one time, the judge had this case against the university, was in effect supervisor of the National Forests in Texas, director of the Texas prison system, superintendent of the state's public schools, and watching over several counties and cities within his realm.

I was the first-named defendant in a First Amendment case: I should not have impeded Si's right to criticize, even if it tore the faculty apart. If I had been the only defendant, I believe I would have roasted alone. Although I owned a house, I had little other that a court could acquire. The president, on the other hand, had considerable wealth.

Si asked to be reinstated, and to receive punitive damages. In the end, he got $200,000, half of it going to Tom Rorie, a local attorney. Later I asked Tom's partner, Bill Petersen, how in the world Tom won that case. Bill's response: "For one thing your lawyer never showed for a pretrial hearing in Tyler." It had to be a poorly executed case, for between the time we lost it in district court and it was heard in the appellate court in New Orleans, not a single person from the Texas Attorney General's Office contacted me. I didn't even know when the case would be retried in the Ninth Circuit. In April, 1982, I outlined in a 5-page letter to President William Johnson (Steen's successor) 18 errors of fact in the district jurist's decision that almost assuredly would have turned the court in our favor, had the appellate trio assigned the case been given a chance to hear them. Following the decision in New Orleans, Dr. Johnson, called me to his office to tell me the university would make the payment—in two checks, lest Tom, the lawyer, not get his fee from his client.

Meanwhile, midway through this litigation which covered several years, Dr. John T. Lewis, vice president for academic affairs, called one afternoon. He wanted my resignation in writing before quitting time. He said the Board of Regents was to meet the next day and the pre-recorded minutes already showed me resigning to accept the Lacy Hunt chaired professorship. Two other deans went down at this, the last, meeting of the Board before President Steen's retirement and all the festivities accompanying that event the following day. The Board would not counter any proposal at that happy time for Dr. Steen.

I had some hint. The president was very cold during meetings in the Tyler court. Dean Gerber seemed to sense it. And Walter Todd, chairman of the Board of Regents, invited me to his ranch on three occasions during the preceding several months. Each time I informed the president of the invitation and, upon return, of the discussion.

When it did happen, Walter called to say the new president would reinstate me as dean. Todd wanted me personally to call upon every regent at his/her home or office. I did so, except for Mrs. Peggy Wright, a friend and a Nacogdoches resident. All treated me warmly. Former board members also tried to persuade Dr. Johnson to send me back to the deanship but, as he told one, "I cannot undo what Dr. Steen has done."

At least three times after Dr. Johnson was in office, the board chairman called me at home to encourage me and to tell me to hold on. He also asked how the new president was working out.

Of course no reason was given for my dismissal; all deans knew we held office at the president's pleasure or displeasure. Word got back to me that it was something fiscal. Si's complaints caused an accountant from the USDA Inspector General's office to spend a week looking for irregularities. He found none. For Si to make a fiscal irregularity complaint was crazy; for all USDA money was Si's responsibility as assistant dean for research. Additionally, at his insistence, the state ran a special audit. Others had fascinating reasons for the reassignment. After reading the previous chapter, I'll let the reader decide.

Two postscripts: Dr. Steen and I often met for coffee after this event. He made no mention of the matter. I sent a full file drawer of the trial and all matters associated with it to the University Archives, asking that they be sealed until after my retirement. President Johnson, however, had them sealed until after my death. Someday they should make fascinating reading for a graduate student of law.

I went into a mild depression following my dismissal, although the administration, by its news releases, wanted it to appear voluntary. The chaired professorship was to make it look like an honorable promotion. The faculty knew better; the students knew better; and soon the townspeople knew better. In truth, during the years that followed I never received any money to subsidize my salary from the Lacy Hunt fund. (Recently I learned some of it was used to pay my salary so that the account allocated for my teaching could be freed for another purpose.)

I'm convinced the faculty played no significant role in this turn of events, even though the president had an open-door policy which made it easy to bellyache about a dean. The only complaint that ever got back to me was that I was "like a sergeant." It was true. I wanted to keep the facilities like new. Isn't conservation the wise use and care of resources? We were conservationists and these were our nonrenewable resources. By secret ballot, the faculty had its say: "The Faculty Council of the School of Forestry does hereby indicate its unanimous support of your continued leadership of the School."

My suggestion to administrators who, with open doors, entertain complainers is that they follow up on the grievances. If they are frivolous or unfounded, the complainant should be disciplined.

The Pine Log, the University's student paper, asked a question in the title of a long editorial: "A change for the better?" I was heartened by the concluding paragraph that followed the listing of the accomplishments of the School. It read, "We regret that the decision was made and hope that Walker will remain with the school and continue the outstanding performance he has maintained." I accepted the challenge.

Stephen Spurr, my college friend and president of the University of Texas, when confronted by dismissal, described the situation with one of his famous puns. "I was hired and fired with enthusiasm," he told the press. And, for me, it seemed so, too.

Twenty some years later, a move was afoot petitioning the Board of Regents to name the Forestry Building in my honor. Among the petitions was a letter signed by 25 faculty members (two omissions grieved me) who thanked me for leadership and School development.

Associated with the depression over a longer period was the exclusion from opportunities to serve in professional or civic activities. I served out my second term as president of the Deep East Texas Development Association, a natural resource conservation organization at the time; served until replaced by the new president on the Chamber of Commerce's University Relations Committee; delivered the several speeches still on the calendar; and faded away like MacArthur's old soldiers. Remember the tune:

> Old soldiers never die,
> They just fade away.

With these lines General Douglas MacArthur concluded his famous speech to the joint session of Congress following his dismissal from the Army by President Truman for insubordination. Our old National Guard anti-aircraft unit sang the song to the band's accompaniment in its finale rendition as weekend drills drew to a close. And today, we mimic the old WWII warrior with

Old foresters never die,
They just pine away,
or
Old deans never die,
They just lose their faculties.

A dean may lose his faculty, but to pine is self-destructive. For this I knew, too, bitterness is a killer. Depression is a disease, for which sometimes there is no remedy, but bitterness we bring on ourselves. So I gritted my teeth, determined not to let the killer get its grip on me. Yet the cloud hung low.

Of course I looked into opportunities elsewhere, in industry, government, and academics. No place appeared as good a town to finish rearing the four kids as Nacogdoches; so we settled in for the long haul, perhaps until retirement and beyond.

Some old cronies looked out for me. J. Walter Myers, Jr., of the Forest Farmers Association, and Douglas Craig, assistant regional forester for state and private forestry in the South, without any encouragement from me, tried to get me appointed state forester for South Carolina. James Hull, former student and always the loyal supporter, worked on the assignment from his position as assistant director of the Texas Forest Service. The interview with the Board opened with the Somberg controversy: that nixed the job.

I was still wondering what to do with the balance of my life when Prof. William Bos of the Communications Department called on me. Bill, a Presbyterian minister, needed some help. The kind of help he sought is the subject of Chapter 16. Dr. Bos' task thrust me out of the doldrums. The assignment, too, convinced me that the best remedy for depression is to help someone in that person's time of need.

John Putnam, "Mr. Hardwood," was introduced to Regional Silviculture students during a field trip to this old stand of black willow in the Mississippi River Delta (source unknown).

15

A FEW COLLEGIANS

Cottonwood plantation, at age 20, visited by silviculture students with author in 1964.

A memoir must not omit some reminiscences of those we endeavored to serve in the classroom, the laboratory, the woods, or at the dean's conference table. However, the slogan ablaze in my wife's kitchen condemns my lack of restraint in telling these tales: "The rooster crows, but it's the hen that delivers the goods."

Van Samford, a large burly man who served as dean of men at SFA, likes to remind me of one senior student. The candidate for the BSF degree, not quite the stature of Dean Samford, showed up for graduation with a few too many beers in his belly. Van learned his intent: to whip the School dean as part of the forthcoming ritual. So the likable Van pulled the student from the processional just before he proceeded to the platform from behind the curtain. We arranged for a private awarding of the degree a few hours later when the libations from the fraternity party had been diluted by copious cups of coffee. No one seems to remember what the aggravation was about. Indeed, I had forgotten the matter when the FBI interviewed me about an Army officer who had been selected for a highly classified task. Our forestry grad received the assignment!

But let's dwell upon happier relationships.

The 1970s, on the heels of the Vietnam conflict, witnessed radical changes on campuses across the country, and SFA was no exception. Streakers, men and women, ran naked through the campus to express their freedom from authoritarian restraint, their disdain for the mores of their elders, and to aggravate administrators and campus police. Like youths of past eras,

> This modern generation seems
> to mock at all our preaching,
> To want to build its house of dreams
> without our wiser teaching.

Meanwhile "Jesus freaks" appeared on campus, gathering into a dozen groups of church-supported and non-denominational organizations. To explain the situation I penned an editorial for the *Presbyterian Journal*. Keith and Gladys Hunt apparently considered it useful for their book, *For Christ and the Campus*, that chronicled the 50-year history of Inter-Varsity Fellowship. Scanning the 1992 book, I found this epigraph at the top of a chapter recounting the 1970s:

> Never mind that the Jesus freaks are scripturally illiterate, that the long-hairs want no part of the organizational church, or that the sandal-clad ones wearing the sign of the fish are insecure in the presence of the short-haired square of the parental generation. They've been motivated by the knowledge...that Jesus is the Christ, the Son of God, Savior of men, and the only way out of the amoral, agnostic and atheistic jungle of the 1970's.

The authors noted my connection with the School of Forestry at Stephen F. Austin State University.

An endnote, after citing the article, recalled that "Walker, a long-time friend of Inter-Varsity, was influenced by its ministry in his student days. He worked as a volunteer, opened his home to students and staff as well as serving on local committees for IVCF and the corporation."

Fellow Forester and Scouter

David Hyink, a South Dakota University football player and guide to Boy Scout groups canoeing on the lakes and rivers of northern Minnesota and neighboring Canada, was a pre-law student. In 1967 we met at the Boundary Canoe Waters Base Camp where, David later noted, I did "a little recruiting." He abandoned the Plains and the projected life of a lawyer to receive a BSF and MSF at SFA. Dave went on for a PhD, and, as I write, leads the Weyerhaeuser Company's Silviculture/Wood Quality Research Team in Tacoma, Washington.

Here are a couple of Dr. Hyink's stories about those days at SFA:

> In the Introduction to Forestry course, Dr. Walker personally conducted a 6-hour field trip every other week. On my first such trip, I found myself sitting directly

> behind Larry, with New Yorker Walter Garasic. As Walt and I were quietly asking one another just "what the heck is an armadillo?"--the bus came to a screeching halt and we soon found ourselves examining "road-kill dillers", and getting the evolutionary history of the "Hoover Hog."
>
> I (and most of my classmates) will never forget the many times (Walker) stepped off the bus, donned his pith helmet, and said: "As they say in the movies...follow me boys!" He would charge off at a joggers pace—students in tow, with Alyse Moore and Mary Thompson at the front of the pack!

I've lost track of Alyse; I last read about Mary as a ranking forester for the State of Montana.

David came all the way from the Cascade Range to participate in honoring me at a Boy Scout ceremony in Nacogdoches. After quoting paragraphs from Sam Cook's "Up North," Dr. Hyink concluded,

> "Feeling the tug...", "making the dream happen...", and "putting the map in the packsack and going...", aptly describes Dr. Walker's approach to everything he takes on. Through his vision, his commitment to excellence, and his leadership by example, he has given countless numbers of us an "Up North Experience."

For the same occasion, Bob Harper, a successful forestry consultant, wrote,

> After three years of less than success in the classroom, I was invited to leave school and enroll in a new venture—the U.S. Army. I returned to S.F.A. in 1972 and came to see you to ask for re-admission to the School of Forestry. You looked at my transcript and looked at me and said, "I'll let you try, but I don't think you will make it." That challenge was all I needed. I graduated in 1974 with a 3.3 gpa in my forestry courses. I have loved my profession ever since.

And from Patrick Angel, of USDI's Office of Surface Mining in Kentucky:

> I left Kentucky for Stephen F. Austin State University in 1970 primarily because of Dr. Walker's encouragement and personal interest in me. When my wife. my two young

> babies, and I stepped off a Greyhound bus onto the streets of Nacogdoches, it was Dr. Walker who was there to greet us. All we had was a suitcase stuffed with a few possessions, a couple of dollars, and faith and trust in a man who laid before us a bright and promising future. And Dr. Walker certainly delivered on his promises! First he took us under his wing, got us a place to stay, a part-time job, financial aid, and a list of tough courses to sign up for at the School of Forestry. During my first four years there he helped my family and me through tough times. My big break was when Dr. Walker helped me acquire a fellowship with Texas Utilities Generating Company in order to work toward a Masters in Forestry. It was the work on the strip mines at Fairfield, Texas, for that Masters Degree that pointed me in the direction of my life career.

Perhaps no student in my memory had more strikes against him than Patrick. That he succeeded so well continues to be an encouragement to me and to others who know him.

Women Enter

Strange it seems today, with women comprising a third of the forestry enrollment, that their entry would be such a concern in the 1960s. These pages now note their influence in the life of the school, the profession, and my family. But at the time the first young lady enrolled, not all agreed the influence would be favorable. Male students were confident female graduates would lessen job opportunities for the men. Faculty complained, wanting me to find a way to exclude women. What the staff predicted soon happened: the mischief occurred at the Field Station. I was relieved only slightly by learning the father was a student (though married) and not a faculty member.

An old Methodist district superintendent in the early days of women preachers, and unfamiliar with strident feminists, called his clerics together. In a speech he frequently referred to his charges as "gentlemen and our lovely lady." Exasperated, the sole female parson rose to say. "I am *not* your lovely lady. I am a woman." Whereupon an oldtimer in the rear bench mumbled, "If I ever had a woman for a preacher, I'd hope she was a lady."

Female forestry students were ladies. Reared in time and place where hugging was considered sensual (I recall never receiving a hug at home, even upon returning from

Early enrollment of women students about 1970 at the Shay locomotive.

war), it took some adjustment in my latter years to interpret a hug as an expression of appreciation and a "glad to see you" gesture.

If among our women scholastics were radical feminists, I was unaware. (Nancy later said I *was* unaware). Students in the general Conservation of Natural Resources course, not open to foresters (they would control it), were another matter. Majors in education and liberal arts would chastise me as a chauvinist. So I chalked this paragraph:

> As one who has championed the cause of women in employment in the natural resource professions, permit a private word to my feminist students concerned about the possibly sexist language in the lectures. How I wish there was an all inclusive term, but the English language we've inherited consists of little that is neuter. To say "man" is no more chauvinistic than to use person, as the bold-faced letters below suggest. Per**son** is obviously masculine; to say men and wo**men** is to add men to wo. Wo**man** is simply womb and man. He is at the beginning of **he**r. He is in s**he**, male is in fe**male**, lad is in **lad**y, Mr. is in **Mr**s. and Adam (Hebrew

for man) is in m**adam**. **Homo** of ***Homo sapiens***, the Latin binomial for the hu**man** race of **man**kind, means man. The race of mankind: that is our concern; the two-footed anthropoids of either gender to which biological nomenclaturists have ascribed ***H. sapiens***, man of wisdom. And, at the risk of overly burdening a point, demo as in **demo**graphic means man. Journals such as *Science* use **man** without apology.

The outline was a mistake. "See," they said, "That's our point. Men control everything. Even the language." Maybe s/he and he/r should be introduced as personal pronouns.

"Tear-drop" Honduran mahogany table and "Gate to the Forest" in the Lacy Hunt Seminar Room.

Western Trip I

One of the western silviculture trips covered much of the Southwest. We stopped the beat-up old school bus at Wagon Wheel Gap, some miles west of South Fork, Colorado. Here, a district forest ranger in the teen years bootlegged research. Rangers often carried out studies of particular interest without authority, a study plan, or financing. The wisdom of the ranger's effort proved worthwhile, for the Rocky Mountain Forest and Range Experiment Station subsequently followed through on the studies in the Gap through which pioneers trekked and wagons lost wheels. The station, with a mite of persistence, found the records of the research in the archives. The file included photographs and map plats that showed photo points. We carried these to the site, located the photo points, and positioned the prints. Hardly anything had changed in the 60 or so years that had elapsed. In this high, dry climate of slow-growing vegetation, the students picked out every tree as they surveyed the distant scene from the depression in the rugged mountains. They then traipsed down the mountain into the valley and then up the far side to locate the watersheds. The young foresters-to-be discussed how each watershed had been treated silviculturally in a specific manner. They located the dams built to measure runoff of soil and water from the treated mountainsides.

The fellows located the site of the cabin where Forest Service workers holed up in deep snows while gathering data in those early years. Only the iron cookstove remained. The students would have brought the whole stove to the second-floor museum in the Forestry Building at Nacogdoches, but its weight and bulkiness required them to settle for the stove door.

That was a close knit class; to prove it they brought back a cross-section slab of a large spruce on which all signed their names. The wooden "memorial," like a limestone-embedded fossil (about which more later), and the stove door for years were sealed in a display case in the main hall of the Forestry Building.

Ed Shirley kept a log of the 1969 Silviculture trip to New Mexico. Ed noted that I required, as the rickety school bus pulled out of the parking lot, each of the 22 students to fill out an emergency identification card "so our loved ones could be notified." Ed's roll listed "The women," three "Yankees," "The Honduran," "The Indian," "The Grad. Student," and "Photographer." Ed considered himself "Tallyman" for what he described a "motley crew."

"Our concentration was disturbed in a (Texas) Tension Zone pine stand by numerous ticks, but didn't faze Doc, who kept on talking" (Shirley wrote). The students dug soil pits in the Post Oak Belt and in the Blacklands, looking for montmorillonite clay and measuring pH, as they journeyed through the East and West Cross Timbers. "Everyone was glad when we got moving again at 6:15. Our gladness soon changed to fear, because Doc took the wheel." Shirley continued his log:

> Reached the "ranch house"...stretched out in the yard or in the bus...washed and skinny dipped in Lampasses River...night's rest ended when Doc stomped through our dew-covered bivouac ringing a cowbell...five miles away to cedar brake lecture...stopped for our mesquite lecture under a live oak, yet!...played with a horned toad (the first our Yankees had seen)...lecture rudely interrupted by little old lady who ordered us off property...lecture completed with dissected mesquite on roadside...12:15am arrived city park in downtown Lubbock for the night (what little remained of it). Mosquitoes terrible...puffy with bites...Nancy and Alyse showered in lawn sprinklers (said they sold spectator tickets)...Dr. Urbanovsky guided us on campus tour of Texas Tech...lecture on parks by Dr. Kitchens...Doc presented with seedling of *Carya illinoiensis* (the seed from President Johnson's Pedernales Ranch)...Nixon asked the common name...shocked us all...Doc threatened to make him walk home...lecture near tributary to Palo Duro Canyon (which included an awful rendering of "When the Swallows Come Back to Capistrano" by Doc...geology lecture as we progressed down the canyon wall...lightning flashes and rolling thunder...Bob Reinhart, park ranger and SFA grad, ordered flash-flood evacuation...Bob took the gals to his home...Doc and guys spent night in a building shell on canyon rim...completed geology lecture in the morning after waters receded...Doc and [five] went to church in Canyon while rest enjoyed leisurely breakfast...stopped for grueling (and boring) shelterbelt studies, wind speeds, etc....Conchos Reservoir, lightning flashing from thunderheads moving toward us. Doc made a deal with the Boy Scout camp on the north side of the lake for us to spend the night there in genuine tents. He says he had to get down on his knees with tears in his eyes and beg. The storm hit and we were in for our second straight wild night. It's getting to Smitty. He wants to post a watch to look out for an Indian attack...Scouts fed us breakfast...Paused for study of lower Sonoran Desert...Pettigrew killed rattlesnake; other one got away...Burch, Farmer, and Minor unearthed a large limestone rock [the size of two footballs] with fossil fern

imprint [100 pounds]. Fossil lodged in "Galloping Goose" stairwell for remainder of trip...stopped at little "Cantina" for gas. Cute little senorita chief attraction. Carlos, our Honduran, wants to jump ship [this was before Latinos had returned to East Texas]...ponderosa pine lecture...pinyon—juniper lecture...Driving on fumes...Taos. Doc made a deal for us to stay the night in a motel...Smitty, our Indian, was glad to see his red brethren getting back at the white man (at last) for socking it to us to visit the pueblo...Field lecture of Rocky Mountain conifers and shrubs in Kit Carson National Forest. Mullally said Doc was meaner than any "bull-of-the-woods." Mount Wheeler, highest peak in Southwest, lies ahead. A real challenge...Beautiful scenery and snow. Entered Wheeler Peak Wild Area [before Congress passed the Wilderness Bill] at about timberline. Some retreated to the bus. We cut low shrubs and made windbreaks in the brush to protect ourselves for the night from the high cold winds. Minor and Weind joined us. Slept well...Nixon turned up at bus in campground, driven in by friendly native of Red River. Seems he got lost, spent the night above in the woods, and wandered into Red River campground. Typical Nixon: blunder & luck & plain old horse sense = survival. Sixteen here. Doc and the rest still up the mountain. Snowy, windy, cold. 3:55 Doc and the last of his party finally made it back. They made it all the way [to the top of the 13,151-foot peak].

Bill Briggs adds this to Ed Shirley's rendition:

When Doc and six others reached the Wilderness Area, they made camp for the night. Doc prepared supper while the 'happy wanderers' with him built a hut for protection. The meal, never to be forgotten, was a mixture of pork and beans, canned stew, vienna sausage, and vegetable soup with french bread and hot chocolate. Briggs and Smith got sick. (Doc insisted it was from the high altitude, not his gourmet meal.) After a quick breakfast of spam and ashy chocolate, the group proceeded up the mountain as soon as daylight came. Three, hiking ahead, climbed the wrong peak. Later, they joined Doc and the others and made it to the top [of the highest mountain in the Southwest]. The hearty souls returned to the campground and bus in a snowstorm. For both days, 11 climbed the peak.

Back to Ed's account:

Doc says that after seeing Taos hippies, our long-haired Yankees aren't bad at all. The Yankees are grudgingly readmitted into the human race...Stop in Albuquerque for bus repairs...trouble going 20 mph downhill. Smitty says we should cremate (the bus)...steep, narrow winding road about bus width. If we meet anything wider than a bike we're in bad trouble...Mountainair: 41 miles. Oh, goody, only 2-1/2 hours to go...Arrived Mountainair Ranger Station...Lecture by Doc. Fine talk and slide show by District Ranger Bill Russell...Arrived Trails End Motel. Buzzards circling already...Our task: find obscure corner of Section 36. Light rain, then hail. Local crews with whom we worked took it in stride—after getting lost trying to locate elusive Sec. 36...Bus slid into logging road ditch, battery died...Fogbound at Tajione work area. Doc, our hero, says we'll have 2 weeks after our return to get our reports in. Doc, during the day, visited each crew. He burst upon Briggs, Mauk, and I just as we were discussing what "fudge factor" to use to hide a 77-link error...Farmer's crew spent most of the day reading magnetic north rather than true north...At supper, Doc told the following joke--an example of his humor (?): The old maid walked by the sewer/And by the sewer she died/And at the coroner's inquest/they called it sewercide...Sunday, headed for Spanish ruins at Grand River National Monument and Quaraii Ruins State Monument...Climbed Capillo Peak to Fire tower at 9500 feet...Dry lightning threatened; soon left. Transfixed by dust devil stirred up by coming storm, until about 10 feet away. We scattered except for fearless Farmer...Doc introduced us to George Proctor, forest supervisor for the Cibola...Lazy day; we finished our timber cruise; Doc lectured on precommercial thinning of ponderosa pine...Ranger Russell showed up with his family for supper, observed our compilation of data, and bid us farewell...200 miles to Carlsbad Caverns...Passed a highway surveying crew. Yelled advice: "Pull tight, stretch that chain! Stick"...Park Superintendent Phillip Van Cleve lectured on NPS employment...Free tour of caverns to the Bottomless Pit. Our two lumberjills were unimpressed by the lecturing of a shapely brunette. Doc suggested Alyse recruit the young

> lady for our School of Forestry. But Alyse showed notable lack of enthusiasm...We unanimously voted to drive straight through [to Nacogdoches]. But at 9:20 am, north of Nacogdoches, Doc pulled a "funny." He pulled over to a dilapidated peckerwood mill and announced the half-hour visit. Greeted with moans and sobs, we were too tired to protest. As he turned the bus back onto the road, he complained he'd never seen such a lack of enthusiasm in forestry students...5,467 miles covered in 14 days, 3 hours, and 34 minutes. Eureka!!!

I didn't edit out the comment about Richard Nixon appearing a few pages back. Richard would not have wanted me to. Perhaps I alone knew of Richard Nixon's dyslexia, the condition that causes reverse images in the brain. No wonder he went down the west, instead of the east, side of the mountain. And one can imagine the reception when he intruded at breakfast at a fishing camp, dressed in a tweed overcoat and carrying a notebook, and introduced himself in the days of the Nixon presidency as Richard Nixon! The young man, to his great credit, completed the degree, dying not long afterward. Richard was a great sport.

Sylvans log cabin on the June Lewis Place.

Sylvans float in Homecoming parades consistently won second place.

Parents' Day at the Sylvans Cabin, 1971.

Students and faculty tune up to entertain parents at the cabin.

Another Western Trip

Terrence Neid provided a more cultural log of a trip to the West with students a few years later:

Colorado Experience

"The Journey"

Up from the lowlands I came,
Through the belts of land man tamed,
To see the heights—find God—the same,
Through miles of brush and sage,
Never aware it was a stage,
God set to show:
Contrast between desert and snow,
Between mountain and plain,
Drought and rain.

"Midway"

Still I traveled blindly—ever seeking,
The Truth—God's speaking,
Through the forest of many shades:
The aspen trees of light greens,
Gray-green meadows where life teems,
The skies of indefinite blue,
And fir and spruce of green-blue too.
The red and yellows of the wildflowers,

Above which the mountains tower,
A panorama of God's design,
Spread before me as though it were mine.

"On Top"
Still higher my travels brought me,
Above the crashing of the sea,
Above the gentle zones of the earth,
Where trees were stunted went my search:
Where the tundra fought for a hold,
With the lichen against the cold.
Among the marmot, pika, and the rock,
Beyond people—only the wind could mock.

"Journey's End"
And there, in solitude, and the peace of the peak,
God spoke in my head and asked—"Why do you seek?
Peace and calm are down below,
And the mountains are just a show,
A reminder of my presence—at best,
But within you lies the test,
The challenge to find calm and peace,
And make your unhappiness cease."

"The Lesson"
"Look about you when you return,
At the oaks and pines and ferns,
Their shadows reflected on the lake,
Marvel in the flight of the white egret.
Look at your friends with appreciation and love,
And you'll find the peace sought here above.
Be loving, and be gentle, and be kind,
For happiness, you see, is within your mind."

"Reentry"
My life has begun anew—now.
I still vary from the "ugh" to the "wow!"
From the valleys I climb to the peaks,
But less frantically do I seek,
And within me a mountain rises..................

Terry Nied continued poetic expression, years later

asking if I could "imagine a South Texas Redneck writing poems?" Yes, I could and do!

On our western trips we stopped by Texas Tech. There E.J. "Urb" Urbanovsky presented us with the aforementioned *Carya illinoiensis* var. *johnsoniana* for planting in our schoolyard. The variety indicates its source: the seed came from the bank of the Pedernales River on the LBJ Ranch. Urb, a horticulturist, was Lady Bird Johnson's personal landscaper. A plaque accompanied the presentation, noting the donor as the Texas Society for Protection of Original Nuts. A colleague at Tech, J.W. Kitchens, signed the plaque as chief of the Selection Committee.

"No camping" sign restored in the morning to its proper position somewhere along the creek between Yellowstone and Grand Teton national parks. The advance party, Jim Dawson and Tommy Findeisen, traveled ahead in a pickup to locate a campgound. Finding none, they created one.

Leaving Yellowstone National Park and heading south on one of these silviculture tours, we found no campground that could accommodate the bus load of weary ones. Tommy Findeisen and Jimmy Dawson, always like two peas in a pod, went ahead in the pick-up to scout out a bivouac. In time the bus caught up to Jimmy standing by the paved road. He directed us to a scenic trout stream which soon, by its cascading melody, sung us to sleep. Well after midnight some burly characters shook us all from our slumbers, commanding our immediate evacuation. That is, until a couple of our even burlier fellows "molted" from their bedrolls. We agreed to be on our way by dawn. Arriving back at the pavement, Tommy signaled the bus to stop and wait. He jogged off, soon to return with a sign, removed with its post and deposited in the brush, that emphatically stated NO CAMPING.

Catherine Stein, a lover of rhyming lines, wrote from her U.S. Forest Service station in Albuquerque. She enclosed Badger Clark's prayer that could have been the invocation of a forester. We recite the opening lines:

Oh Lord, I've never lived where churches grow.
I loved creation better as it stood
That day You finished it so long ago
And looked upon Your work and called it good.
I know that others find You in the light
That's sifted down through tinted window panes,
In this dim, quiet starlight on the plains.

Potpourri of Students

Some students we haven't seen in years, though scattered far afield they remain in our extended family. With the Randolphs we have agonized over the tragic loss of a child. Paul wrote, " On Thursday evening as we were watching TV, Katie turned to Mary and asked, 'Mommy, How do I die?' Her question came as such a surprise that I answered, 'You just go to sleep and when you awake, you are in the arms of Jesus in heaven.' She said, 'O.K.'...Friday she was as busy as ever...During the evening, Katie suffered a brain hemorrhage, so she merely 'went to sleep'...On Monday, she died." And we also have witnessed from afar the Randolphs' joys: December 1993:

> Many changes have taken place with the Forest Service this year and along with those changes both of us took on new positions. Paul has taken a job as Geographic Information Systems/Database Coordinator (which means he is working

with computers and loves it!!!), and I (Paula) took on Land Management Planning Staff Officer along with Public Affairs and now have a staff of 6 people. Just supervising them has been a full-time job for me! We love Sheridan, Wyoming, and feel fortunate we were able to change jobs and stay in the same community."

Forestry's weather station continuously maintained by students since the 1950s.

Paul kept me posted on his Peace Corps' work in South America, advising Chileans or developing multiple-use plans. He vented frustrations and acknowledged accomplishments in long letters, detailing every event. Paul knew I'd be interested.

Another Peace Corpsman, Brian Riley, worked with the Saraguro Indians in Ecuador. Brian noted how "minga" works. The community gathers socially to carry out a task. That's how he got trees planted. "Foresters," he wrote, "like missionaries, go into all the world planting good seeds." Indeed they do.

Douglass Peabody writes friendly letters. In 1977, no doubt exaggerating, it was about my "capacity to think of little people, like me, when your day is filled with..." Little people? How should *I* take that? Doug underestimated his abilities. He sang professionally with the Dallas Symphony Chorus and in a synagogue (though he is a gentile), and

Durward Erminger made this picture for a photojournalism course in 1971. (l-r) unknown, Joe Ackeroid, and Ken Bragg.

Mohammed Abido, graduate student from the Mid-East (r), receiving help on his thesis dealing with the physiology of aleppo pine.

went on to doctoral studies in Tennessee. I recall he once said he wanted to be a sociologist. As a recreation-oriented forester, he *is* a sociologist. It is that which Regional Forester Jim Vessey called *socio-forestry*. As this goes to press, I've learned Doug gave his life a ransom to a robber in January, 1995.

As for my size, an "unknown admirer" among the student body sent a 1974 *Wizard of Id* comic strip. The royal guard, spear in hand, approaches the little king: "A group of midgets wish to perform for you, sire." The royal runt grumpily responds, "I guess I can spare a few minutes." In the last box, perspiration flows from the guard's face as the king, witnessing the performance, asks, "Where are the midgets?" They're all taller than the king.

These lines about many students may give the reader a taste of the great variety of fine young folks with whom I have been associated over the years. Sue Tortello, our first woman to receive a Texas Forestry Association scholarship, stood on a box for the TFA photo op, like I sometimes do. Judy Schutzer received the MS and has published forest history material in *American Forests*. Eric Beard advanced to an industry PR task in Washington, D.C. Billy Lemons, a consulting forester, did a professional job editing my book, *Farming the Small Forest*. The student knew more than the teacher! Ann Holland used her forestry knowledge as a Volunteer in Mission for the Episcopal Church in Kenya. She is now studying for ordination in a Virginia seminary. Ralph Weitz, a church worker in Virginia, currently edits a devotional book from a forester's perspective: it relates to trees mentioned in Scripture. We recruited Ralph for SFA from Pennsylvania's 2-year technical school. Kenneth Hayman, another forestry student, went into the ministry.

Jeff Grimes worked up Environmental Impact Statements for the Texas Highway Department before going into public relations. Ed Schmidt uses material from the introductory course I taught with his Boy Scouts. Sara Jane Lee, who taught me how precious a person's name can be, is an industrial area forester in Virginia. Marc Goldberg, who showed up at my lectures in Israel, wrote from Ilanot in 1978 to protest the treatment of a classmate who did poorly in Summer Camp surveying. Henry Palmer, while a graduate student, with his wife built their own house, and went to work for DuPont in Delaware.

"Summer Camp" became the "Field Station" to avoid conflict with Vietnam vets who'd had a bellyful of camps.

Dendrology students in the herbarium.

Jerry deBin, a hunting buddy, took over management of a South Georgia plantation where the hands always addressed him as "Mistah Jerry. Anythin' else, Mistah Jerry." Jerry added, "I make sure they are treated well too." An Oklahoma city boy would have to learn true Dixie culture to practice forestry in goober country. Jerry really touched me with this: "I regret you weren't my grandfather so that I could have grown up around you."

Dr. Lamar Teate, earlier my graduate student, became the head of Louisiana Tech's forestry department. Tom Hartz intermingled consulting forestry opportunities with foreign mission service. As I write he is managing an experimental forest in Nepal. Years after graduating Jerry Don Harris wrote from his managerial desk with an oil exploration company to express appreciation for "one gem of wisdom you gave.... It went 'Whose bread I eat, his song I sing.'" Jerry found it vital for surviving in management. I think I was trying to get across the idea that if what the employer wants offends your conscience, look elsewhere for employment. Jerry Don added:

> Equating profit $ with good forest and land management is the only language most executives can understand. Balancing the fiduciary responsibilities to the stockholders and the stewardship of land can create exciting problems to be solved.

Bill Sandeen, a frequent corespondent, was responsible for environmental matters at U.S. military bases in the Philippines until the early 1990s withdrawal of Americans from the Pacific islands. Students called him "Roadrunner" because he took long strides and always was in a hurry. That served him well when he edited perhaps the finest forestry school yearbook, *The Texas Tally*. He's now a Nevada Certified Environmental Manager at Nellis Air Force Base.

Linda Ann Howes, bless her heart even if she did marry a forestry professor, was honored with a "Profile" in the local paper. In our town she had succeeded in a variety of fascinating industry and government tasks. When asked in an interview "Who has been the most influential person in your life?", she gave the name of this chronicler: "He convinced me I could accomplish anything I set my mind to."

James Hull became the Associate Director of the Texas Forest Service. Twice he was almost elected president of the Society of American Foresters, having served with distinction on the SAF Council. Bob Bloom, also with the Texas Forest Service, helped with the introductory course while a graduate student, freeing me up a bit.

Fellow scouter David Knotts had been in graduate classes that led to the DF degree. After his graduation, he directed a conservation camp in the Rocky Mountains. About that time he nominated me for the Sportsmen's Conservationists of Texas and the Wildlife Society's Conservationist of the Year award for forestry. I gratefully traveled to Austin to receive the statuette of a wild turkey. Similarly, John Norris, earlier a congressman's land manager and who now handles tasks for the Texas Forest Service, nominated me for the Syracuse College of Environmental Science and Forestry Alumni Association's Achievement Award for 1993.

Correspondence continues with many grads. Some who come to mind are Frank Fredenburg, an industry senior forester; David Hacker, an SFA DF recipient; Harry Faulkner, faithful Sunday School student; Robert Haggard from his California office; Industry's Terry Burrow; Don Baun in Ohio; Larry Holers; Richard Cole and his wife Karen Tucker; Hunter Mullikin in the swamps of

Louisiana's Red River; Dan Alsides from the Oregon Cascade Range; Brad Barber; Charles Ray Douglas, a Duke doctorate; Susan Roe Howarth, now an Oregonian; Finis Prendergast, Jr., who often brought his father to visit the school and who now handles wood exports for a major supplier; Chris Miller of the Wild Turkey Federation; and H.C. "Brad" Bradbury, now an environmental consultant after breaking ground for the profession as a food giant's director of environmental affairs.

Bill Terry, PR forester with the Texas Forest Service and later in a similar job with the feds in the Northeast, frequently cheered me on with exaggerated expressions of appreciation. On the back of a print of a photo he made of an old conifer on Grand Canyon's south rim he described the bent, gnarled, wind-blasted, stunted tree as "an inspiration because it shows how adversity can be overcome." Then he testified of his faith.

In Anchorage on a 1966 Forest Service assignment, I'm shown the town by Bill and Sarah Wimmer. Soon I find in the mail box a copy of the Bible translated from the Aramaic (rather than Hebrew and Greek), the language spoken by the Jews in Jesus' day. Bill wanted me to grasp the Russian influence still present in Alaska as evidenced by the use of Lamsa's translation from ancient eastern manuscripts.

Anne and I enjoyed caring for the 8-year-old son of Gail Denkhaus so that she could attend the summer Field Station studies at the rustic facility on the Sabine National Forest. I introduced Gail to Forest Service chief Ed Cliff as "the first female regional forester." Alas, she went with industry, and there became a manager with responsibilities over more than a dozen wood-using plants across the eastern United States.

Most readers of this narrative will know the art, if not the name, of Bruce Cunningham. Learned in forestry lore from our classrooms and field trips, he became the unofficial artist of the Society of American Foresters and for many other groups.

While not an SFA student, Ann Ross reminded me that I had aided her in a graduate thesis at Southern Methodist University. The study dealt with timberland. You bet, if we can win an elitist (which many Texans would consider most SMU students) over to a way foresters see land management, it's worth the effort.

Students from abroad whose sojourn among us is now recalled include Elli and Florida Luhat from Malaysia. Their children, by virtue of their births in America are US citizens. They were foreigners in Borneo when their parents returned home. Elli left behind an ajat, "a basket of ethical values." Handwoven by his mother from fine splits of rattan palm, Elli carried one of these backpacks as he hiked from the village to school, many days distant. The traditional handcraft would remind him of his responsibilities, he said.

A student from Damascus lived in fear that Syrian spies might discover one of his professors was a Jew. For his thesis we had to hide the citations from Israeli literature by referring to the abstract journals in which those articles were noted. To suggest only Tel Aviv researchers had knowledge about allepo pine unknown to Damascus would be grounds for immediate departure to the homeland.

Pilis Malim came from Malaysia, Oumar Balde journeyed from Senegal, and Feme Olaleye crossed the sea from Africa. Franklin Adekiya's given name was attributed to English-speaking missionaries in his African homeland. Kun Ho returned to Korea. All became became high-ranking officials in resource development.

Lord Baden-Powell's "Kin"

Garry Davies, an English lad, was among the students when I arrived at SFA. An International Carryall (forerunner of the van) load of students, Garry among them, went with me at spring break to hike the Big Bend National Park. The plan was to take five days to cross the dry, rocky wilderness from the Basin to the Rio Grande River. Our map, though dated 1906 by the Coast and Geodetic Survey, showed the water holes upon which we depended. But by 1965 the springs were all dry. So, we holed up in the rocks for shade on the second day and in one night we traipsed through the desert shrub and scrub back to the Basin. A parched crew learned at the Park headquarters how little the federal rangers knew about the Big Bend. It would take a burro to carry the water for us and another burro to carry water for the burros. Garry (and his wife) now "chases dreams and rainbows, skiing the incredible glaciers" of Alaska. On the occasion of my receiving the Hornaday Medal for Conservation, this British Scout "visualized Teddy (Roosevelt) and Bill Hornaday...discussing the conservation of the American Bison."

Wedding Bells

A special reward comes with performing the weddings of students. (How this could come about we'll tell in the next chapter.) They then become like family. Among those foresters taking matrimonial vows: Dianne Kwiatkowski and Charles Black, IV; Sean Harper and Kyra Sanders; Suzanne Birmingham and Michael Walker; Shelley Arnold and William Scott Jackson; Christopher Crowley and Elin Ferrell. As the last couple was united on the occasion of the 45th wedding anniversary for Anne and me, Elin later wrote that "It really meant a lot to us."

Charlie Laman appeared at the office door, a bit discouraged. The graduate student from Malaysia, half-way around the world, had not seen Flora, his fiancee, for two years. The US State Department provides no visa for entry into America by "intended" wives, no matter how strong the intentions. Legal proof of marriage is essential.

My task was to advise Charlie. Should he go home? Go home for a visit? Interrupt his studies to go home to marry Flora? Or spend $3000 for a round-trip air fare? Another option seemed possible. Why not marry by radio-telephone via satellite?

When Flora and the parents of the couple had agreed to this, the County Clerk went into action. Hope Skipper, the enthusiastic and helpful clerk, promptly checked out the law for such a unique wedding. No such marriage, with only one of the principals present, had ever been conducted in Texas' Nacogdoches County. Once the laws on both sides of the Pacific Ocean were satisfied, a time was set, and Charlie and his friends gathered at our home. A speakerphone was in place.

Alas, the ceremony would be delayed. Hour-after-hour through the evening, we tried in vain to call Flora on the tropical South Pacific isle. A storm, we later learned, had hammered the village on Borneo. Ironically, only two phones were knocked out—those of Charlie's parents and his sister, the phones designated for this urgent telecommunication. Charlie, again discouraged, said, "Maybe the Lord doesn't want us to be wed." Another time was agreed upon. And this time the Lord's blessing was added. With Charlie and his fellow countrymen gathered around the speakerphone, I read the vows. Both said "I will" at appropriate times. (Flora had been mailed a copy of the ritual.)

Flora and Charlie will have a matrimonial problem: they can never celebrate their wedding anniversary on the same day. For Charlie, the wedding took place on Saturday evening on the 28th. For Flora, on the other side of the globe, the time was in the morning on Sunday, the 29th.

Charlie's and Flora's names are those of Anglican missionaries who, decades past, led the forebears of these young people to Christian faith.

Student Deaths

The loss of a student may be traumatic to faculty and fellow students. Scott Yunk's death at the SFA Forestry Field Station on Lake Sam Rayburn had us grieving. In a race to determine the crew that could finish a surveying task the first, Scott's team took a shortcut through a slough rather than offsetting around the swale. The strong ex-Marine seemed held under the water. Divers later found Scott entrapped in the hydrophytic vegetation. Faculty and students asked me to conduct a memorial service at the Field Station nearby the Lake. It was a sad moment. We continue to hear from the Woodwards whose son, Pat, died while a student. We buried others too.

The Prisoners

The dean of the School of Applied Arts and Sciences needed an instructor for the several state penitentiaries about 80 miles from Nacogdoches on the Trinity River. I obliged; this would be a new experience.

I taught a general conservation course and tailored regional silviculture to forest geography. Here's what one of the incarcerated wrote:

> I've been locked up over 5 years, and my people are all from New Hampshire... My dad came down with some good news, Excellent News! I'm being transferred to New Hampshire State Prison for the remaining of my time... I want you to know (that) as stupid as this letter you are reading may be to you, this class was not a total waste. I learned a lot.....

I really believed that if every sixth-grade student spent a half-day behind these walls and wire, we could clear the prisons in a generation. "Not so," a warden assured me, "The men we house haven't got it made back home 'til they've been in here." One student had been a guard at the classroom door the year before. He'd sold drugs to the inmates. I learned many are smart (though only 200 of the 8,000 inmates were eligible for upper-level courses) and expert at conning the instructor. I wrote a final examination at home; then personally carried, distributed, proctored, collected, and carried it home to grade. Yet men sitting on opposite sides of the room would have identical essay answers on the test, even to the errors in thought and spelling.

Special Course Requests

When asked to hurriedly put together a 6-week course for five graduate students with a variety of interests, we reviewed my silvicultural efforts of 40 years previously on the Sabine National Forest. Going through the old atlases where I had inked in every fire, thinning, pulpwood or sawlog harvest, and TSI task, we then visited the sites. The students saw one thing abundantly clearly: we wasted our energy on selection harvesting of pine stands. The hardwoods had thoroughly captured the site. The regional forester correctly directed these timber types to be managed on an evenaged basis.

Always ready to travel, I took a van full of students to a meeting of the Society of American Foresters in St. Louis. Enroute, we scouted the Womble Ranger District and stayed at Camp Clear Fork in the Ouachita Mountains; visited the watershed management plots on the Ozark National Forest; then proceeded on a self-guided tour of the Mark Twain National Forest in Missouri. The students reviewed coordination of timber management for scenic values. Students were amused by seed-orchard machinations; all four major southern pines producing cones and seeds from a single parent tree to which branches had been grafted.

The students came away with a better understanding of hardwood silviculture because of their trip through the Salem Ranger District and Sinkin Experimental Forest. In the 1975 trip, the management objective was to grow full yields of highest value products the site can produce in the shortest time. In these oak (*Quercus* spp.) stands, intermediate cuts removed damaged and less vigorously growing stems. Here the oaks are more shade-intolerant than elsewhere, requiring evenaged management. Seedlings require full sunlight. An effort to restore shortleaf pine (*Pinus echinata*) to its importance in Missouri's cutover forest is underway. As Missouri is mapped by the Public Land Survey System, students had a much needed refresher on this method of laying out the land. It sure beats the metes and bounds, hoot 'n' holler, method of Texas' Spanish land grants.

Student Organizations

SFA's charter for Xi Sigma Pi, the forestry honor society, was issued in 1968. Application had to wait accreditation. An earlier SFA honor group, Phi Eta Sigma, with the same requirements for membership, made the application. My old notes say I spoke to the charter members as "honorable friends of the trees," recalling my initiation at Penn State's chapter of the fraternity for forestry scholars.

One notes with satisfaction that it was SFA's Sylvans Club that petitioned the Society of American Foresters to establish student chapters. SAF's Constitution was soon changed to make this possible.

Forestry students took over leadership of a campus high-adventure Explorer Post of the Boy Scouts of America. So many Eagle Scouts were among the student body, it was a "natural." The Post brought Walt Wenzel to campus. A conservation specialist for the BSA, he put on quite a show in survival skills. Example: you can light a safety match by striking it on a U.S. dollar bill.

I suppose the relationship of Christmas trees to forestry

was responsible for Karol Spedale, chair of the tree-lighting event for the University Center Hospitality Committee, lining me up to speak on the occasion. A crowd of students gathered.

One of the special problem courses provided a woodland nature trail through the grounds of the Tucker Estate. The seniors designed it for youths of the community, enticed local scouts to assist, and celebrated its completion with a well-attended dedication. The program says Dr. David Kulhavy introduced me as the speaker for the event.

A quotation from President Theodore Roosevelt's 1903 address to the Society of American Foresters reminds students of their responsibility to society.

New Dean

R. Scott Beasley reminded me that he was in my first regional silviculture class at SFA, the weekend field trip among his earliest experiences beyond his hometown, nearby the college. When Scotty returned as dean 25 years later, a reporter asked what he remembered about his academic career. His response was to name this biographer because of a "deep commitment to forestry and to students." Had I given this fellow an easy A? He wouldn't tell me. The registrar squealed: it was a B.

Scott made possible the typing of this manuscript.

And from Mrs. Theodore Fox, in 1968: "Since 1964, on my utility room blackboard there has been hanging a

Christmas greeting to Fellow Foresters of SFA written by you. Ted, a future Navy officer, was so moved by this that he brought it home for us to read...most grateful to have Ted instructed by men who have a strong faith in God and a keen interest teaching about the wonderful resources the Creator has so graciously bestowed on human beings." I think this is the greeting to which Ted's mother referred:

Note to forestry students:
It was night when they tried to find a bed in the hotel.
It was night when the star shone, and when they followed.
It was night when they tried Him.
It suddenly was night when He was executed.
It was still night when He walked from the rocky cave.

But Then it was light!
The Sun rose.
He said, "I am the Light."
He said of those who follow Him,
"You are lights..."

So, for a world in dark despair,
minor skirmishes here,
bloody coups there,
cosmic disintegration possible everywhere,
The Light shines in the darkness.

Your faculty joins me in wishing you a joyful Christmas season and a New Year full of good things.

A parent teams with the dean for the cross-cut sawing contest. Invariably students whipped the faculty. Soon mothers joined daughters in this annual sporting event.

As older folks nurture youth with ideas, so these old-growth shade-tolerant birch and maple trees nurture the seedlings and saplings in this northern hardwood stand (USFS photo).

16

"THE NEEDED TRUTH TO SPEAK"

Douglas-fir planted in the Scottish Highlands. Heather of the extensive heath cover the foreground.

A distinguished professor of constitutional law at the University of Georgia, in addressing a faculty gathering, showed clearly that one, regardless of that person's faith, cannot park that faith outside the schoolroom door. What we believe is part of us, like hair and hands. To abandon that which we hold dear when stepping to the teacher's podium would be like severing a hand for the sake of satisfying a critical class or a complaining student. We are what we are.

The same is true for the researcher. Much as he or she utilizes randomized sampling in order to avoid bias in an experiment, we enter the laboratory or lay out the field trials with a bias dependent upon our past. From whence we each have come plays a role in guiding our scientific studies.

I've never preached to a forestry class, nor even spoken what some would call a testimony. Objectivity, not subjectivity, was my aim. Yet objective instruction may lead students to know where the professor stands on life's great issues. Visits to my office and home, even years after retirement and by alumni, attest to this.

However, the concluding lecture I gave for a general university course dealing with all the natural resources—water, air, grass, soil, minerals, forests, aesthetics, and so forth—concerned *the* most important natural resource: people. To encourage future educators, chemists, and entrepreneurs among the students, I spoke on the nature of mankind. I suggested that the answer to why we do what we do to use and abuse the earth's bounty begins with The Fall. The lecture was listed, "The Solution to Pollution." The following class period was given over to the students for expressing other concepts about our behavior, some attributing self-centeredness to the social environment, to legislative edicts, and industry's greed. Occasionally students would insist that self-interest was not inherited by himself/herself. They believed sincerely that they never had done anything objectionable to anyone; they had never sinned. Yet students expressed appreciation, if not agreement, for the view held by St. Augustin of the early A.D. years, Martin Luther and John Calvin of the Reformation period, and a host of theologians today. The *Journal of the American Scientific Affiliation* published that lecture in 1986.

The "Call"

So, it should not be surprising that shortly after leaving the deanship, Dr. William Bos, an SFA professor of speech communication, knocked on my office door. Bill, a theological seminary graduate and occasionally a seminary lecturer, had a problem. A country Presbyterian preacher had died, leaving two flocks abandoned. Bill, being ordained, would provide ministerial services for the larger congregation and provide guidance for me if I, as a layman, would help out with the smaller assembly.

I had no Pauline experience on a Damascus Road, no angelic voices, no meeting with a Captain Cornelius. It was the simple "Come over and help us" asked of a first century apostle by the followers of The Way in Macedonia. So in 1976, I took on this weekly assignment. The good folks at Rock Springs, their church claimed by some to be the oldest Protestant church in Texas, endured my ministrations. Rock Springs is also likely the place where Richard Overton Watkins was ordained in 1840, the first Protestant to receive the "laying on of hands" in the new Republic.

Prior to the Texas Revolution, legend has it that the Protestants, more-or-less wards of Stephen F. Austin's colony, worshipped in secret in a log house nearby the site of the present church. The house, recently moved to the historic park at Millard's Crossing—Nacogdoches' answer to Williamsburg—appeared in courthouse tax records in 1834.

Enjoying the new freedom to worship, citizens of Watkins Settlement, which I have located on old maps, built a church. It was a crude six-sided log structure with port holes on each wall through which sentinels watched for Indian raiders. The second sanctuary, erected in 1868, had pews of split logs and peg legs. The present structure of

Rock Springs Cumberland Presbyterian Church, site of the author's ordination, in rural Nacogdoches County, Texas. The boards were sawn from logs harvested from nearby virgin pine forests about 1902.

locally sawn boards harvested from nearby forests dates to 1902.

In time I came "under the care of Presbytery" with an aim toward ordination so that these parishioners could receive the sacraments—baptism and the Lord's Supper—and have a clergyman available to perform weddings.

As a nonseminarian, Presbytery (composed of all the ministers in a region and an elder from each church) didn't know what to do with me. They passed the buck to the dean of the seminary in Memphis. Dr. Hubert Morrow, the dean, assigned the text readings; read, critiqued, and graded my papers. After five years he was satisfied that I was ready to be examined for ordination under the church's extraordinary clause (for those not holding seminary degrees). It was now presbytery's turn to pose the questions, for that is the body that ordains. I am probably the most thoroughly interrogated candidate ever. The vote was not unanimous. A scientist and a faculty member of a state school were reasons enough for suspicion.

In 1981, Bill Bos retired from the University, and he and Doris began a new work in Oregon. What now for the little church in Trawick, once a significant town with a railroad station? Even the tracks are gone. The congregation's patriarch and a few other folks checked me over and asked for my help. Five miles of logging road ruts through southern pineries, or 15 miles of pavement, separate the two churches. With an early morning worship at Trawick, I could serve both. Never officially installed, 14 years later I'm still helping them out. The task, as Whittier saw it: "Free, the needed truth to speak. Right the wrong and raise the weak."

The Lord's Funny Bone

The Lord must have had a sense of humor in creating the Church. And preachers better have it, too. An example: Some Roman Catholic visitors, unknown to me, were present for an infant baptism. Following the benediction, a visiting lad asked, "Are you the father?" I tried to explain that Mr. Stevens is the father of little James Keith. But the child adamantly insisted that I was the father. I suppose my face was beet red 'til a helpful soul rescued us both. The lad equated me to his parish priest. That I do not wear clerical garb probably confused the child.

Or the time a man died and was cremated in a distant place. He left instructions that his ashes were to be scattered on a certain hill in a Nacogdoches cemetery. His legal will also directed his heirs to secure a Presbyterian minister for the ritual that involved disposal of the ashes. The undertaker, enlisting me, said, "Here, you take these flowers out to the graveyard and do your thing." I took the floral spray and the box of ashes, about the size of a cigar box, brought by the family from the West Coast. The funeral director said the family would follow me to the cemetery shortly. Realizing how long it would take to distribute the ashes, a handful at a time, and that I'd run out of scripture verses committed to memory, I decided to partially empty the box before the family and friends arrived. I poured out some of the contents just as a gust of wind blew from the west. My black suit and shoes were suddenly a shade grayer than ash-white. Quickly, I dusted myself, wiped my shoes with a handkerchief, and hurried to another hill for the solemn occasion just as the cars approached. Alighting from the cars, someone said, "That's not the right hill.' My heart sunk. Would they go back to the mound I had just left covered with the calcareous dust of a human skeleton? "Over there is the family plot," someone said. Fortunately it was in the opposite direction of the disposed ashes.

Even so, I ran out of scripture to quote. Fifteen minutes had gone by as I paraded back and forth quoting Jesus, Paul, Peter, John, and Mark. Feigning an empty box and sensing the mourners were satisfied, I closed the box and pronounced the benediction. My science-teaching daughter used the leftover salts for class experiments.

As George Bernard Shaw said, "My way of joking is to tell the truth. It's the funniest joke in the world." Ministers regularly witness this verity.

Folks often say what they don't mean, or say what they mean unintentionally. One dear woman: "I stayed awake to hear your *whole* sermon today." She'd been present every Sunday for years.

Another: "You shouldn't apologize for being a poor preacher. You have other talents." And after a long pause in which to dredge up the other talents from an exhausted memory bank, "You're good with the children!" And a man on the street stops me to say, "Walker, you ought to learn to preach like Jimmy Swaggart. I sent him 20 bucks last week." Never mind; keep your twenty.

Two Drs. Larry Walker work in our town. Besides myself with Larry as a nickname, the *real* Dr. Larry Walker is a surgeon. He cared for Mary Stevens, an elderly parishioner who underwent surgery for an aneurysm.

Visiting Mrs. Stevens in the hospital's Intensive Care Unit, her eyes closed and she only minimally alert, I introduced myself as Larry Walker and asked if she would like prayer. Her hours on earth were indeed numbered, she thought, if her surgeon was suggesting prayer. An hour later the surgeon made his rounds, gave his name to the subconscious women, and proceeded to remove the sheet to examine her. Startled, she quickly wakened, thinking the preacher was getting fresh. Mrs. Stevens, in gales of laughter, told the tale more often than I.

Twice my trousers tore while in the pulpit. Once I looked down to see, after kneeling, that the knee was gone from the twill material. On another occasion, while bending over for the children's talk, I heard the seam of the seat give way; the ripping noise was clearly audible. I backed to the pulpit. Now I know why clerics wear robes.

'Twas in the mountains of North Georgia to which the presbytery's superintendent of home missions sent me early on a sabbath to supply a pulpit. Midway through the service, an elder of the rural church rose to speak. His concern was the kinds of folks the superintendent sent to the village. "They don't belong here," he exclaimed after each category in a long litany listing English professors, seminary students, seminary instructors, big-city retirees, and college students. Finally the old man arrived at his point, even before the sermon. "Today," he said, "Mr. Clements sent us a forester, and he belongs in the woods."

I thought the same techniques used to "grow" a forestry school could serve the church. How lacking in understanding was I! Small, rural congregations are family. In each, one likely finds a patriarch and a matriarch, not necessarily spouses, who manage matters. These churches need no pastor, translated shepherd and defined as leader. Minister, meaning servant, is the appropriate title for the task; the ordinand often not unlike the chaplain to a fiefdom as in the days of feudal estates.

Could Shakespeare have known about these *wee kirks* when he exclaimed, "What tedious homily of love have you wearied your parishioners withal!"

Upon being "honorably retired" by the Presbytery from the work at Rock Springs, and serving the Trawick church only on alternate Sundays, I thought I'd take it easy, yet being available now and then to supply vacant pulpits. That wasn't to be. The folks in Tennessee, once a thriving community to which the beauty of the church attests, asked for our help. As I write, Anne and I journey to the Shelby County community, not far from the boundary of the Tenaha Ranger District, to lead worship in the Presbyterian Church.

Land of the Thistle

"*I am* a forester." That is what I thought I heard the caller say in a 2:30 a.m. telephone conversation. The strange accent at that hour immediately brought to mind one of our foreign students in trouble with the police. The call, however, was from a minister of the Church of Scotland, the church the colonialists brought to North America and here called Presbyterian, simply meaning rule by elders. The cleric's name turned out to be *Ian* (Scottish for John) Forrester, so you see my confusion. He wanted to know if we would serve his parish of four churches on the North Sea while he "went on holiday" during the summer. Indeed, we'd be delighted to do that.

The Church at Inverkeilor, one of five served by the author in Scotland. The Scot Calvinists built the sanctuary in 1635, only about 70 years after the death of the Reformer John Knox.

The Inverkeilor Church, one of the congregations, dates to 1635, its granite walls four feet thick. The manse was "modern," built in 1785. I climbed the attic rafters to discover huge hand-hewn oak timbers supporting the roof of slate slabs. Bells still hang in the kitchen where a servant, hearing the tinkle of a particular bell, would know to which room he or she was called. Ministers, addressed as "*The* Reverend," are treated with unexpected courtesy. They are, by virtue of ordination, notaries public. That's because in the middle ages they were among the few people who could read and write.

When I baptized wee Jamie Stuart, he was the thirty-sixth child to wear the same gown. A book accompanied the dress, logging the recipients of the sacrament on both sides of the Atlantic Ocean. The first was this child's great great uncle in 1891.

At the conclusion of a funeral, the undertaker, dressed in tophat, striped trousers, and tails, asked me to announce that the mourners are all invited to tea at the hotel. The tea, financed by the family, was a full banquet-style meal. But to be asked to propose a toast to the deceased was a bit much!

We served another summer at Coalburn, a mining town as the name suggests. Reclaimed surface mines filled the landscape across the road from the large old manse, once the mining superintendent's residence.

Surface coal mining and reclamation on government lands across the road from the manse in the town of Coalburn, Scotland.

In Lanark Presbytery's parishes, they're still fighting the excesses of Queen Elizabeth's armies of the 1500s. Nearby in a farmer's field stands a monument to David Steele, shot at the site by Dragoon Lt. Creighton for secretly assembling to worship with other Convenanters in the fields and forests. A scroll chiseled in the granite obelisk reads the Book of Revelation's admonition, "Be thou faithful unto death." Scots never forget! One of the descendants of the Covenanters, a shepherd watching his flock out in the hills, curses under his breath as a whaup rises screaming from the bent grass at the herdman's feet. Why the oath? The migratory bird, a curlew to North Americans, betrayed the hidden and hunted Covenanters during the killing times. When disturbed, the whaup's cries of alarm guided soldiers to the spot where Covenanting Christians lay low among the heather of the heath and the mosses of the moor.

Gravestones record the events, often in rhyme; this one telling that only heads were buried:

Stay passersby, take notice what thou reads.
At Edinburgh lie our bodies, here our heads.
Our right hand stayed at Lannark,
These we want,
Because with them we signed the
Covenant.

For weddings the men wear dress kilts and Bonnie Prince Charlie jackets. The law no longer permits wearing the dirk, a small dagger, tucked in a stocking. And ministers are subject to prosecution for violating a long list of statutes spelled out in the Marriage Act: a man may not marry his grandfather's former wife; A man may not marry his former adopted daughter, etc. I needed to be cautious!

On such exotic excursions I visited foresters, heard about their problems and was called upon to lecture. I spoke to foresters at the University of Aberdeen as I had done on a previous trip at Oxford. At Oxford, the dean scheduled me at the same hour in which Prince Charles arrived to receive an honor. In spite of this conflict, the forestry students attended, especially interested in the silviculture of the fascinating longleaf pine.

An original macadam road nearby the North Sea. Gifford Pinchot captured Pennsylvania's governor's office by promising to get the farmers out of the mud with macadam roads to every mailbox. MacAdam, a Scot, developed the gravel and tar combination for modern blacktops. Pennsylvania's "Pinchot roads" were as narrow as this one traveled by "the minister" on the borrowed bike.

Hamish (Scottish for James) on a grouse-hunter's drywall—perhaps several hundred years old—on royal lands in the Highlands.

As I wore two hats on these trips, so too my forestry ventures abroad usually included a study of the Church. Hence, while lecturing on silviculture at universities in Egypt, Jordan, Israel, and Greece, I could report, in a series of articles, on the problems and successes of those pilgrims. My observations of the church in Costa Rica, Indonesia, China, Siberia, and Brazil also were published.

The end of the road, what then?? Second-growth red-spruce at the end of a train in West Virginia's mountains (USFS photo).

17

MANURING EUROPE'S FORESTS

Nine-year-old stand of poplar in Italy's Po Valley. Dr. Prevosto, tree geneticist, hybridized the trees from America's eastern cottonwood and Europe's black cottonwood.

Anne had me figured out: "You're happiest when the wheels are turning." From scouting days and maneuvers along the St. Lawrence River as a 15-year-old national guardsman, venture challenged me. Anne, always contented (read "preferring") to stay at home and nurture the kids, never complained.

Nutrient Needs

As the consulting forester for the National Plant Food Institute, the educational arm of the fertilizer industry, I traveled to a half-dozen western European countries to view the practical and applied research underway as well as the applications of nutrient amendments to various species of commercially valuable forest trees. The director of the W.R. Grace Company's research group accompanied me on such a venture in 1962. The company was interested in diversification and, perhaps, the movement of fertilizer freight on its ships. Dr. Ray Stirostka, Peter Grace's man, opened for us the doors of the best hotels and restaurants in Europe, while I lined up the folks who knew the most about forest fertilization.

The world renowned Drs. C. Olaf Tamm of the Swedish Forest Institute and W. Wittich of the University of Gottingen, Germany, each spent several days with us traversing their respective countries. Prof. Wittich had a pre-war VW bug, but it served us well in those days prior to rental cars. With the Swede, we traveled by train and then to the woods in government cars. In Finland, our host, Prof. P.J. Viro of the Finnish Forest Research Institute, took us to the far north, to the woodlands where the Laplanders herd their reindeer. Our children still have slippers made from the hide and fur of these draft animals, both sexes of which display gigantic antlers.

At Helsinki's airport we spotted a Buick: What's it doing

here? In every respect save one, it was an American Roadmaster. A five-letter word in Russian gave it another name. Did the Soviets even steal U.S. automakers' engineering blueprints?

We met Fleming Juncker at his office in downtown Copenhagen on the Isle of Sjaelland (Zealand). This tall, stocky Dane instructed us to meet him at a certain dock at the waterfront that afternoon. We did. And when he arrived the bells rang, whistles blew and, to our surprise, our host was piped aboard. He owned the shipping line! Our staterooms on the captain's deck and the meals in the master's cabin for the overnight journey displayed unencumbered opulence.

A chauffeured Volvo met us at the dock on Jylland, which some maps call Jutland, west of Zealand. The car pulled into a circular drive to let us out at the entrance to a castle, the moat long ago filled with soil. My accommodations, high in a round tower at a corner of the castle, were the Forester's Quarters. The room, specifically for visiting foresters, looked out in three directions onto stands of conifers and broadleaf trees. The forests, both planted and natural, grew on drained bogs and mineral soil.

The host's history slowly unrolled over meals and field trips in a vehicle in which he could, by tripping a lever, raise the vehicle's body so the car could pass over high centers. He'd lower the frame when we returned to pavement. Was it a Saab or a Volvo?

As a lad, Juncker had walked past this castle on the way to school. Someday, he dreamed, that would be his home. Eventually he graduated from the Danish Forestry Academy, got involved in the pulp and paper industry, and by the time of our visit was the owner of a pulp mill and a flooring plant that fabricated beech (*Fagus* spp.) into sheets of beautiful boards. Juncker pulped Denmark's trees and shipped the baled fiber in his vessels to England for paper production.

As a student of less than moderate means, Juncker turned to a particular faculty member as his mentor. That professor, Karl Mar:Müller, at the time of our visit lived in a home on the castle grounds and managed Juncker's forests on the Overgaard, meaning Upper Farm.

The forest manager told us of his former student's exploits during the Second World War. Wanted by the Germans, who had captured Denmark, for his underground activities, Juncker fled for his life. So large a man could not possibly avoid capture at the docks on Zealand nor on the train that rolled on to ships with accommodating railroad tracks that carried passengers to Sweden. Juncker's wife escaped alone aboard the train-boat that ran from Helsingor to Halsingborg. The giant of a man swam to freedom across the Oresund Channel that separates the two nations. Throughout the war he was sought by Hitler's hordes for his guerrilla tactics.

The castle, although used as a regimental headquarters by the Germans in the war, was neither looted nor damaged. Shelves in the library still held the books of the estate's financial and biographical records dating to the 1500s. In the hall gallery hung life-size portraits of every lord and duke or prince that had ruled from this spacious citadel.

Overgaard, owned by Juncker since 1954, receives about 18 inches of precipitation per year. *Juniperus communis* captures much of the heath. Then Norway spruce (*Picea abies*) seeds in. European beach (*Fagus sylvatica*) grows on 40-year rotations. Lodgepole pine (*Pinus contorta*) from Puget Sound and mountain pine (*P. montana*) from the Alps and the Pyrenees replaces the heather and the spruce. For ash species (*Fraxinus* spp.), provenance is important even in this small land. Both height and diameter growth vary by parental origin of seeds used in reforestation.

The forester explained his use of the ammonia form of nitrogen (NH_3) only, never nitrate (NO_3), to correct for deficiencies of this element. Nitrate, he noted, buffers the soil pH.

As the tide in the Katteskat Sea of the North Atlantic Ocean only amounts to about 6 inches, draining and diking is easily done. The waters here contain just 2 percent salt as NaCl. Prof. H. Holstener-Jorgensen of the Danish Forestry Institute introduced us to tile drains that lower the water table in a fjord originally ditched and drained a thousand years ago. Ecologists of 1900 considered the beech trees virgin. Not until 1805, however, did the government establish forest land by law. None now can be used otherwise.

In Germany, Dr. Wittich took us to Luneburger Heide, a site cleared in 1200, now covered for 100 years with a heather, *Calluna vulgaris,* leaving the plain deficient in nitrogen for the growth of trees. Also in North Germany, we found the Syke Forest, clearcut for WWII reparations but now planted to broadleaf trees. Records on the Syke date to 1833 when Scots pine (*P. sylvestris*) was planted. *Fomes annosus* attacked the stands as they were thinned, even a hundred years later. These acidic soils were being limed with calcium carbonate at government expense, the granules distributed by blower machinery.

In the Netherlands, Engineer (as a forester is sometimes called in many nations) C.P. VanGoor showed us the hybrid *Populus euroamericana* on the polders, land claimed from the sea by dikes. Inland dunes developed in the Middle Ages and subsequently afforested with Scots pine exhibited magnesium deficiencies. Applications of Epsom salts ($MgSO_4$) then enhanced the color of the chlorotic foliage.

In Sweden we observed severe moose damage in peat bogs, called Lagg. Fences that exclude roe deer appeared ineffective for controlling moose. Both moose and hare prefer nitrogen-fertilized stands for the protein component in the forage.

Farmers abandoned vast areas between 1800 and 1900 when, due to starvation, some 10 to 15 percent of the Swedish population migrated to the United States.

The Swedish State Church owns much of the forest land. Old pine and spruce trees growing on the glacial outwash of the Bottnaryds Parish Forest, when cored, exhibited ages of 300 and 200 years, respectively.

Tamm took us to fertility plots his father, also C. Olaf, established in 1918 using wood ashes. Deep soil occurs here on hilltops, the mineral matter derived from glacial moraine deposits of a variety of minerals with adequate essential elements for tree growth. In the lowlands, "miners" harvest peat for fuel. Frost-heaving of seedlings is especially serious on these sites. Woodland pastures are fertilized and grazed throughout the year, even in this subarctic climate.

In northern Finland, mor humus layers exceed a foot in thickness due to the cold, high-humidity climate. For the nation, some one-third of the woodlands are on wet sites, the peat always at least 6 inches deep. When drained, trees grow well. Until WWII, ditches were dug by hand; now dynamite sticks placed in holes a foot apart, especially where stony land lies beneath the peat, speeds the task. Poles lying in trenches and covered with peat also provide for subterranean drainage.

Visits with French foresters reminded us of the famous forestry school at Nancy, attended for a year by Gifford Pinchot. France, too, is where plant nutritionists developed *foliar diagnostique* as a means of quickly recognizing nutrient deficiencies. Forests there still had not recovered from the fires, bombing, tank intrusions, and military harvests of WWII. Prof. M. Bonneau of the Nancy School and Engineer (Ingr.) R. Keijean, a forester of the Ramouillet Forest, discussed their reforestation efforts.

A sojourn in the Black Forest for inspiration, if not for education, was essential. Oberforstmeister Dr. F. Querengaesser showed us problems with roe deer, their numbers about 20 per 100 hectare where the carrying capacity called for a limit of 7. In this forest, treasured by the Germans, timber markers decided and flagged the direction of fall of each tree to be harvested in order to avoid damage to residual stems. We counted eight species originating in the Southern Appalachians and planted on the obermeister's district. Could that somehow be related to the establishment of the Biltmore Forest School in North Carolinas's Pisgah Forest and the origin of its chief, Carl Alwin Schenck, the German forester friend of Pinchot's?

Two articles about the forests in Europe, resulting from this trip, appeared in *Better Crops with Plant Food*, an industry journal in 1963. J. Stoeckler, whom the U.S. Forest Service sent on a similar fact-finding mission, co-authored these papers, one on the use of lime and legumes, the other on commercial fertilizers (including inorganic salts) the Europeans call *manure*. Another article on "The Source of Nitrogen" appeared in *Agricultural Ammonia News*.

Cottonwood, the European Fiber

I stepped off the Athens-bound plane in Rome in the

middle of the night with a few others destined for the City on the Seven Hills. An airport bench was a safe haven until dawn, a phenomenon here confounded by the smog that engulfs the city. Once at a hotel and my passport surrendered, I hurriedly visited the great attractions of what was, less than 1500 years ago, the capital of the world. Here I found the Palatine Hill, now covered with the ruins of ancient buildings; checked on the Swiss guards and the Pope's gold at Vatican City; and ventured into the Coliseum built in A.D. 75. Once seating 80,000 spectators of the gladiators' murderous sport, the grand structure is now eaten away by the sulfuric and nitric acids emitted from vehicular exhausts.

Satisfied after two days of jostling with traffic and small-car drivers who do not distinguish sidewalks from boulevards, I headed for the catacombs that tunnel the Eternal City. One still detects the scratches in the rock walls made by clawed fingernails as Christian martyrs and others abhorred by the Caesars etched in the dark their testimonies in the stone of the dank walls to which they were bound and held captive.

But my task in Rome in 1966 was to learn of the work of the Forestry Directorate of the United Nations headquartered there. The international bureaucrats, later for a while headed by one of our forestry graduates from Honduras, pointed me toward Casale Monferrato in the fertile Po Valley. Along the coast of the Ligurian Sea, the train steamed by Pisa. I thought the locomotive's vibration would bring down the Leaning Tower.

In the deep, rich, alluvial soil, accumulated over many millennia from the erosive Alps Mountains of Switzerland and Austria, I'd find the Institute of Poplar Culture. There the silt and clay texture of the soil encourages production of highest quality wheat and other grains. But it also provides the nutrients and soil moisture regimes that enable rapid growth of trees, especially cottonwoods of the genus *Populus*.

By the conclusion of the Second World War, the forests of much of Europe were in shambles. Rambling tanks, the engineers' crawler tractors, shelling, bombing, the fires that flared and spread from the cannons' targets, and the demand for wood for replacing demolished bridges and for temporary housing consumed vast volumes. For example, at the conclusion of hostilities, 80 percent of German housing had been destroyed. However, within ten years, every citizen of the country was adequately housed; hence 80 percent of the population by 1955 lived in homes less than ten years old. And erecting all of those houses, along with other essential buildings and infrastructures, required wood. With the armistice, wood famine faced the FAO leadership.

The UN's Food and Agriculture Organization placed agronomists at the helm of the project to find species or to develop hybrids that would rapidly grow to replenish the wood supply. Foresters, I was told, ignored the opportunity. They considered working in a greenhouse or nursery bed somehow beneath their dignity. One aspired to be a forstmeister, a leader. I've known European foresters too elitist to use an increment borer or to dig a hole to examine soil; a subprofessional accompanies the forester for this kind of manual labor.

Agronomists concluded that poplar would serve best because of the genus *Populus'* widespread occurrence, its ability to grow well on a variety of soils, and while short-lived, it is fairly immune to insect and disease attacks. Poplar wood also is versatile, suitable for lumber, plywood, particle board, and paper of various grades. Holding in my hand a sample of plywood, 1-5/8 (4 centimeters) inches thick, I count 53 laminated layers. Architects use this material for structural beams. Much better than steel, the wood may char but never collapse in a fire.

The scientists searched the world for various species, races of species, and provenances. They grew seedlings from seeds and sprouts from cuttings, planting them in greenhouses and nursery beds.

Finally a particular hybrid, I476, showed promise. It is the cross between *P. deltoides*, the eastern cottonwood of the southern United States and *P. nigra*, the ubiquitous black cottonwood of Europe. Now, dendrologists call the hybrid *P. euroamericana*. From a single highly desirable tree like this, cuttings can be taken and from those cuttings new trees arise of the same genetic code. It's like cutting off a finger, planting it, and getting a person exactly like the one from which the finger was removed. I have since observed the hybrid clones in other European nations, in northern Africa, and in countries of the Near-East. Perhaps it is I476 that presently grows far up the Nile River to the south of Egypt.

The deltaic Po Valley now would not grow spaghetti's raw material. The Italians import wheat grown in the alluvium of Egypt's Nile River overflows while growing *P. euroamericana* in short rotations at home. Twelve-year-old stems from cuttings tally 16 and more inches dbh, but at about that time certain poplar-loving insects, especially *Crytarinous*, begin to feed on the trees. Chickens, carried by laborers in box-like roosts from place to place within the forest, feed upon the bugs. In the Po Valley, insects may destroy 20 percent of the wood produced. A fungus called *Venturia populina* also attacks the trees. Professor Ettore Castellani of the Institute, established by the cellulose industry in 1930, took me to a 5-year-old stand, R2F2, that averaged 11 inches dbh. Nine inches in 5 years, 18 inches in 9 years, and 75 feet of merchantable height at age 9 are not unusual growth rates. Poplars were first successfully hybridized by a Professor Yakometki in the 19th century.

Following the final harvests, seldom at more than 14 years, bulldozers push the stumps and roots from the ground. The effort is not to prepare the site for a new stand, which obviously could be established by sprouting. The ground is cleared for wheat, as the law requires that an agricultural crop must follow a tree crop for at least two years.

During a logging operation I picked up a slab showing some growth rings exceeding two inches in radius. Back on the same logging job two days later, dozers had completed pushing up and piling the stumps, with roots attached, to the edge of the tract. A creek alongside served as an irrigation canal. Next day, the stumps were gone. What happened? In the night, I was told, they were dumped into the stream.

Long before Mussolini messed up Italy, the Mediterranean Populetum had been established near Rome as the *Instituto di Sperimentazion per la Pioppicoltura*. Scientists there had shown that hybrids of *P. nigra* and *P. deltoides* vary greatly by the parents selected. Here, to enhance growth of these poplars, workers cultivate in the spring, June, and September. Farmers interplant the cottonwoods with food crops for three years. After that they plant forage crops. "Factory farms," villages of over a hundred people, work in these poplar plantations.

Nursery-grown trees with both tops and roots one year old are called barbatelles. The name derives from the appearance of the roots and branches, when held in the hand together, looking like a beard. When tops are two years old and roots three years old (due to cutting back after one year), the hybridizers call them pupelles.

It's great to be a man of stature, as I was in the village of Casale Monferrato. Everyone looked up to me. For a change, this runt was taller than most of the citizens of the town.

In a restaurant there I ordered spaghetti. What else? Across the table, a fellow was eating what looked to me to be a delicious slab of pie. Gesturing, I asked the waiter for the same. He waved his arms, "No, No." Not understanding the objection, I persisted to his negativism. He brought the "dessert": it was a thick piece of real Italian pizza, 2 or 3 inches deep.

Champion Paper and Fibre Company tried to duplicate the FAO Institute's effort in Texas. Company foresters attributed their failure in pine—hardwood sites to the inability to control competition, even though site preparation prior to out-planting included picking up by hand every piece of twig that might sprout.

Swiss Forestry Oversell

Prof. H. Artur Meyer convinced his students in our undergraduate Penn State days of the excellent management of the forests of his native Switzerland. Once before I was determined to see the majestic stands of conifers about which he bragged. And once before I was stood up by the Zurich University professor, a Dr. F.R., who promised to escort me through the spruce and fir stands described in Dr. Meyer's mensurational classes. This time I was to meet a forester in canton employ at a train station in Moritz.

As all European trains run on precise schedules, I arrived at the appointed place and at the appointed minute.

Several hours later it dawned on me these Swiss don't want to show off their much-touted managerial skills so often reported in the technical literature. With the help of a map, I'll see these intensively managed woodlands on my own!

I studied the map for the most remote part of the little mountainous land of French-, German-, and Italian-speaking peoples. No roads led to my destination; one could get there only by train, including cable-drawn cars to the town where I'd settle in. Here, certainly, I'd find the little elves who assemble watches in tiny snow-covered cottages in winter months as well as the elusive intensively managed forests. By train to Andermatt, transferring to another rail car to Brig, passing by 11,657-foot Mt. Leone, and finally a cable-drawn coach took me to the end of the line. An almost empty carriage assured me I would find the remote forests I sought.

Stepping from the cable car on a perfectly clear early spring day, I gazed upon the spectacular *der Matterhorn*, the mother of mountains. I was in the town of Zermatt, one of Europe's principal holiday sites, both summer and winter. To my good fortune, the time of my visit was between seasons, and local people were now on their "holiday." Hotels line the streets, but only one was open for business. Horse-drawn carriages transport people and goods.

The breath-taking view of the 14,679-foot mountain from Zermatt's vantage point several thousand feet below left me awestruck. This was better than studying tree-plot tallies. So, next day early, I began at the trail head that led upward. When I slipped on the ice, a camera slung across my chest breaking the fall, yet banging my skull, wisdom suggested I abandon the trek and return to Zermatt. Already I'd noted the graveyard reserved for mountain climbers. Attached to the markers were the gear that accompanied them on their last attempt to scale the alp.

World Forestry Congress

Scheduled to present a paper at the World Forestry Congress in Madrid in 1966, I entrained for a scenic ride through the Swiss Alps. Transferring to a French train, I was on a route that paralleled the Mediterranean coast at the Gulf of Lions. I then boarded a Spanish train that took me to Barcelona *via* the eastern edge of the high wall of the Pyrenees Mountains, home of the sheep-herding Basques who became shepherds for America's western sheep ranchers.

I knew I had bought a second-class rail ticket to Madrid, never imagining Spain's railway has three classes. I looked for the least inviting coach, assumed it was second class for this deprived Iberian nation, and hopped aboard. From Barcelona to Madrid, for some dozen hours, I rode with the peasants and their livestock on wooden slats, looking at the cross-ties (Europeans and Asians call them stringers) through the gaps in the floor, and holding on for dear life to the seats in the "Toonerville Trolley."

I delivered my paper on "Forest Fertilization in North America," and there met Dr. Krishnaswami Kadambi, chief conservator of forests for Bangalore State in India. Dr. Kadambi wanted to come to the US for a couple of years, having retired from the Indian Forestry Commission and having established, under UN sponsorship, a national forestry school in Nigeria. I felt we needed some cosmopolitan influence on the faculty, and he was willing to come as a technician. The low wage set aside for that position was the only money available. Constantly with students, day and night, Dr. "K" served well. He wrote a bulletin on teak which the school published in order to gain some favorable international recognition. The publication may be found from Indonesia to India, and points in between. An Indian publisher has reprinted bound copies.

Various side trips from the convention site in Madrid took the conferees to historic places and to exhibits of forestry and grazing practices. At Toledo, the bull fight and the reenactment of the Military Academy's defense in the 1936 revolution entertained us. Here we saw the Alcazar where for 71 days the youths held on until Franco's forces relieved them.

Discouraged by the lack of any real intensive forest management, I set off—this time really on a second-class train—to see how abusive have been the land-use practices over the centuries and the results of the most-recent civil war. The whole Iberian Peninsula, Portugal included, has succumbed to the ravages of mankind's greed and laziness. Poplar stands are limited to 2000 hectares and the law requires two years of vegetable gardening between cottonwood crops. Cork oak (*Quercus suber*) provides a valuable commodity. The bark is stripped for its commercial use every 8 years.

I crossed the border into Great Britain's Gibraltar at the Bay of Algeciras and, from there, went by a small launch across the narrow neck of the Mediterranean Sea to Tangier, Morocco, on the north coast of Africa. Once again I come upon the rows of planted poplar trees that hold the sand in place. Are they *P. euroamericana*? I could not learn the answer. Moving along to Spanish Centa, I headed back to Spain on my first ride on an air boat. They glide along inches above the water, leaving almost no wake. Now, at Gibraltar, I had a bit of trouble getting back into Spain. Finally I was admitted. From there to the gardens of Lisbon and home.

Cork oak from which the bark is removed at 8-year intervals for bottle stoppers and other commercial purposes. Sheep graze these southern Spain woodlands.

Lodgepole pine from Canada's western provinces has become a favorite reforestation species in Scotland.

18

PACIFIC AND ASIAN ESCAPADES

Rain forest of dipterocarps in Indonesia.

The 727 stopped at Honolulu on its way to New Zealand. Other passengers bunked for a short night's rest in local hotels. But I had never been to Hawaii and, not knowing if opportunities would ever return me that way again, I observed all I could of the local scene until closing eyelids demanded rest. A bench by the not-yet-so-famous site of the bombing that initiated the U.S. entry into WWII, on the island of Oahu not far from Honolulu, served well until time to return to the plane.

From Auckland on in New Zealand, I was the guest of New Zealand Forest Products, Ltd. In this socialist nation, an entrepreneur, several decades before my visit, had the audacity to sell stock for a few cents a share in an industry that would grow trees, cut them into boards, and sell the wood. No trees of commercial value even grew in these islands when he went door-to-door to peddle the certificates. Housewives bought his "paper" to get rid of the pest. They are now rich.

The businessman had observed the dramatically rapid growth of Monterey pine (*Pinus radiata*) in Australia. He believed the same vigorous production of fiber could occur in the volcanic soils of New Zealand.

The success of the venture resulted in the government of this South Pacific nation of islands creating a competing enterprise. Could socialism compete with free enterprise in a land where egality plays such an important role? Here, the chief forester lives in a house identical to that of his loggers, even the pastel colors of the house not distinguishing executive from laborer.

A mild earthquake interrupted my first breakfast in the hotel in Rotorua, the forestry capital of the nation. A common event, all was silent until the rumble ceased. Geologists claim the islands were spewed from the ocean floor within the last 100,000 years. Hot springs abound, pipes running here and there to carry the warm water to homes and hotels. In the woods on a cold day, a hole dug into the bank of a road releases enough heat to warm one's frozen hands.

By the time of my visit in 1972, New Zealand Forest Products, Ltd., was competing with Australian and North American companies for sales of wood to Japan. Exports picked up as environmental groups insisted on limiting wood shipments abroad from the US Pacific Northwest. I would meet up with the company again, in the Fiji Islands.

The Aussie Visit

Next stop, Sydney, Australia, and then to Canberra to participate in a forestry conference to investigate the cause of a "second-generation malady" of Monterey pine, the tree introduced from a California peninsula in the early days of the twentieth century. Rapid growth of the planted trees of the first rotation seemed unattainable in the regenerated stands. Was the difference due to a fungus pathogen, an insect, nematodes, climate change, diminished soil fertility, or something undreamed of? Had some unknown factor that kept the growth of *P. radiata* checked on the US Monterey Peninsula lately arrived in Australia? Should New Zealand anticipate a similar ecological and economic catastrophe? Our group did not arrive at even a suggestion of an answer.

Foresters tally about 700 species of *Eucalyptus* in Australia. Some, like *E. grandis,* are especially useful for construction timbers. Some serve solely as food for koala bears, while others provide oils for medicinals and industrial ingredients. Most, because of the highly flammable oils in the broadleaf foliage, are susceptible to ready ignition. Holocausts quickly consume thousands of acres, as the pyrrhic disaster of January, 1994, west of Sydney in the Great Dividing Range attests. Near Melbourne, our study group visited the site of such a burn, now regenerated to a fine stand of eucalyptus. Many species of the genus depend on fire—or the absence of fire—for regeneration or stand survival. Ecological succession or regression relates to fire history.

In eucalyptus forests, fire towers abound; so too do fire shelters. In these dugouts in the ground into which woods-workers or travelers escape from fires that easily outrun an Olympic track star, the first "refugee" to arrive finds a drum of water and blankets. Quickly that person soaks the blankets and hangs them over both the openings to the shelter and an inner doorway that serves as a sort of a vestibule. Early-day shelters had a pipe for ventilation that extended through the soil-covered roof, as some 20 or more persons may jam into the 8x8-foot "room." Alas, the suction generated by one raging fire pulled from the shelter all of the oxygen. Rescuers found only a stack of human bodies.

In the group of Yankees on the Australian venture was George Staebler, Weyerhauser company's research manager, and his family. His two little girls became good travelling companions. One, now the talented editor of the *Journal of Forestry*, recently reminded me that she titled me "Tex" on that long-ago excursion. Rebecca is not a professional forester, but George's enthusiasm for his life's work netted the profession a valuable prize.

Eucalyptus species have been introduced to many of the world's forest lands. An excellent stand of perhaps 40-year-old trees on the University of California campus at Berkeley tallied, I suppose, more than 500 square feet per acre basal area. Recall of the visit in the 1950s suggests these stems exceeded 3 feet d.b.h.

We'll write elsewhere of the species importation into Israel by the Israelis, there called by the Palestinians the Jew tree.

One of the South's major wood-using companies, Buckeye Cellulose, under the direction of chief forester Don T. Stevenson, introduced certain species of eucalyptus into the Florida Panhandle. And SFA's Dr. Kadambi initiated experiments in cooperation with Southland Paper in the Rio Grande Valley, near Edinburg. Two years' results for the latter study were sufficient; trees commonly exceeded 4 inches in diameter at the ground line. But this land is too valuable for fruit production to utilize for fiber. Eucalyptus trials on the campus at Nacogdoches continue to be killed back each year by frost, only to sprout again to the present time. They serve as a living memorial to Dr. Kadambi.

Chinese Gold and Fijians

Next stop, the Fijian Islands, paradise place of film and

story. We'll see. Here, too, New Zealand Forest Products, Ltd., harvested timbers, but these were not solely stands of Monterey pine.

I located a peckerwood mill perched on a mountain ridge. At the lumber dock end of the sawmill, a Fijian laborer carried freshly planed 1x6-inch boards to two sawhorses. Laying the boards across the scaffolds, he carefully measured the designated length and, with a 15-inch carpenter's saw, removed the excess few inches. I watched this simple procedure for a while, eventually inquiring, in slowly spoken English, as I picked up a remnant from the ground, "Do-you-know-the-name-of-this-wood?" Quickly came the reply in Oxfordian English of a species of which I was not familiar. He noticed my puzzled expression and asked, "Should I spell it for you?" Since I was still not grasping his English pronunciation, he took from my hand the slab of lumber on which I was trying to record the species' name and wrote in precise lettering, "Kauri." "Would you also like it in the Fijian tongue?" To my affirmative answer, he wrote some heiroglyphics.

Why do I tell you this story? The Fijian, with a command of English as though a graduate of Oxford and doing the menial chore I've described, was dressed only in undershorts and a hat woven from the fronds of a palm. The sleeves to protect his upper arms from the blistering sun outside the tin-roofed mill were also woven from the green foliage of the tropical isle. Why would one so educated—many schools would have been happy to have him teach lettering to drafting students—earn a living as he did? Is this not a waste of our most important natural resource? (Foresters often speak of water, wood, wildlife, range lands, and recreation as resources for which they are responsible. I'd put people at the head of the list.) I returned to the little island nation's capital city, really a village, to learn the answer to this riddle.

Asking the question in the city of Fiji, I was told to walk down the main street, visit the shops, note the ethnicity of their managers and owners, and the answer will come. It did. When the Chinese businessmen and women fled the mainland of their homelands as the Peoples Republic emerged as a Communist dictatorship, the entrepreneurs left with their gold. One can't eat gold, but it is useful for buying up businesses in Fiji, Indonesia, the Philippines, Hawaii, and many other islands of the Pacific Ocean. Fijians, greedy for gold, became the servants of the migrants who bought them out.

Plantations of Monterey pine replace the kauri and other forest trees on these isles.

Another renewable natural resource subject to exploitation is coral. The living organisms, rich in lime secreted by the animals and colored by the mineral impurities of the seawater in which they grow, became the tourists' favorite souvenir. Many species abound, each with its peculiar configuration as its colony rises from the peaceful ocean floor. Coral lines the shelves of gift shops. The compound animals, resembling small sea anemones, unite into branching of more or less solid colonies that enlarge by asexual reproduction.

I succumbed to the temptation, put on my "tennies," and sampled the trophies in the sea. Some appeared like many tined, but lace-like, reindeer antlers, others more solidly structured. The fishy smell soon filled my little hotel room. Perhaps a clorox dip from a nearby store would counteract the odor. It did, and in so doing bleached out all the beautiful colors of the skeletal deposits produced by the anthozoan polyps, the minute animals of the sea. So back to the shore, this time at high tide, I ventured. I'd simply endure the odor to protect the reds, greens, violets, and other hues. But the tennis shoes! Torn to shreds by the sharp barbs of the coral strands, they remained in Fiji.

To the Great Wall

A call from a medical doctor in Hong Kong inquired of my interest in lecturing in the Peoples Republic of China. Will travel! It would be the trip of my dreams. Not only was Taps played over my older brother's first interment in Kunming (though we doubt if any of his remains were recovered in the WWII crash in the Himalayan jungle), I had other professional concerns and questions.

How one of every four people in the world were willing to surrender their freedom so readily and to live so long in bondage intrigued me. (Actually, when the large Chinese settlements in North and South America, Indonesia, and the South Pacific isles are included, the tally of Chinese ethnics approaches one in three of the world's people.) The trip took place only a few months after the 1989 Tiannamin Square uprising.

The United Nations sponsored this trip. Dr. Chung (a pseudonym for his protection) visits China to see what kind of specialists they want him to recruit from abroad for short-term tasks in the Communist land. With the list of experts needed, he cranks out on his computer Chinese-born people in the United States and Canada who are also Christians. (The UN cares nothing about separation of church and state.) No foresters' names appeared on his screen so he searched elsewhere and—how I do not know—located me.

Others in the team were wizards in jet propulsion, computer design, artificial intelligence, rocketry, mathematics, and medicine. All the others were young folks, one young lady among us, and all holding doctorates. The MD team leader did a residency in California.

If the PRC was to have our services, the government would have to agree to a short self-introduction that would include a few words about our faith. We carefully chose our words, all translated into Mandarin, to say in effect that we were in China to share our knowledge because of a love, generated by Christian faith, for the Chinese people. As Bibles cannot be bought, and new ones can't be legally given away, we marked up a few personal copies we could rightfully carry and give to curious inquirers. Hungry for any books, one man got my daughter's copy of C.S. Lewis' *Miracles* while another pleaded for Dr. "Lings" treasured personal Chinese-language Bible.

I delivered my lectures on various silvicultural subjects to the staff of the chief forester, to the forestry college at Beijing, and to forestry groups in the Kwangtung (Canton) and Kwangsi provinces that stretch across southern China.

Foresters in the Guangzhow District established thousands of acres of slash pine (*P. elliottii*) plantations in recent years. These trees, native to the US Deep South, are believed to have been introduced into China in the 1930s by a school teacher who wanted something by which to remember his home in Dixie. The vast plantings were ordered by a bureaucrat in the Communist regime, a nonforester who liked what he saw. So ordered!

The foresters described a needle cast that is first noted when the trees are about 4 years old. The cast seems to diminish at age 8. The silviculturists sprayed for insects and for unseen fungi. But when they took me to the field, the trees looked healthy except for a minor aphid. The falling needles was simply a natural condition. Why did the needles not fall much after age 8? They do, but not so noticeably, and the trees were not yet 9 years old. I pointed out that the "problem" didn't appear until the trees were 3 or 4 years old because the evergreen slash pine holds its foliage that long.

Readers should recall that all of the forestry books in the Peoples Republic were burned during the Cultural Revolution. Foresters must start over. Chen Pei Dong promised to translate my 113-page SFA bulletin on the *Silviculture of Slash Pine*. After several tries, a copy eventually got through the censors to the forest entomologist at the Forestry University in Beijing.

I wrote out a management plan that itemized planting container mixes of soil texture classes; plantation spacing; thinning regimes for fence posts (almost no fences here), pulpwood, poles (they are of concrete and steel here), railroad crossties (also of concrete and steel), sawlogs for lumber; and harvests at rotation ages of 40 to 50 years when the final cut would provide large poles, plywood bolts, and sawlogs. I told them about preservatives like creosote and chromium-copper-arsenate, how to treat posts locally in heated drums, and the opportunity to sell pulpwood sticks or chips to Japan when this reforestation crop is thinned. Gum naval stores operations were described and industrial uses outlined.

In an effort to improve growth of a plantation by fertilization, laborers had strewn about a bushel of garbage around each sapling. With the garbage were broken glass jars, rusty cans, aluminum pans, and plastics. I suggested the practice be discontinued lest loggers be injured by the sharp objects in the soil. For politeness, I didn't mention the unsightly litter.

Nowhere in China could I find any forester who had heard of chestnut blight, caused by *Endothea parasitica* and believed to have originated in China. Subsequent to its discovery in North America and its destructive effect on American chestnut (*Castanea dentata*), our most valuable tree, American foresters journeyed deep into China to obtain plants and seeds of Chinese chestnut, believed to have an immunity to *E. parasitica*. World War II ended these explorations. The old accounts tell of how the Chinese far in the backwoods would bow to the Anglo researchers, treating them as gods. 'Tain't so. That was Chinese courtesy.

Only during debriefing back in opulent Hong Kong did my new friends tell me of the foods I had eaten in the 4-star hotels in which the UN booked us. (The hotels by the standards of other nations would be 2- or at best 3-star.) Rat, dog, cat, snake, and monkey were included in the menus. Not able to read a menu, they all tasted delicious to this round-eye. But scorpion, a delicacy, can't be camouflaged in bowls of fancy sauces. As a courtesy, I ate one: "In Rome do as the Romans do."

The team traveled by air and train together, sometimes heard each other's lectures, bunked on the same floor—the one with the bugged rooms—in the hotels, and parted only for our speaking engagements. Always the equivalent of a Chamber of Commerce-type host accompanied each of us. Mine was so obviously a CIA-equivalent, making notes she thought went unnoticed. That the young professionals were so kind to me I attributed to their knowledge of my ill-health which I had explained only to Dr. Chung. No, he had not divulged the matter. They treated me kindly because I'm not young. An example: as we walked a sidewalk and approached an icy spot, I'd have an arm on each of my elbows to make sure I wouldn't slip. Only I required a translator.

Dr. Chung made certain we understood the purpose of the trip. We agreed to four items, one of which I cite:

> We need to have a humble and realistic view of what can be gained...our visit is merely one step in a long-term learning process. We hope to learn to understand China—its people and their environment, to understand Chinese intellectuals, their needs and their struggles. We must observe and listen in order to serve

> and share meaningfully...so as to address the educational, professional and spiritual struggles and aspirations of China's modern intelligentia.

Renewed Relationships

Shortly after President Richard M. Nixon made his unexpected and highly publicized trip to China and subsequently appointed George H. W. Bush to be the USA emissary, a Chinese-born businessman from Dallas approached me. His interest involved exporting wood from Texas to the homeland from which he had fled three decades earlier. He had the contacts, both in Hong Kong and in the PRC.

DeSheng and Company retained me to do a feasibility study. Could pines and hardwoods be sent from the ports of Orange, Port Arthur, Beaumont, or Houston to Shanghai at a better price than the Chinese pay for wood from Australia, New Zealand, or even Brazil? Does East Texas have enough large-diameter timber to make the venture worthwhile? How about the attitudes of local wood buyers seeing the price of stumpage rise as foreign companies compete? How much will it cost to peel and dip the logs in a fungicide to avoid infection by blue stain or rot-causing fungi? Must an insecticide be added to the dip to be sure the Chinese wouldn't quarantine the logs at a PRC port? Would "under-the-table" gratuities be expected? Who has the available wood and how near to the three Texas ports does it grow?

As the China trade mission of the American wood-using industry indicated, China needs wood. Its neighbors' forests are depleted, as are its own, due to rapidly increasing populations and 50 years of war-caused destruction of both forests and the structures built from materials grown in forests. The PRC consumes 80 percent as much wood as Japan and as much as France and Germany.

The answer to these questions are of little value now due to changes in US wood availability, the construction of new mills in Texas, and the competition by private timberland owners in the Pacific states whose shiploads need not pass through the Panama Canal and thus can deliver logs more reasonably than can marketers of Texas-grown wood. We can argue about the quality of second-growth logs leaving West Coast ports, but not about the price. At that time, companies owning ships lying idle would practically give away stumpage to maintain a profitable fleet. So my report, distributed widely in English and Mandarin, was shelved.

At the time of this study, no Texas mills belonged to the National Lumber Exporters Association. As a consequence, Conference Line ships cannot dock at Texas' ports; wood hence moves from Canadian ports to foreign wharves more cheaply than from U.S. ports.

Indonesia Beckons

We stepped from the dense canopy of dipterocarps into an opening astride a spring. From the spring, a trough of split bamboo fed water to a small platform, perhaps a meter square, woven of the same ubiquitous fiber as the aqueduct. There, on the little shelf, just above the pond into which the water emptied from the spring, kneeled a Javanese maiden. She showed no shame, nor would she have reason to know of the mores of another world. She must be clean, and to bathe in the water from the mountain stream was the only way she knew to be cleansed. Her majestic form was neither misshapened by the inventions of the French nor contorted by the practices of the Victorians. The young woman's tanned skin blazed golden in the tropical sun.

Johannes A. Hoesada, a former graduate student, would introduce me to this South Pacific tropical paradise in 1984. It is an archipelago of over 3000 islands in which the nation's president manages its forests by decree. As I wrote in the *Journal of Forestry* in 1984, creeping expropriation discouraged foreign investors. Weyerhaeuser and other US wood-using industries, Indonesia's third largest national group of investors, sold out to those with more patience for an archaic bureaucracy. (No citizen of the US has room to talk. Jakarta's lunacy preceded ours, by two decades, in the micromanagement of the nation's woodlands according to the persuasive politics of so-called public-interest groups.) US companies and their Indonesian partners controlled over 2.5 million acres of tropical forests until the pull-out.

Prosperity was still to be anticipated as the US government, through the Agency for International Development (AID), poured money into the country for forest research, giving the local professionals an incentive to carry on. AID sent me to plan a research laboratory and to design experiments which could be implemented by foresters trained in the several forestry schools there. (Continued correspondance fails to reveal if a single plot was ever established, though the Ministry of Forestry did build the experiment station laboratory in East Kalimantan on the Island of Borneo.)

East Kalimantan, a state about the size of Colorado, had a single paved road. It led from the airport to the capital city of Balikpapan. The Indonesians assigned to me a newly hired driver and a jeep off-loaded from an American ship the week before my arrival. The young fellow had never driven on pavement nor understood how blacktop becomes slippery when showers and sunshine, occurring at five-minute intervals, coat the surface with lateritic silt. (Here one is always wet from either rain or sweat.) The road was the color and texture of melted butter.

Hoesada had warned me that Indonesian culture does not allow criticism. Children are never scolded nor employees reprimanded. Even my counterpart, Johannes Sudasrip, a Borneo forester, sitting in the back seat of the jeep with me said nothing to the driver about his reckless ways and the speed by which he propelled us along.

The inevitable happened. On the first roll-over, the metal screws that held the locally installed sheet-metal cover apparently ripped out. On the second rollover, the top flew off and Sudasrip and I involuntarily ejected. On the third gyration, the jeep must have missed us by inches, for it landed right side up. Front wheels rested on one side of a deep stonewall drainage ditch, the rear wheels on the other.

An off-duty policeman happened along as Johannes and I were regaining consciousness. He took us to a nearby river, commandeered a boat passing by, and directed that we be taken to a Roman Catholic hospital some miles upstream near Samarinda. The kindly missionary Brother charged me $13 for the five days of loving care. To look upon one of the missionaries, a man from the Netherlands, was to look at myself in the mirror. Stature, hair, skin, and face resembled mine. Momma was right; the Colton in my name was earlier that of a Dutchman who settled in Pennsylvania.

The accident rated an expense-paid trip to Bali, the sandy beaches a favorite of French and other European vacationers who must have forgotten to bring the tops to their bathing suits. Eve's fig leaves, I'm confident, covered more than the Parisian styles displayed here.

Ministry personnel phoned to Bali about my arrival by plane. The forester assigned showed me the isle, including its aboriginal head-hunting tribal site (from a far hill) at Batur Volcano and the lake formed by the eruption. We travelled along the timbered ridges and down the cultivated slopes to the water's edge along which grand hotels cater to wealthy tourists. So embarrassed was the Ministry of Forestry by the jeep rollover that they booked me in one of these.

Upon flying back to Jakarta from Borneo weeks later, the first question asked of me at the Ministry office was, "Did the roll-bar hold?" It did. But why the interest? Like the sheet-metal cover, the roll bar was installed in Indonesia, not by American Motors. And Indonesians are well aware of their lack of mechanical ability. Yet no one will argue with the native ability of these people as craftsmen in wood-carving, silver-smithing, and textile-dying (their batik is world famous).

Before leaving the scene of the accident, readers who believe in angels may be encouraged by this episode. The wreck occurred at about 2:30 p.m. on a certain day. At 2:30 a.m. on that same day, though exactly 12 hours away on the other side of the world, Anne was aroused from her sleep, slipped from the covers to the bed's side to pray for me, believing I was in trouble in the Enemy's territory, as C.S. Lewis, the Oxford don, referred to the Evil One. So significant to my wife was the event that she marked it on a calendar. Not until three days after I returned home did she learn, reluctantly from me, about the accident.

I never learned if Kosin, the driver, survived. When I asked about him, my hosts would just shrug their shoulders. Although still grasping the steering wheel when Sudasrip and I left the scene, Kosin was unconscious.

Nevertheless, even in that condition and in spite of Indonesian culture, he got a severe chewing out for wrecking the new jeep. Sudasrip needed only a day at the hospital to recover. Oh, no seat belts there.

From East Kalimantin we flew on a rickety airplane to the end of the run, then by handmade boat with a Japanese kerosene motor up rivers to mangrove swamps and jungle timbers of dipterocarps. We chose not to ride on the rafts of logs on which men build shanties in which to live through drenching rain and sunshine for the 10-day trip to the coast. Back at the airstrip we waited two days for seats, gaining them only upon surrendering $40 in "grease." While waiting we chased monkeys in the woods.

Research Assignment

Samarinda, the town nearby the jungle mission hospital, seemed like a good place to build the laboratory. The site the government owned, called Sampaja, however lay beyond a swamp near the Karang River. I wore boots and clothes appropriate for looking over the area. Several foresters joined me, all in office clothes. I was learning: Indonesian foresters don't go to the woods; they are all bureaucrats. Nevertheless, these good sports took off their shoes, rolled up their slacks, and followed me across the swale.

I learned too that many foresters are tax collectors, assessing values on timber harvested and exported. Just being a "Matthew" brings disrespect, but money under the table encourages disgust for the whole profession. Most, of course, are honest.

A visit to Itci, accessible only by boat two hours upstream on the Kutai River, enabled me to understand this cultural problem. Itci, an acronym for International Timber Company of Indonesia, was a Weyerhaeuser Company town. Well-planned with schools, three churches, and garden plots for employees, this was home for expatriates from the States along with Indonesian citizens. Nearby were some well-designed silvicultural experimental plots. Indeed, the reputation of the Weyerhaeusers, from founder Frederick, to the present, the Company "does all things well." The world should be grateful to E.N. Johnson, later a Weyerhaeuser vice president, for establishing these plots. (I knew personally C. Davis Weyerhaeuser from membership on the corporation of Inter-Varsity Christian Fellowship.)

When the Company sold to a consortium of Chinese entrepreneurs, the new owners hired a Japanese executive to manage the Itci complex that involved logging, manufacturing, exporting timbers, and research. I needed badly to see the experimental plots and the accumulated data involving species, planting, and regeneration techniques which the Tacoma-based company had left for the Chinese owners. Accompanied by a local forester, we made the long boat trip to meet with the Japanese executive. The meeting was not cordial. Finally I broke in to explain our mission. Just that quickly the manager's attitude reversed, we were shown to the best rooms in the guest house, and assigned his jeep and driver. All data were now available for our study.

Why the 180-degree turn? As foresters, the Japanese gentleman had assumed we were tax collectors. When he truly understood our mission, he realized that our efforts as researchers would benefit his company.

I designed studies involving species selection, fertilization, water infiltration, thinning, and tree improvement (genetics). I left behind plans for a seed orchard and seed-production areas for various species.

Along with the Federal Republic of Germany, my AID assignment involved evaluating the largest forest fire on record. A year of drought, extending through the monsoon season of 1983, dried the soil so much that the converging fires in Kalimantan covered an area twice the size of New Jersey. We found low-value ecological pioneer species replacing valuable dipterocarps. Ground fires, fed by peat and smoldering in the bogs, erupted as surface fires to kill seedlings. Fires crowned and ran rapidly with the wind, killing large trees of the climax forest. Diseases and long-horned beetles beat salvage loggers to the stands in roadless areas. This pyrological catastrophe provides foresters with the opportunity to manage woodlands in their ecological infancy and to expedite succession by mechanical and chemical manipulation. Hopefully, the folks stationed at the Samarinda research center will enthusiastically participate in such studies.

Forest products from this chain of almost 14,000 islands, stretching a distance equivalent to that of Los Angeles to New York City, go to developed countries of the Orient and from there to the world. Designers prize the woods from some 400 species of dipterocarps, marketed as Philippine or lauan mahogany, species of *Shorea* called meranti, teak, and rosewood. We watched the monkeys play in the mangrove trees that grow on their own stilts along the river courses. There, too, one finds a rainforest conifer called agathis. Although a forester named Kartawinata tallied 122 genera of trees in 45 families, US industries introduced exotics like eucalyptus (*Eucalyptus deglupta*) from Australia, true mahogany of the Meliaceae family, *Pinus caribaea* from Central America, and Monterey pine (*P. radiata*) from California. And again we chance upon the near ubiquitous cadam from the Malaysian peninsula.

Rivers remain the principal "roads" for moving logs to coastal ports where rusty warships of the US Victory and Liberty class carry them to mills in Japan, Taiwan, and South Korea. In those places the false lauan mahogany becomes panel and door veneers for homes in Europe and North America.

Contractor for the AID project was Resources Management International, doing business world-wide. However, they had no resource management people in Indonesia. I was on my own.

Politics Continues

Indonesia was a necessary stepping stone for the intended Nipponese invasion of Australia, as the notorious Great Barrier coral reef protected potential landing sites on that continent from intrusion. Dutch foresters became slave laborers for the building of a railroad the length of Sumatra, second-largest of the Indonesian isles. And East Indies oil was important for fueling machines of war.

Following the Japanese surrender in 1945, the Indonesians began a 5-year war for independence from the Dutch government at The Hague. Even while the war raged, a Dutch company contracted for cutting rights on over 3 million acres. When hostilities ceased in 1950, the Indonesians (formerly the Dutch East Indies) adapted forestry decrees that had been issued from Holland. With establishment of a forestry bureaucracy, the folks in Jakarta would issue the edicts. One after another these proclamations discouraged investments from abroad. Alas, a 1961 decree declared the concessionaire a government company. Soon a lack of financial and industrial leadership took its toll; the nationalized industry declared bankruptcy.

A final blow clobbered relationships in 1982. After 1985, log exports would be prohibited. Expatriates would either build plywood mills on the 1300 inhabited islands or go home. The decrees, designed to enhance employment for Indonesians, backfired. Capital now flows from Indonesia to industrialized nations for machinery and for salaries and wages of mill managers and maintenance men. Products coming out of the mills are inferior and, if not, they are sold to customers of former owners. Vigorous salesmanship from abroad could leave the output of 160 new plywood plants stockpiled in national warehouses.

I learned the importance of nationalism in Indonesia. The Scouting movement is encouraged; even the president, along with all boys, wears a Boy Scout uniform on Fridays. Missionaries of anti-patriotic groups are unwelcomed. This nationalism does more than promote the use of the relatively new Indonesian language. It affects forestry, for it discourages the encroachment of foreign investment on sovereign soil, in spite of edicts to the contrary.

Let me conclude this discourse on the lovely Spice Islands by relaying several fascinating observations: Estate forestry plays a leading role in the nation's economy, for this is the world's fifth largest country in population. (It seemed to me that every woman of child-bearing age had one youngster pouched externally on her stomach and one cradled inside. Indeed, a poor man's pleasures are simple ones.) On these plantations grow coffee, tea, and rubber. On the isle of Java, large plant leaves serve as troughs down which latex flows into coconut shells. Throughout the land, loggers harvest rattan vines and dry them for shipment. I watched home craftsmen and women weave and bend, using blow torches, these into wicker furniture that brings high prices in the western world.

I discovered foresters mapping the region into land

units based upon natural features and land-use planners then determining a site's highest use. Low soil fertility, lands prone to flooding, and rough terrain earmark much of the timbered country for continuous wood production. Some lands may be designated as potentially *estate* sites, where rubber plantations would be appropriate; while small tracts may satisfy requirements for farms for families living in transmigrant communities away from the land they till.

Transmigration, a governmental program, resettles people from the slums of the cities to the countryside in islands with few people and only river transportation. The migrants establish new villages of tidy two-room cabins built at government expense. Each family receives five acres of land, a livable stipend, and a year's supply of seed and fertilizer. An agricultural county agent-type accompanies the group recruited by social workers. How does the system work? A few become loggers. Some destroy the lateritic soil as they convert the land to shifting agriculture, often referred to as slash-and-burn farming practices. All is not peaceful. Riots have occurred where resettled migrants, unable to cope with wilderness life and physically rugged occupations, can only plead, "Show me the way to go home." Even if it is to the old haunts in city slums.

You ask me how they escape the islands? I suppose as stowaways; not even Johnny Weismuller could swim across the Flores Sea from Kintap to Surabaja.

While visiting small towns, I'd walk to the local Ministry office from the hotel or rooming house. Always groups of kids would soon forget their games and follow me, giggling. At a corner I'd buy a bag of candies from the ubiquitous vendor, distribute the sweets, and send the children on their way. Another group would soon follow and receive the same treat. It happened in every town.

I could never learn the reason for the laughing. No Indonesian would tell. "Was I the only roundeye in Indonesia?" I'd ask. "Or the only blue eye in Borneo?" Or the only bald man in East Kalimantan? "Or the only foreigner to ever come to Samarinda?" Finally, Sudasrip confided, "It's because you walk funny!" Indeed I do. The army always bellowed about my bouncing. But, then, the

Bible says, "A man is known by his walk."

In my 121-page report, I complained about foresters "persisting in going over terrain in jeeps where I wouldn't take a horse, simply to avoid walking." To get the faculty at Mulawarman University to carry out field studies, special pay per day provides the incentive.

A little aggravated, I noted my job rated combat pay. "Old planes without airworthy certificates, hand-made small boats for river travel, dangerous roads, reckless drivers, bad water, risky food in a cholera-epidemic area, and primitive hospitals in a hinterland are but a few of the deprivations."

Many Indonesians, like Dr. Suarso and Dr. Sadam, dean of the forestry program at the National University at Bogar, and Ir. (forest engineer) Mandong have but one name. Running down genealogies must be impossible. On the other hand, the new Indonesian language, developed by edict, is easy to learn. For instance if *hotti* at a road construction site means caution, then *hotti hotti* means extreme caution.

When visiting these woods, take plenty of peanut butter for sustenance and chlorine phosphate tablets for water purification. I lost over 20 pounds in the two months assigned there. That in spite of the big dinners with which you are greeted by government brass at each new place along the way. Foreign public servants don't understand why we don't entertain them lavishly when they come to America. Most other nations provide budgets for the purpose; even the lowest bureaucrat can be expected to take advantage of the occasion to have a many-course dinner at the finest hotel.

En route home I visited Somsak Sukwong, dean of the Faculty of Forestry, at Kosetsart University in Bangkok, Thailand. How kindly he and his students treated me, always greeting with two hands folded at the chest and bowing respectfully.

Mountains of Magadan

“Get diphtheria and hepatitis shots, don’t drink the water, and you can’t use credit cards.” With these admonitions I joined a group of foresters to study the vast boreal forest of Siberia’s Far East. The Aeroflot TU 154 carried us from Anchorage, at that city's latitude westward through 3 time zones in as many hours for the 3000-mile trip, to the rugged Magadan province in a Third-World country.

Those familiar with Alaska's taiga of black and white spruces (*Picea mariana* and *P. glauca*), found respectively on the north- and south-facing slopes of The Great Land, may expect similar vegetation on glaciated sites of the Asian continent. However, erosion of the land bridge across the Bering Sea apparently separated vegetation types as well as peoples. Here, the hills are covered with larches (*Larix cajanderi* and *L. dahurica*) unless bogs, shallow soils on stony slopes, tree line, and bare rock interrupt these rather pure stands of second-growth seedlings and saplings and a few poles. *Larix's* frost-resistance and ability to grow in permafrost and there to establish adventitious roots encourages its success.

Foresters who scouted Siberia in 1994. Government boat and helicopter carried us to remote forests in this roadless territory.

Here, too, one finds the 5-needle dwarf Siberian pine (*Pinus pumila*). This white pine grows erect but stunted; its lateral branches also lie prostrate. Burdened by the weight of snow, the branches produce new trees by layering. Anatomically the wood structure on the upper and lower sides of the laterals differs, compression wood forming on the lower sides. As these wood cells shrink with cold temperatures, the branch bends downward to make contact with the soil. Rooting then takes place.

From the vantage point of an ancient 18-place rappellers' helicopter's windows, we flew into the roadless interior to see old-growth. These stems, according to our estimates, were as large as 20 inches dbh and 75 feet tall. Most of the old-growth was harvested in the 1930s to accommodate the development of a huge gulag and farming cooperatives in the early Stalinist era. Many of the trees found use as pole timbers for corduroy wood roads as well as wood for prison walls, many still standing. Our bunk house was once the officers' quarters for the gulag. It still had no showers, but a community sauna sufficed.

I called this trip to Russia's Far East in the fall of 1994 a leg of "My Last Great Adventure." With this title, I honor my oncologist whose name is Ventura, *Italienne* for venture.

Ten-year-old Russian forester's son in a larch stand of about the same age.

19
JOURNEYS TO THE NEAR EAST

Lombardy poplars, like these 4-year-old trees grown from cuttings, provide living fences for farms in the countries of the Near East.

While I was still on a retainer as chief forester for the National Plant Food Institute, the Washington office of the educational arm of the fertilizer industry dispatched me to Israel to look into the drip-irrigation technique that made the desert bloom with vegetables, fruit, and trees like eucalyptus. So, on this journey I looked into forestry education, lectured to foresters and students, and visited the woodlands and reclamation sites of the region.

Grecian Visit

I flew to Athens, saw cactus growing in the Acropolis, visited the majestic Parthenon, the Temple of Athena, the Aeropagus on Mars Hill, and many other two-plus-millennia-old structures and statues now eaten away by the acids in the air of the industrialized region. I visited with the director general of Hellenistic Forests, Napoleon Papalexious, a Yale Forest School graduate, who introduced me to Mr. Arivantinos, my host in the ancient city. We looked at groves of pistachio and olive trees (some stems of the latter claimed to be over 2500 years old), now useful for lumber. Planters converted poor stands of oak to olive groves for furniture wood. Foresters were growing *Pinus euroa* No. 214, a hybrid of California's *P. radiata* and the region's *P. halepensis*. Interest derives from the drought resistance of *P. halepensis* and the growth rate and quality of *P. radiata*. But the latter, in this climate, consumes five times its needle weight in water every day.

Mr. Arivantinos put me on the Pullman for the night journey to Thessalonika. Upon arrival in the ancient city, Lampros Romanas, head of the Silviculture Section of the Forest Research Institute, took me in tow.

On April 23 (St. George's Day) each year, all sheep move to summer range in the mountains. Shepherds return them, like clockwork, on October 26. Foresters thus need to be range managers; and watershed managers, too, constructing concrete waddles to protect streams from erosion siltation.

In Greece, the law requires management plans for all forests. That, apparently, hasn't solved the problem of the lack of jobs for graduates of the "Saloniki" school. Some of these foresters, however, at the time of my visit were establishing plantations on lands near Athens, lands lain bare since the time of Plato. Others were restoring forests to sites surface-mined for lignite.

Among the products of these Hellenic forests are parquet flooring from small oak logs and turpentine from pines (*P. halepensis* and *P. brutia*) for export (some is added to wine). One boat builder cuts *P. halepensis* trees himself, then cuts the logs from which he searches for boards naturally bent to the desired shapes for the knees, keels, and keelsons of the vessels. The task is reminiscent of the days of the live-oakers in America's southern second bottoms. Genetics has facilitated the effort; the last straight trees of this species were harvested a thousand years ago, each successive harvest taking the straightest boles and leaving the "deformed" to produce seeds for subsequent generations.

Millmen cut chestnut (*Castanea* spp.) trees, seemingly not noticeably susceptible to the *Endothia* blight fungus, for many uses. And all around the woodlands, I spot Smokey Bear, its forepaws hugging a tree as the symbolic bruin stands on its hind legs. The Greek legend, deciphered from my recall of fraternity letters, says, "I love the tree." While the bear hugs and loves the trees, Athenians and their kin stand on street corners fingering worry beads as they wait for lights to change.

Environmentalists have their day, too, in Greece, even during this visit. My host showed me the foundation for a new building in downtown Athens. Alas, construction workers unearthed an old aquaduct. Further work was put on hold.

Dr. George Tsoumis, also a Yale Forest School graduate, met me at the bus stop in Thessalonica (English spelling), site of the principal Greek university. Its name honors Aristotle. The Apostle Paul and his friend, Silas, may have lectured there. George, dean of the forestry school, apologized, as we entered the campus, for the bold graffiti and desecration on every wall, and even on ornate statuary. I awakened to a civilization, once the most advanced in the world, gone berserk. Police are not allowed within the campus' wrought-iron gate, a tradition of separation of state and education. Consequently, every so many years, the university shuts down for repairs necessitated by vandals.

George and his lovely wife, Alexandra, were gracious hosts at their condominium on the dazzling Aegean Sea. We drove to the Turkish border, but not a foot beyond, visiting reforestation and reclamation efforts along the way. The Moslem Turks, as I understood, during the 400-year occupation of Greece severely abused the range. Their control over Greece encouraged black goats to overgraze the land. Hellenists insist the Turkish destruction was intentional so that, if ever again Greece was free, the land would be totally unproductive. Although that occupation ceased in 1912, not until the late 1940s did an American agricultural mission encourage the protection of watersheds and the planting of pines and olive trees.

Red poppies abound, particularly in the zone near the Turkish border. To grow or possess yellow poppies is to invite severe penalties. At the time of my 1978 visit, fear precluded use and trading in illicit drugs.

I had forgotten the New Testament account of the Apostle Paul's treatment in "Saloniki." When asked if there is a monument in the city to the missionary, Dr. Tsoumis reminded me, "Paul was not well received here; Demetrius is our patron saint."

I learned in attending an Eastern Orthodox Church that, even if Paul was not well-received in Saloniki, his words to the young church endured: "Be at peace among yourselves." And again he wrote "...speak not to please men, but to please God who tests our hearts." (Was *Sylvanus*, who traveled to the region with Paul, a forester?)

Egypt's Nile

At the Cairo airport, I asked a taxi driver to take me to a hotel where German and British businessmen stay. They aren't on fat American-industry expense accounts; thus they know the best deals. The cabby misunderstood, taking me in the evening's dusk through dark alleys and dingy streets where multitudes gathered outdoors to avoid the heat inside their small squalid quarters, a practice observed in many tropical lands. With each passing block in that speeding, swerving cab, I had every reason to pray. (Before the morrow ended, I had learned that's the only way to drive in that city of 40 million people.) We arrived at the German Hotel. That's its name, but certainly not where a "Jerry" salesman would spend the night. The driver motioned to wait and he'd see if a room was available. Fortunately none was. He then took me to just the kind of hotel I sought, carried my bag to the desk, gave instruction for my care, and left.

The only disappointment: bottled water was not factory-filled. Pleasant surprise: a servant, sitting at the end of the hall to promptly attend to your needs, was aroused by the ding of a bell.

The British, controlling this land from 1918 to 1940, made traveling easy. Everyone, seemingly, knows some English. But Cairo became an unmanageable metropolis when, as a result of the 1967 war, hundreds of thousands fled to the safety of the city or were forced by the government to do so.

A cabby couldn't find his way to the Maadi Yacht Club, where Rotarians meet in a Cairo suburb. The driver, frustrated, stopped a car on a little-traveled street. In English, its driver responded, "Oh yes, I know. My father is a member. Have your passenger ride with me." That's how I was introduced to the attractive daughter of a professor of ophthalmology at Cairo University. He got his residency experience in Chicago. The woman left her Mercedes to introduce me to the club's manager who, in turn, cared for my needs until the eye doctor arrived. From then on, I was his guest, sitting where he could quietly translate the speaker's message. The Rotarians thoroughly quizzed the socialist gentleman, chairman of the Nassar Bank. The bank pays no interest on deposits!

Natural resource managers have much to wonder about in this old land. Egypt was old even when King Tutankhamen and his kin were buried at the valleys of the Kings and Queens, near Thebes, astride the Nile River upstream and a great distance south of Cairo. Slaves somehow transported limestone quarried far off in the Libyan desert for use in these monuments, discovered in

1922. Sycamore wood joined large pieces of stone, the wood soaked with water to expand to hold the stones firmly in place. The city of Karnack lay under 35 feet of sand when Frenchmen at the time of Napoleon discovered the ancient city.

Archaeologists ponder how the heavy stone tombs were transported deep into the dark caves of quarried rock. How, too, did artists paint murals and sculptors chisel rock? Apparently crystal mirrors were used to reflect light from the sun for the working slaves. No wood was available for burning for light; and if it were, its combustion would have exhausted the oxygen in the dungeon-like caverns.

Oh, yes, slaves buried gems and gold with the royal family members. Then, to protect the treasures from thieves, the slaves sealed and camouflaged the tombs with sand. Final precaution: the slaves' heads were removed. But what about the fellow who sliced off the head of the last surviving slave? This last decapitation occurred well away from the treasure cache, so he did not know its whereabouts.

Some old monuments carved in stone cliffs were carefully cut away and moved 150 miles to the south when the second Aswan Dam was built. Else these treasures would be under water today.

Flooding of the Great Delta at the Nile's mouth on the Mediterranean Sea ceased with the building of the second, lower Aswan dam, financed by the Soviet Union. While farmers can now plant and harvest three crops annually, the lack of fresh sediment carried into the sea each year destroyed the sardine fishing industry. The first dam, erected in 1907, restrains the crocodiles. They no longer come downstream to the north to feed. While shelterbelts of cottonwoods line the Nile, goats graze its banks.

If it can be done inefficiently, in Egypt it will. Floor tiles laid using the thumb for a trowel and a drinking cup for pouring cement exemplify the waste of the human resource. Here I learned the word, *bukecheeze*. That is the request for a tip for the least assistance. To say "Hello" may be met with "bukecheeze."

AID folks lined up my visit to the agricultural faculty and students at Cairo's University. Among the graduates are those who plant the poplar trees upstream on either side of the Nile River, for hundreds of miles south of the city to the Valley of the Kings and the Valley of the Queens near Karnack. These *Populus* species also grow throughout the Delta. No one could tell me the source of the cottonwood cuttings. Were they *Populus euroamericana* (the cross listed as *P. deltoides* x *P. nigra*), another hybrid, or a similar species?

In my journey upriver to the South, I learned from a guide a lot about Moslems we typical Americans are not likely to know but, by knowing, will help us to grasp something of the hostility felt by some Egyptians to others. For instance, special courts handle Christian legal matters, the law more lenient on divorce. And statutes allow hotels catering to westerners to serve alcohol, forbidden to a Moslem. Loudspeakers mounted in the minarets that call a Moslem to pray do disturb a fellow's sleep. A Moslem is not "complete" until he (always *he*) has journeyed to Mecca in Saudi Arabia. That pilgrimage guarantees entry into paradise, and allows a special pastel coat of paint on the man's little clay hut.

The scarab, to the Egyptian a sign of immortality, is called the "son of the sun." The beetle appears from the Nile mud to endure the dry sand of the desert. My guide continually pointed out to me its likeness, everywhere displayed in carvings in stone.

The Trinity of Thebes, my new friend emphasized, represents father, mother, and son which, in that culture, signifies sun, earth, and moon. Another account: The Trinity of Thebes refers to the three towns to the south!

Taking a shortcut to the Cairo Airport, the driver took me through the city dump. Yes, I wondered if that would be my destination. Equally squalid to Mexico City's refuse pile is that of Cairo. Hundreds of people live in the dump, among wrecked cars and garbage, and eat from its "bounty."

The occasional cross contrasts with the crescent mounted atop the sanctuaries of the land. That distinction needs to be recognized in the wake of the Camp David accords of the Carter presidential years and the recently signed (1993) Palestinian-Israeli recognition act. I found nonpolitical professional people want that peace dearly. Why do I think so? The Moslem, finding me a stranger, took me in.

On to Jordan

"You can't get there from here," as the old saying goes, applies to travellers from Egypt with an Israeli destination in mind. The 747 plane, packed with Egyptians on their way to work in the Hashemite Kingdom of Jordan, Saudi Arabia, or Kuwait, was forced to fly north of Israel through Lebanese and Syrian air space to touch down in Amman, quadrupling the distance of a direct route. I was the only nonpeasant laborer among the passengers, all carrying their lunches. How'd I know? When a uniformed airline employee saw me at the end of the check-in line at the gate, he grabbed my arm, took my bag, and led me to the front of the line.

The brother of the Jordanian prime minister, a forester with a PhD in land-use planning from Colorado State University, hosted me in his beautiful marble-floored home. Tropical woods paneled the walls and exquisite furniture, brought by van through ten or so countries from the galleries of Rome's marts, decorated the interior of Salem Allozi's palatial residence.

The lecture over at the University of Amman where Dr. Allozi serves on the Agriculture faculty, three young men piled with me into a beat-up jeep to head for Petra. That is the ancient city made famous by the Queen of Sheba. Rooms in the royal quarters were not built; they were hollowed out of hills of solid rock by Nabataeans in the distant past. Some folks believe, because of its inaccessibility except by one narrow rock-enclosed trail, that the 144,000 Jews mentioned in Revelation 14 will escape to hide there in the last days. One organization has stashed Bibles in air-tight containers for those who will gather in the innards of the excavated mountain. But I have news for them: Mahmoud Juneidi, forester with the Ministry of Natural Resources, and his colleagues took that jeep over many trails a tourist wouldn't see, trails long used by Bedouin goatherds to escape the wrath of government overseers. Juneida, forestry class of '58 at Utah State, showed me remnant pines believed to be 500 years old. He said Prof. Ted Daniel (my good friend and fellow silviculturist) would remember him as the "dendro dropout."

I wanted to study land reclamation. Perhaps something here would be useful for the work of surface mine-site restoration by Texas Utilities on whose environmental steering committee I served. Jordanians remain angry at King Solomon for cutting the tall cedars for the Temple construction in Jerusalem. Though the harvest was in Lebanon to the northwest, tradition holds that it started the clearfelling of Jordan's stately timbers, too. Only in recent decades have foresters been reclaiming these lands.

Bedouin goatherds now care for their flocks on these lands in need of reclamation. Called the "black locust," the goat crops all that grows, even where it snows upon the mountains. Cattle leave a few inches of grass, sheep an inch, while goats nibble to the ground line and below.

The desert nomads depend upon the goat for milk, meat, clothes made from its hides, and their tent houses made from the skins. Black absorbs the sun's heat for winter warmth. St. Paul, as a tentmaker, probably used these skins in his trade. (My kids say I should raise goats; then I'd have mo' hair.) Bedouin nomads are laws to themselves. They herd their black goats from one nation to another, avoiding check points as they cross over unmarked desert boundaries of deep, blowing sands.

Land-value inflation moves many black goat herders to government-built farm communities. From these new permanent residences, many comprised of animal skins hanging on poles, trucks carry workers to plant trees. I was amused by the feeble, but costly (in time), attempts to imitate on the east side of the River Jordan the success of the Jews to the west. In Jordan, dozers intermittently level benches on terraced land. Poplar, eucalyptuses, and Mediterranean pine are then planted in deep holes on the benches. As in Israel, water may be provided by allowing it to drip through holes punched into plastic tubing only at the locale of the trees. Workers also pour water from buckets onto these trees to protect them from drought.

A few other notes on Jordan. Trees line the roads; among them locusts (*Robinia* sp.) Arizona cypress, (*Cupressus arizonica*), eucalyptus (*Eucalyptus* spp.), alleppo pine (*P. halepensis*), the leguminous acacia (*Acacia* spp.) from which gum Arabic is obtained, and

Lebanon cedar (*Cedrus lebani*).

The Al Faisal Tree Nursery at Jaresh was producing a million seedlings per year. Four other nurseries each grow about the same number, all for planting high in the mountains. Nursery workers place five seeds in a tubular plastic bag; thinning to one or two plants, the containerized stock is then planted. At the time of my visit, one percent of Jordan was forested. Plans call for re- or afforesting eight percent of the country.

One needs no kidneys in this climate. Rainfall in much of the land averages 5 minutes per year. Perspiration takes care of the need to eliminate human waste water!

The Rev. Albert Hashwey, an Israeli Palestinian refugee, joined me for dinner to tell another side of the on-going saga of national rivalries in this part of the world. The octogenarian minister, with congregations on both sides of the Jordan River, seemed to be everybody's elder statesman and patriarch.

King Hussein, who married the daughter of a wealthy American because his former queen seemed destined not to provide a male heir for this patriarchal society, has an interest in forestry. Trees may be a more apt term as, at his direction, small parks and gardens sprout here and there. That is the king for whom the rickety bridge spanning the narrow Jordan River to Israel is named. An insult to his name, the wood structure may yet be the "bridge" to peace in a part of the world where there seems to be no peace.

Israel

On the Israeli side, the bridge bears the name of Allenby, a British Army officer assigned to Palestine when the nation was, until 1948, under England's flag. I arrived with the first busload from the Jordanian checkpoint, a mile or two east of the bridge. The cab driver could bring me no closer. Who were all the people on the bus, packed within like the proverbial sardine can, some hanging on to the luggage rack on the roof? They were Jordanians going to work in Israel's fields and orchards. They would return the same way that night. They cannot walk across the bridge before I and an attractive well-dressed woman do; and we can't go until a fleet of trucks from Jordan's port at the north end of the Gulf of Aqaba pass by. The trucks are empty. The driver's seats are not upholstered, only canvas suspended from galvanized pipes. The vehicles have no dashboards and no spare tires. Windows have been installed in the gas tanks. All of this is for convenient inspection, thus to assure that terrorist munitions don't get into Israel from Jordan.

Security is tight. You must snap several pictures, of anything, even the young lady guard, to prove the camera is not loaded with anything other than film. I was tactfully informed, a cupped hand to the speaker's mouth, to cease taking notes and to put the tablet away.

Finally, I'm in Israel. Later after more questioning, including my knowledge of Hebrew and Arabic, I meet up with the olive-complexioned woman I saw on the bus to the bridge. A Lebanese with a US passport, she was visiting relatives in Jordan and parents in Israel. She was strip-searched, supposedly as harassment. (In those years, an Israeli visa stamped in a passport prevented the bearer from entering Arab countries. So one concludes one's visit to the Near East in Israel. Now the Israelis don't stamp the passport, but slip a card into the passport. Bearers then remove the insert when entering an Arab nation.)

I jumped into a limo-style taxi with Jerico marked on its window. When filled, it moved on to the oldest continually inhabited city in the world. After checking on the famous wall that tumbled down at the sound of Joshua's trumpeters, then being excavated by archaeologists, and noting the vegetation of the River Jordan flowing nearby, a cabby took me "to a hotel where English and German businessmen stay" in Jerusalem. The Ritz, a Palestinian hotel, a block from the old city wall and adjacent to the reel of barbed wire that reminded visitors of the Six-Day War of 1967, was the ideal place for this weary traveler. An important lesson was learned promptly: all Palestinians are not Moslems. This hotel, displaying Gideon Bibles, had more scriptural placards and banners than a Billy Graham crusade.

My lectures at Tel Aviv University went well. A botany professor, Danni Eisokovitch, insisted that I check out of the hotel and stay with his family in their home near Lod. I watched as he touched the marker on the side of the front door, a tradition reminding the devout that where the blood was sprinkled on the door post, the avenging angel would pass over that house. "It is the Lord's Passover." Raviv, his wife, cooked delicious meals.

Commerce continues as Israeli West Bank farms produce food exported through Jordan to other nations on the trucks described earlier. Commerce is active, too, on the large plaza at the Dome of the Rock. A Palestinian tour guide who wanted $8 to show me the gold-domed mosque constantly pestered me as I walked. I continually rejected his offer of aid. Finally, exasperated, he cupped his hands and called so loudly the folks in Bethlehem could hear, "You cheap Israeli Jew." I feared an international incident.

On the road south to Bethlehem from Jerusalem, a pleasant four-mile walk from the wall of the old city, the Israelis have established one of their finest forests. The trees memorialize individuals, including many Americans. The Peace Forest was set aside in about 1958.

My best views of Israel's forests, planned to cover 7 percent of the little nation's land, were orchestrated by Marc Goldberg, an SFA student of a few years earlier. Marc had migrated "to learn what it is to be a Jew.' The Tel Aviv University folks, planning for my lecture, notified forestry groups of the speech, and then I saw Marc in the audience. What a surprise!

Marc's problem, and that of other foresters, involved protecting the plantations from the Bedouin black goat herds and herdsmen. Security guards must be posted. Even at night, if left unattended, the animals will be driven to the woods to feed. These nomads also burn the woods to improve grazing. Marc told me how the Bedouin work for the Druze, a non-Arabic tribe with a secret religion. They are dangerous, he says.

Jealousy over Israel's success in reclaiming land and irrigating crops with drip irrigation has its political overtones. Mosquito-infested swamps and desert sands were bought cheaply from Palestinians. Maybe the sons of Ishmael chuckled as deeds were signed, prices for seemingly worthless land so high. But now, long-fibered cotton exported to the US; forests of eucalyptus, and fish in large farm ponds grow so well that environmentalists complain of the destruction of nature by farm and forest planners. Radicals among the nature lovers also object to tree planting because forests reduce the habitat available for rare plants (iris and tulip) while increasing habitat for the deer-like gazelle (*Dorcas* in Greek and *Tabitha* in Hebrew for those recalling the account of the Sewing Society for the Poor.)

Siviculturists leave openings for herbaceous plants. Mark showed me white pine (*P. strobus*) and Douglas-fir (*Pseudotsuga menziesii*) introduced from Canada, beech (*Fagus* sp.) from Scandinavia, and tropical woods from Nigeria.

We chuckled over a letter found in a leather pouch in a clay jar along with the famous Biblical scrolls in the cave near the Dead Sea. The letter, dated *circa* AD 100 and translated, read, "Bring your myrtle and willow. See that they are tithed." The young transplanted forester also noted that oleander seeds go to Africa where tribesmen smear arrows with the poison oil extracted from the fruit.

Over a thousand species, including 120 eucalypts, were being tested at the Forest Experiment Station at Ilinot. The Aussie trees get their name from the Greek, the seed being "well covered." Eucalyptus species, introduced into Israel about 1900, are important for more than wood. Apiarists maintain bee hives in stands of these trees. Rotations for this species run around 12 years. Meanwhile allepo pine escapes to cover the land.

Every foreign venture has for me, it seems, its adrenaline thrills. Here, at night on a street in the old walled city of Jerusalem, I ask directions. I was led by three youths courteously enough to my destination. They wouldn't give directions, but would only go with me to the Joffa Gate. When I thanked them, that wasn't sufficient. Pinning me to a wall, they wanted to be paid for their effort. Of course they could have my whole wallet. We settled for two US dollars.

In Bethlehem, I caught an intercity bus to the South for Hebron, site of Sarah's death and Abraham's family burial. At the end of the line, I rose to get off. "Where are you going?," the driver asked. "To see this old city, to see the cave, to just walk around" was not an adequate answer. "I can't let you. It's too dangerous," he said. Then, to my amazement, he shut the door, put the bus in gear, and for 15 minutes, until the schedule called for his departure to

Jerusalem, he showed me Hebron, a Palestinian-controlled city.

And then there is Mt. Gilboa near Afula, the site of King Saul's suicidal battle. King David later cursed the land. Nothing would ever grow here again! Until, that is, a forester tried. And on the side of a hill a Jew points with good humor and a bit of pride to the stately forest he attributes to the will of God.

20

GRINGOS AND TICOS

"Don Pepe" Jose Ferrer-Figueres (l), president of Costa Rica and the author's host at *La Lucha*, a high-mountain *hacienda*. The diminutive prexy chose the higher ground.

The Rich Coast of Costa Rica

My 1972 task in Costa Rica for the US Peace Corps was to evaluate the work of the 90 American young people assigned there to assist in developing parks and managing forests. The puddle-jumping Pan Am flight to San José was uneventful except for the Salvadorean exiles, including a university rector, who wished to enter the southern Central American country for protection. Their presence on the 707 delayed its clearance for take-off from Nicaragua's capital.

I visited 18 young folks, scattered throughout the little nation, to observe their work. Juan Castro, economist with a degree from Cornell, and earlier in charge of parks and forests, now administered Peace Corps activities in the *Rich Coast.*

Columbus named the country, the size of West Virginia, when, on his last voyage, he saw gold adorning the necks of the natives. Bananas, coffee, cocoa, and prized woods like mahogany and rosewood were yet to add to its wealth. The world in time would come to know about the beautiful *nazareno*, so named because of its rich purple wood resembling the robe of the Nazarene Christ. *Guapinon*, from the word, "pretty," and a blond mulberry tree of soft texture eventually would provide *colones* for the treasury.

Two potential national parks, the only ones accessible by road, received much of my attention: Volcan Poas and Santa Rosa. Another two, reached only by narrow-gauge rail, trail, and boat remained low priority. I was told the parks would not be created were it not for the enthusiasm and imagination of Senora Figueres, first lady of the land. The trim blonde lady of Danish parents and Yankee birth proved to be a delightful hostess at the president's *hacienda* he called a farm.

Invited to the office of the nation's president for a brief courtesy visit, I was permitted to leave only after 45 minutes, coffee and snacks served at 15-minute intervals (the time-table standard), and a promise to visit his farm in the mountains above San José on the following Saturday. Jorgé Blanco, Minister of Forestry, accompanied me to the farm of President José Ferrer-Figueres. Jorgé was not sure whether we were to arrive at 8:00 *Gringo* time or *Tico* time. The former refers to the punctuality of Americans, the latter to the "time to smell the roses" of the Latinos. That President Figueres was an MIT graduate in engineering suggested we should arrive on Gringo time. However, the place looked deserted at that hour, and no one stood guard in the little "pill box" at the entrance. A few toots of the horn brought a small man (my size!), with a pistol tucked into his trousers, trotting to the gate. He carried the message of our arrival to the Big House, returning to say the president was waiting for us for breakfast. But the little fellow couldn't find the key to the big iron heavy steel barrier. Ah, climbing the rungs of the gate, he found the key safe in its hiding place at the top of the concrete post that supported the gate. Passing through the portal, I glanced back to see him return the key to its "proper" niche.

This head of state, Don Pepe as the peasantry affectionately called him, had professional and business interests in forestry. Indeed he wrote a small forestry book, a copy of which he pulled from the top left drawer of his massive desk of beautiful hand-carved mahogany and autographed for me. When I requested the pen with which the president signed his name on *Cipres Con SAL* by J.F., he handed it over. Inscribed on the pen are the words, "U.S. Government"!

Cypress, introduced from Nicaragua to the north, has long interested J.F., the initials needing no elaboration in Costa Rica; they are as well known as LBJ hereabouts. The pamphlet opens with "On my farm today at La Lucha (meaning *The Fight*, site of the opening battle of Costa Rica's revolution of 1948) is the record of 40 years of errors in growing cypress (*Cupressus lusitanica*)." In planting windbreaks on the hacienda, he noted the effect of the tree position in relation to the sun at this place on the globe, just 10 degrees from the equator. Subsequent plantings and even nursery beds are laid out with an astronomer's precision so that trees are aligned on the terraces of the hilly landscape to receive maximum light and, hence, exhibit maximum growth.

L in the booklet's title is for *luz*, light, *S* is for *seleccion*, a reference to choosing wisely the preferred genetically inherited characteristics of the species, while *A* represents *abono*, Spanish for fertilizer and a reference to soil.

We rode with Don Pepe to his woods in a BMW with loaded submachine guns in three positions. We dug holes to look at latisols, checked on a variety of introduced species, especially cadam (*Anthocephalus cadam*). The president was intrigued by this species, the silvical characteristics worked out by the late Dr. K. Kadambi of the SFA faculty, prior to his becoming chief conservator of forests in India. Dr. "K" discovered in Southeast Asia the tree that grows with a square trunk and whose age is measured in months. The reader has seen *A. cadam* or *kadambia* earlier in these pages.

Back at *La Lucha*, we pored over maps while the First Lady and two young daughters served coffee made from beans picked from the bushy trees interplanted with tropical timber species. Hot soup, tasty beef, and ice cream from local cows completed the meal as we sat around the great fireplace on a cold, rainy day.

When Jorge, the Oregon-trained Minister of Forestry, approached a comfortable chair, Mrs. Figueres, with great tact, exclaimed, "May I ask you to take another seat. That greasy one is Don Pepe's, and not even ambassadors are permitted to sit there."

As we rode through the hacienda and stopped to look at a stand of trees, no one would be in sight. Within five minutes, 10 or 20 people would gather, old women and little girls taking their turns for a hug from the 67-year-old Don Pepe.

I never learned the size of the hacienda, but some 1700 people lived and worked there. For instance, at a pre-fab house factory, we found 2x2-inch studs supporting walls of thin slabs, each side with a window. These homes could be erected in a day or two, thereby providing housing for the poor nation's abundance of peasants. How does one furnish a low-cost house when the cheapest furniture would cost more than the dwelling? To use typical furniture would defeat the president's purpose for providing modest homes.

He solved the problem by building a factory on the hacienda to make cheap, but serviceable, furniture. For a bed and mattress, a plastic bag filled with native vegetable matter would suffice.

I was shown with pride the farm's new school where 450 children are required to learn to read and write and where all wear shoes, the latter specifically noted, for I had observed their bare-foot grandparents picking beans in these fields.

During one of our stops along the way in a forest of quapinol and mora, the battery on the heavy BMW died. Shoulder to shoulder with the chief of state, we pushed it to a steep decline where the driver got it cranked.

Don Pepe escorted me to his inner office at the hacienda, really an inner sanctum. There nailed to a wall were a pair of boots, a hole shot through the ankle of one. That explained the limp of the leader of the revolution.

Bob was my interpreter on a trip to Santa Rosa National Park, near the Nicaragua border, and on a spectacularly beautiful Pacific coastal beach. A hard-to-reach zone, where man has not yet fouled the land or sea, waves of turquoise roll on the gray white sand and lap the jagged mountains that dip precariously into the surf. Property maps showed that already New York developers had discovered this wilderness. Would the Gulf of Papagayo become a playground for the wealthy?

I was impressed with the Peace Corps' people. Curtis Freeze did research on the shy and petite white-face monkeys and two other species of primates. Margaret, his wife, fixed beans and rice, the national dish, in a dozen ways. Lack of refrigeration makes innovation a necessity.

Sent in an ancient 4-place Cessna—pieces of wire substituting for knobs on panel instruments—the mission was to look at the work on Cahuita's coral strands and *Lianura de Tortuguero*, the turtle everglades, on the Atlantic coast. Not long after leaving San José, the plane's electrical system failed. Three times the pilot of the little plane aimed for passes in the mountains that surround the capital, only to be driven back by rapidly descending clouds. Finally a pass was clear and we were on our way. En route toward the Caribbean, Antonio Chavez navigated by the rivers.

Studies of giant green turtles on the Atlantic shore charted their emergence from the sea to lay their eggs. After the long swim from Florida's feeding grounds, the reptiles move inland scarcely 30 feet from high tide to dig a hole, deposit eggs, cover them up, and return to the Gulf. This occurs in the brief span of a night. We could not help but erase the deep wide trails they carved in the sand as the plane landed on the beach.

Traveling by dugout, we paddled upriver to some shacks which housed the Peace Corpsmen and in which we'd bunk. A swarm of bees in the dense jungle, brushed by the canoe, let you know of their presence. Walt Auffenberg took eight mean stings before we got untangled. I settled for three and a swollen head.

Hand-cranking the Cessna propeller, we were off to the Coral. There a crude landing strip cut by machetes along a littered beach provided a temporary haven. Running low on fuel, Antonio taxied down a dirt road to a country store, put a handkerchief across the tank's opening, and filled it with gasoline. Leaving the landing strip, we flew over banana country to see from a low altitude the red coral growths of the shallow coastal sea.

Already I had seen the crystal blue lake of Volcan Poaz on the ground. Now, I'd look into its throat, dipping a thousand feet below its rim. The water gurgles as it turns red from the release of heat in the bowels of the earth. Looking down, we saw the plane where four American foresters, employed by industry, perished. They apparently didn't realize they were inside the cone when they made the descent on the west of the *Cordillera Central*.

Not until we touched down back at San José did I understand that Antonio's flying buddies never expected to see us again. They raced to embrace him. No search parties, no telephones, and no radios for transmission over the mountains was the way of life here.

As I left the Rich Coast, my new and cherished friend, the president, said, "Professor, don't buy our wood; buy our furniture." Mr. Figueres, while acknowledging that he understood the frustrations of America's youths at work in his country, urged upon me the need for a continued and strengthened Peace Corps program. Could one ask for more assurance of a job well done? I documented the Costa Rica account in *Explorer* magazine.

Juan Castro, the Peace Corps' Costa Rican director, and I headed to the Pacific in a 4-wheel drive carryall. We'd need the power in the front wheels as we traversed swamps and swales and dug ruts in deep sands and quick sands. We viewed likely locations for a new national park visiting center, cooked a pot of beans, and began the drive back. In the middle of a swamp just inland from the sandy beach, the transmission quit. Without tools, but recognizing a problem prohibitive of solution in the remote locale, we, too, quit. We hiked over the rugged terrain for a day, chanced upon a peasant's cabin where we were fed more delicious beans (the hungrier one is, the more delicious they are) and lay upon the dirt floor for the night. Up early the next day we continued inland. Mid-afternoon we came to a well-traveled road through the jungle, hailed an oxcart, and for a few colons, rode in the wagon to a highway where we flagged a bus.

The Organization for Tropical Studies and the Interamerican Agricultural Science Institute at Turrialba, the Tropical Science Center at San José, and the Caribbean Conservation Corporation in Philadelphia have plenty to think about when it comes to natural resources in the Rich Coast. They join the Peace Corps' volunteers with questions. Is there any one month in this tropical climate when planted trees survive best? How do you convince ranchers to remove cattle from certain woodlands? Should watershed management override other resource considerations? Can *Ticos* develop markets for wood to compete with *Gringo* exploiters? Is it wise to use valuable figured hardwoods for fence posts? Should parks be administered by the forestry ministry? How do the *Gringo* volunteers resolve their frustrations over the slowness of *Tico* work habits in a land, a culture, and a language with which they are unfamiliar and without much oversight from either Washington or San José? To the Americans, Latin American administration was chaotic. But, I could not help but note that some of the cabins in which the volunteers resided resembled pigstyes; a few of the Corspmen could use a G.I. scrub-brush bath.

"Ticos," the diminutive these Latins call themselves, may refer to their stature (though even I, to quote the short, but long-time mayor of New York, did not think of myself as "a dime among pennies"). Ticos may also be the trademark of a courteous and lovable people.

AID in Honduras

When we at the SFA School of Forestry learned the Agency for International Development was planning to assist Honduras in establishing a forestry school, we joined the competition to be selected for the task of training a cadre. Little unknown SFA beat out Louisiana State University, the University of Florida, and the University of Washington for the privilege. Once we were selected, President Ralph Steen and I set off for Tegucigalpa, the capital of Honduras and, of course, the site of the American embassy. Dr. Steen, never having had a passport, was issued a special paper that would, on short notice, permit him to leave the US and enter Honduras and return. After two days, the prexy had had a belly-full of the third-world impoverished country, even though our accommodations were the best the Central American nation could offer. While in "Tegu," we stayed in the hotel where O. Henry, the novelist, bunked when on the run from the law. The hotel stood at an intersection in the capital that is the equivalent of Times Square in New York City.

AID selected the students, the finest the country could offer, for this training. Unfortunately, we almost had a murder on our hands in Nacogdoches, for AID interrogators hadn't recognized they'd sent to Texas the sons of two long-time feuding political enemies. Tipped off in time by a peer, we flipped a coin to determine which one to send home.

Unfortunately, too, we hadn't prepared the town of Nacogdoches to receive the brown-skinned youths. The influx of immigrants from Mexico had not yet reached "the oldest town in Texas" which once was a part of the empire of brown-skinned people. Some news stories and an editorial or two by the esteemed editor and publisher of the Nacogdoches *Sentinel*, Victor Fain, quieted things down. We also met with the Honduran students to explain their responsibilities in this culture.

Oh, you ask, why was I interested in the AID program? I could envision our anglo graduates finding good jobs in

Latin America from contacts made with the visitors. Our students needed the association with people from afar. And a boost in enrollment would provide funding that would, in turn, allow us to employ more specialists in the many fields of forestry; i.e., forest entomology, forest pathology, and game management.

All of these students were required to return to Honduras upon graduation, principally to work on the Olanchito project, a vast concession of timbered acreage leased to the US International Paper Company. Some worked for the government in Tegucigalpa. One eventually rose to the FAO's forestry directorate position.

To get to the Olanchito country, I traveled in an antique Piper Cub. Prior to landing, the pilot circled the strip to shoo away the cattle, circle again and buzz if necessary to clear the path, and eventually land. By the time we were on the ground, always seemingly unexpected, the forest caretaker had left his rustic cabin to meet us with a jeep. The embassy would not allow me to travel overland, lest I would be killed for the gold in my teeth. The *bandidos* first kill and then check for fillings.

Pinus caribaea was the tree of choice for the Olanchito. IP anticipated building a kraft paper mill to utilize this resource, a forest that extended from the Caribbean Sea on the north far into the interior mountains that lie to the south.

Meanwhile IP gave up on the Olanchito. Company executives told me the Honduran Ministry of Forestry kept adding new requirements to the contract until, finally exasperated, the company utilized the stock-holders' money, set aside for the Central American project, to build a mill in Northeast Texas. Kraft paper boxes for shipping pineapples and bananas now come from North America. It is the production equivalent of the mill designed for the Olanchito.

The Hondurans came to SFA with new boots for field work. On a summer field trip to the Rocky Mountains in which three of them enrolled, I learned how they hated those boots. The thirty or so students and I planned to camp at a place called the "Parting of the Waters" in a wilderness area of the Teton National Forest, east of the Grand Teton National Park. At the Parting of the Waters, a rock divides a small stream, some water flowing eastward and some flowing westward. The water that flows east theoretically goes into the Gulf of Mexico while the westerly aimed stream heads for the Colorado River and then the Gulf of California and on into the Pacific Ocean. For the Hondurans, seeing the snow on the peaks above, this point on the Continental Divide was not to be the end of their day's journey. "No, professor, we must touch the snow." I couldn't let them go alone in this wilderness, so I trudged along. The Americans, exhausted from the day's ecological studies, contentedly remained at the improvised camp. Not far up the trail, the Central Americans sat down, took off their boots, tied them over their shoulders and renewed the trek. They had natural soles on the bottoms of their feet at least an inch thick. Their bare feet carried them over rocks and into the snow, to roll, play, and fight. Late that night, the anglo students, now uneasy because of our long absence, prepared a search-and-rescue just as the weary Hondurans and their "guide" stumbled into camp.

After a few years we had provided a faculty for a five-year bachelor-degree program and a three-year technician school. Several of the Honduran foresters returned to the states to work for industry. A couple had married North American gringos. Living conditions in the wilderness of that land encouraged these women to come home to Texas.

South America's Red Wood County

On assignment for the International Executive Service Corps in Brazil in 1989, I was sent to help *Companhia Vale do Rio Doce* (CVRD) and its subsidiary *Florestas Rio Doce* establish forest management research programs at several sites and to lecture to company foresters and nonforester executives at the company's headquarters in Rio de Janeiro and Belo Horizante. The Valley of the Doce (Sweetwater) River Company, one of the nation's industrial giants, was a gracious host.

Before we leave for the woods and trek into the rainforests of the Amazon basin, let me tell of my dramatic introduction to the country named by explorers for the exquisite fiery brilliance of several tropical red-colored woods. We get brazier and braze in English from the same root word the Portuguese used to name the country. Traders in the 16th century, even before the continent was mapped and still thought to be an island, carried brazilwood to Europe for dye production.

Met at the airport by an IESC representative, I was taken to the Hotel Capacabana, the best in Rio. Indeed while I was harbored in the 5-star hotel, so too was the President of the People's Republic of China. (Everyone wants in the act where natural resources are available.) The IESC country manager gave these instructions: "Don't go out of the front door alone, day or night. Only in the daytime can you go out the back of the hotel." So after a brief nap from the long journey and a quick shower, I went out the back door that opens onto a beautiful sandy beach of the Gwanabara Bay. Enjoying the beach and mingling among the people all around and not many yards from the hotel entrance, the blow from a fist reached my cheek. Simultaneously I was tackled from behind and just as quickly a fist full of sand was jammed into my mouth. Quickly, and in broad daylight, knives slashed my pockets to expedite the search for valuables. The trousers in shreds, my glasses missing, a cheap watch stolen, $50 gone, and no wallet for identification, I could have been in a "fix." Promptly a policeman arrived, summoned by onlookers. I suppose he was patrolling not a hundred yards away. Many hastened to help find my glasses, brush me off, wipe the blood they could see on my face and hands, and profusely apologize for the inhospitality of their nation. A police car shortly plowed through the sand, picked me up to look for the culprits, in vain of course as I never saw them, and took me to the hotel for a bit of first-aid. Soon I learned the fist knocked the crown off a tooth. The hotel summoned my IESC host who, from then on, made sure I was never alone; a CVRD man was always by my side, always in the hotel room next to mine, always from Rio to Brasilia at my beck and call. He was a wonderful companion, a forester, and a Christian of Roman Catholic persuasion, yet knew well reformed theology. To learn of the faith of others is required in the parochial school he attended.

With that traumatic experience I was to address the assembled CVRD people and Rio Doce Forest Company executives and forestry staff for three hours the next morning. Throughout the month and even after returning stateside, I suffered "gunshyness," always looking behind me, stepping aside to let a follower pass.

CVRD deals principally in minerals. Its iron ore provides bartering capital for forestry research. Highest-quality hematite iron ore moves to Japan in exchange for financial assistance with which to implement forestry and forest-product research. CVRD's 2000-plus square miles of forest in five areas, including lands reclaimed after mining, provide pulp and paper for a Japanese affiliate. Much of the exotic caribbean pine (*Pinus caribaea* var. *honduriensis*) and eucalyptus will eventually be utilized in the Land of the Rising Sun. The company annually shipped short-fiber bleached wood pulp worth about $100 million to Europe, Japan, the U.S.A. and its neighbors in Latin America.

Brazilian pulp purchased by Lake State s paper mills impacts the economy of Michigan, Wisconsin, Minnesota, and the adjacent Canadian province of Ontario. The South American supply competes with the super-abundant volumes of trembling aspen (*Populus tremuloides*) available for pulp. High labor costs in the US sideline the Lake States' easily regenerated aspen for the cheaper fiber imported from 10,000 miles away. Moved from the forest by raft, rail, boat, barge, and finally ship, the Brazilian pulp costs less delivered to the mills of the North-Central states than locally harvested aspen.

The world worries about the Amazon rainforest. It disappears for three reasons. (1) Slash-and-burn agriculture removes much of the timber. What is the alternative? Starvation for the Indian peasants. (Brazil vies with Indonesia for being the fourth largest nation in population after the break-up of the Soviet Union). A couple years farming these lateritic soils leaves the silicates leached and a clay crust on the surface. (Laterite derives from the Latin word for brick.) The impoverished people move on to again slash-and-burn a few acres, else to go hungry.

(2) Conversion of forests to grazing lands that supported cattle—which, when fattened, went by air to Japan—obliterated much of the rainforest. The government

no longer provides financial subsidies for these ranchers, essentially halting exploitation for beef production. The sparsely populated Rondonia State in the southwest corner of the Amazon Basin, transected by a single road, loses its ranching peons to Rio and San Paulo slums, there to maim and steal and kill in order to survive.

(3) Not all of the trees of the Amazon Basin produce quality timber. Good stands may have but a half-dozen trees per acre worth harvesting. One cruise in these woodlands tallied 1200 tree species in a 12.5-acre plot. No more than ten species in the sample were merchantable. Of some 3000 species in the Basin, timbermen utilize but 230. The mills cut 80 percent of the lumber from about 50 species. Rainforest rotations may require 50 to 100 years under intensive management.

The government protects 30 million of Brazil's 907 million acres in the Amazon. That 3.4 percent compares well with the 3.6 percent of the USA's 70 million acres in national parks.

The Rio Doce Forestry Company staff emphasized the need to plan research for species useful for charcoal manufacture. Ore refining and steel production require heat. In a land rich in iron but poor in oil and gas, the carbon must come from wood. Charcoal made from tropical hardwoods burns long at high temperatures; when dropped, the coal doesn't break but clangs like a band's cymbal.

The world's mightiest cascade of water roars deafeningly as it pours over Iguaçu (pronounced E-waá-su) Falls, a most beautiful sight I shall forever treasure. Now protected in a three-nation international park, the sight and sound could lure travelers as Niagara Falls did a century ago. Security rangers in the Park guarantee a safe haven for a mugged North American to begin to overcome gunshyness.

Iguaçu Falls consists of 300 separate cataracts that come together to dump water over 300 feet downward in a two-mile wide sweep of the River. The Falls shares its glory with Argentina on the other side of the river. Nearby lies Paraguay, its Parana River of similar flow to the Iguaçu powering the largest hydro turbines in the world. They produce 10-fold the electrical current of the Grand Coulee station on the Pacific Northwest's Columbia River. Here discussion revolves around watershed protection for the Itaipu Dam. The artificial lake above the dam covers 520 square miles. Foresters protect its banks by planting various tree species, 20 million seedlings having been planted by the time of my visit.

As I reported in *Forest World*, both industry and government have natural areas set aside for future ecological study. Though these preserves are presently without adequate protection, I recommended they be used as ecological benchmarks. Good data need to be collected now.

Seven distinctly different forest types cover about one-half of the nation's land area: babassu palm, dende palm, caatinga (a semi-arid scrub land of cactus-infested thorn forest), cerado brush that replaces much cutover forest, *Araucaria*, pampa cerrado (rangeland forest), and interior tropical hardwoods. These latter, of course, are of the greatest economic value.

As herbicides had not yet been introduced, I designed experiments for testing them, especially for controlling palm weed trees. However, extracts of the spiny papunha palm kill vermin, its trunk can be used for wood, its fronds for baskets and roof thatch, and its fruit for "cabbage" (reminiscent of the cabbage palm of South Carolina which provided human sustenance in the famine times of the Civil War).

I was fascinated by the use of "foxtail" slash pine for wall panelling. Boards sawn from these trees show on their faces minute knots where needles once grasped the boles. The defect is now considered an enhancing figure-mark. Remember, dear reader, when knotty white pine was the rage. Until an imaginative salesman came up with its use for panelling, the second-growth stems of inferior quality in northern forests were a trash wood.

The practice of forestry in this land of tropical and subtropical forests seemed much like that of The United States in the 1930s. The forestry profession emerges dramatically as timber famine threatens and economics makes at least extensive management feasible. Industry and government encourage silvicultural practices, reminding Brazilians that the green field of their country's flag symbolizes woodland wealth.

U.S.'s Tropical Forests

A show-me tour of the forests of Puerto Rico, the USA island colony off the mainland's Atlantic coast, guided by Frank Wadsworth, the director of the US Forest Service's research center on the island, showed me for the first time the cadam tree. When I told Frank about our Professor Kadambi, we returned to the station library to learn, indeed, that *Anthocephalus cadam* was named by and for our illustrious forester from India. Planted throughout the tropics, cadam becomes nearly ubiquitous.

Cadam, named for SFA's Dr. Kadambi, in a nursery in Costa Rica.

Mexican Trips

As a member of the Council of the American Association for the Advancement of Science, I met with the group in Mexico City. On this trip I learned from a cabbie something of the US monetary system. The taxi-driver wanted to know if I wanted to buy some US silver certificates. "Why," I asked, "should I be interested in that?" He explained what I should have known: Congress quietly replaced silver currency with paper money for which the Treasury had no obligation to secure with silver valued at so much an ounce. Thus, during the Johnson years, paper money could be printed at will in order to finance the Great Society and its welfare programs. I paid $2 in US Federal Reserve Notes (the new money) for $1 in silver certificates. The certificates are now worthless for currency!

Another trip, this time to lecture at the Mexican College of Agriculture a few miles beyond the border of Mexico City, rates mention. One can travel to the campus, once a wealthy hacienda, on pleasant roads; but the shortest way takes the traveler by the city's garbage dumps. There thousands work as scavengers, living in cardboard shanties or auto body parts. Children play among the rat-infested heaps and vegetables grow among the rubble. I thought, "Are those the strawberries, cabbages, and carrots we buy in Texas' markets?"

A bigger surprise awaited me during a tour of the campus. Handles on the doors of what was once the hacienda's chapel were carvings of hammers and sickles, emblems of the international Communist party. My host said these replaced symbols of Christian faith, the cross and crown. Inside, where the altar had been, a huge mural now displayed a much more than life-size reclining nude, painted by the world-famous Mexican artist Diego Rivera. I inquired as to why the government should desecrate a holy place? When the great farm village was nationalized in the years following the revolution of 1913, it fell under communist rule. My host in 1971 said, without apology, "It remains so." How little we know of our neighbor, writing this page as I do on the day the Congress of the US debates the North American Free Trade Agreement (NAFTA).

On another occasion, a Mexican army colonel called me from Mexico City on a forestry matter involving his extensive holdings near Victoria City on the east face of the *Sierra Madre Oriental*. His two sons met me at the Nacogdoches airport, flew me in their plane to the site, spent several days with me and others in the woods, and deposited me at Laredo, on the US side of the Rio Grande. Alleging plane trouble, the young Mexicans put me on the bus for home. I wrote the report, suffered a case of Montezuma's revenge, talked and corresponded with the Colonel, but was never paid.

Anne, Janet, Wendy, Jean and I took the sleeper train from Nuevo Laredo to Mexico City. What a ride! Upon emerging from a tunnel in the mountains, the engine bashed into a huge boulder that had fallen onto the tracks. The engine jumped the tracks but continued on, the wheels crunching the concrete crossties as it proceded, stopping on the far side of a trestle just as the last car cleared the tunnel. We waited in that high country desert sun for the rest of the day and into the night for a wrecker to come to wench the locomotive back onto the tracks. Without power from the engine to cook, the chefs built fires from mesquite wood and fed us well. The fires in the cold night air also kept us warm. To see the extent of the damage to the engine, as we were in the last car, I hiked down the valley and up the far side. The cow catcher and locomotive "grill" had been battered severely by the now well-splintered rock.

Earlier we waited for several hours on the main line, not knowing why. When finally the train moved down the track we saw we were waiting for an overturned engine to be hoisted back onto the rails. A similar accident occurred on our way home. I thought I'd never get Anne on a Mexican train again. But I did.

Another trip took us by train to the Gulf of California. *En route* we spent a few days at the rim of the Copper Canyon of the Yaqui River, Mexico's secret competitor to the Grand Canyon of the Colorado River, and one of the most spectacular scenes on earth. Down in the canyon we bartered for a handmade *tesiopas* (violin), in which I count six different woods for various parts of the fiddle. There we learned Mexican Olympic marathon runners train for the Big Event. The *tarahanuras* "run like deer." Some day the little cabin in which we kept warm by the fire while it snowed beyond our door, and in which we ate beans and more beans, will be replaced by giant lodges. People will come from afar to view the *Posado Barranca.*

Windbreaks of a cypress species enclose the pastures of an Amish dairy community, migrants from Pennsylvania, in Costa Rica.

The pre-Columbian vessel, a gift of Peace Corps volunteers, and the Indian hand-crafted violin.

21

"WOODS AND TEMPLED HILLS"

Giant sequoias (the bigtree) of the moist west slope, near the summit, in the Sierra Nevada Mountains (USFS photo).

Opportunities to visit forests in all 50 states came my way. Yes, forest trees do grow in Nevada, in the high reaches of its mountains; and Delaware and Rhode Island have farm woodlots. Hawaiians, too, care for stands of wild stems among the orchards of fruit-bearing trees. While I described many fascinating forests in *Forests*, other accounts of intriguing trees growing in the woods and among the templed hlls of Katherine Lee Bates' 1910 description of beautiful America bear recording here.

The Great Land

No more spectacular scenery has been engraved by the Creator's hand than that of the Inside Passage, the "Interstate (Marine) Highway" where ferry boats connect Seattle with Alaska. The state's name means Great Land in the Native-American tongue. On one trip I made to the Far North to study soils, the boat at the dock in Seattle was already full. A helpful gent suggested taking the bus to Vancouver, BC, then the 90-minute ferry ride to Swartz Bay near Victoria on Vancouver Island, thence an intercity bus from Victoria via Nanaimo on Canada's "Interstate" 1 to the end of the gravel road at Port Hardy, and another ferry to BC's Prince Rupert. By that time and place many passengers would have disembarked from the ferry and there'd be room for the straggler. And there was. In that 1960s' venture, I was warned to carry a bedroll. The deck served as an open-air stateroom.

An alternative was to sign up for a swabby job on a freighter and jump ship at Ketchikan. The union boss at the dock nixed that idea, and the labor fraternity's dues were a bit steep.

En route northward in the Inside Passage, I noted king, sockeye, and silver salmon playing among the seals and porpoises. Bald eagles, diving to claw a meal from among the school of fish, struggled to carry the heavy catch to a nest high in a seaside tree. Now and then a whale appeared, spouted, and submerged.

Ketchikan, the south-most town in Alaska, was built on the side of a mountain, the wood-decked streets supported by scaffolding on the lower, seaward sides. When I asked the Forest Service ranger how many vehicles he had for managing the half-million-acre district of the Tongass National Forest, the answer was, "one." For clarification he said, "I only have eight miles of road. One sedan is all we need to get to the dock from which we take a boat to another island to board a plane for the ride to whatever island we will work on that week." When told there are no grizzlies on the islands, I solo hiked a trail in an early morning fog when, not a mile from town, a black bear crossed my path. No cub was in tow.

Glaciers galore and glorious! We traveled a fjord upstream in a Forest Service boat, the pilot ever vigilant in the fog for icebergs, readily seen on the radar screen. When one was spotted, the pilot suddenly withdrew the throttle, reversed the engine, and turned the boat about. It now raced a fast-moving mountain of ice broken from the LeConte Glacier face not many miles up-stream. Later the sound of calving glaciers drowned the drone of the motors on a Twin Otter plane; huge chunks of ice broken loose by the vibrations of the pilot's skillful throttling nearby the wall of ice.

We learned fireplaces don't exist in Alaska. When contractors built modern homes in Anchorage following the Good Friday earthquake of 1964, they didn't know better. The flues not only sucked up the chimney the heat from the fire, wind siphoned out the heat within the home. But we, in July, stood around pot-bellied stoves "warming" our hands and backs after a day in the field. Eventually, someone would look inside the fire box to find it empty!

The aforementioned earthquake, an above-9 tremor, released energy exceeding 50,000 times that of an atom bomb. Huge seawaves, called *tsunami*, rolled onto the coast as far away as California and raised or lowered ground level up to 10 feet over 200,000 square miles.

Muskeg holes of acidic peat and muck 12 or more feet

deep interrupt the level terrain of taiga, the "land of little sticks" of black spruce. Foresters fall into the holes in the spring and fall when thin coats of snow on the ice hide the treacherous traps. Smoldering fires burn for weeks in the extensive peat bogs and muskegs.

Loess soils, windblown from ancient alluvial fans, and glacial flour, the finely pulverized mantle of rock left at the face of the mountains of ice, underlie forest trees. Red alder (*Alnus rubra*) quickly covers the gray silt-like material, nitrogen-fixing actinomycetes in nodules on the trees' roots providing up to 150 pounds per acre of usable nitrogen each year. Nutrition on this outwash, adequate for sustaining growth of Sitka spruce, allows succession within a few years to the high-value conifer.

Rocks in incipient soils, those formed within 50 years of a glacier's "retreat," are clean on the bottom and silty on top. In time, the fine granules leach from the upper surfaces and the fragments become clean on all sides. Thus one may gauge where in time along the scale of soil genesis the site may be.

In these cold, wet soils, true podzols—displaying ash-white surface layers—appear similar to granite weathering. The latter is apparent in the many terminal moraines. Here, too, shallow soils result in windthrow, suggesting shelterwood silvicultural systems for many conifers. And, too, the bare mineral soil associated with clearcutting provides a hostile seedbed: frost-heaving of seedlings takes a heavy toll, especially for western hemlock (*Tsuga heterophylla*).

Indicator plants signal site quality extremes in the southeastern Panhandle of Alaska. I've seen pure Sitka spruce (*Picea sitchensis*) 185 years old growing on Site Index (SI) 50 lands and, not far distant, *Vaccinium* spp. denoting SI 50 for the same species. On such sites dogwood (*Cornus* sp.) as a ground cover grows to a height of 5 inches and white spruce 3 years old is all of 3 inches tall. Skunk cabbage serves as an indicator of wet sites on which one may tally 30 species of willow (*Salix* spp.).

A few other ecological observations about this fascinating frontier: porcupines kill white spruce (*Picea glauca*) by girdling; tamarac (*Larix laricina*), though called *eastern* larch, does well on permafrost: duff is a good seedbed for western redcedar regeneration; and slumps—especially on clearcut areas—bury A horizons, ofttimes many feet.

On another Forest Service expedition, we looked at the land the Service sought under the Alaska Statehood Act of 1958. Forest Service folks knew it would be a hard sell, for the Secretary of the Interior, never in any administration a friend of the Department of Agriculture's agency, was given the authority to recommend to the president and the Congress the lands for national parks, scenic rivers, wilderness areas, and wildlife refuges, as well as national forests.

We traveled by plane, in tandem. One of the Twin Otters flew low so that we could get a good view of timber types; the other plane's occupants watched the ship below. In time, we reversed the procedure. If the FS appeared edgy, it could be attributed to Louisiana's Congressman Wade Boggs, his Alaskan colleague, and their pilot disappearing in the proposed Wrangell Mountain National Forest shortly before our visit. When a plane with the heat of friction and its engine plows into a glacier or mountain of snow, the plane and its contents are quickly swallowed and hidden from view by the fluid snow. And pilots get vertigo, unable to discern sky from water. Not trusting the instruments, they dive into a bay (as one of our pilots did, sadly, six weeks after our venture). Truly said, Alaskans seldom die natural deaths; plane crashes and drownings, usually associated, take their toll. The survival record in summertime coastal water is about 30 minutes, seldom long enough for rescue when a pilot dives or ditches a plane into the sea. Native Alaskans know nothing of swimming; new hotels with pools in Anchorage may change that.

Our team reviewed proposals for 500,000 acres of additions to the Chugach National Forest, and new units nearby Mt. McKinley (the Yukon-Kuskokwim) and in the Wrangel Mountains. Another of these proposals was the vast area of the Porcupine River. We landed at Ft. Yukon, a US Air Force DEW Line base, and from there the military carried us by van a short journey through stands of black and white spruce (*Picea mariana* and *P. glauca*) to the village. In the town of 300 people, we learned lessons in sociology. The Bureau of Indian Affairs (BIA) had a white man handling the native-American business, selling almost $1200 worth of booze a day to the natives. The 30 airmen didn't buy the suds; they'd get it cheap at the PX. And the three Wycliffe Bible translating missionaries weren't likely to imbibe. Meanwhile, the US Department of Housing and Urban Development shipped in prefabricated houses from the "Lower 48" while a vast forest of trees just the right size for log houses—and in which the Indians had lived for perhaps a millennia—go unharvested. The new 2-room houses cost $35,000 in 1976. The Hudson Bay Co. established what became Ft. Yukon, a wooden grave "stone," marked 1847, telling a bit of the story. Wood doesn't rot in this climate.

FS folks didn't spell out their interest in this land of the midnight sun at the Arctic Circle and up against the Canadian border on the Yukon River. Harvesting these trees in the vast river alluvial plain could take place only for a few weeks in spring and fall. The soil is deep muck in summer; in winter deep snow prevents timber harvesting. In the two short spring and fall seasons, logs could be skidded to the frozen rivers, rafted in summer to the pipeline bridge over the Yukon River (the only other bridge over the Yukon is at Dawson, in Canada), and from there chipped and/or shipped down the River for pulping abroad. Old shorelines in the timber indicated the undependability of the rivers retaining their courses.

The area of the proposed forest, below an elevation of 600 feet, just a few years earlier had been sought for the gigantic Rampart Dam hydropower generating plant. The dam would have backed up water for 400 miles on North America's third-longest river. Congress finally nixed the Federal Power Administration idea when no one could suggest who would use all the electricity, and where. Some say waterfowl supporters defeated the project on the Yukon flats.

Only the curvature of the Earth prevents one from seeing Russia from Kotzebue on the shore of the Chukchi Sea. At this village above the Arctic Circle, black insulation board clearly stenciled "Dierks" provided siding for houses along the single lane at the waterfront. Dierks, dear reader, was once the principal wood-using industry in Arkansas and Oklahoma. Those 4x8-foot sheets originated in the pine forests of the mid-South.

Forget plane schedules. When passengers deplaned at Kotzebue, the pilot issued to everyone a colorful green jacket. When the captain was ready to take off, he walked the town, locating his passengers by the colorful garb. Twice we left for Nome, not because of plane trouble but, arriving in Nome, the Weems dispatcher decided deliveries back to Kotzebue were more important than taking passengers on to Fairbanks.

Kotzetue had its racial turmoil. BIA moved Aleuts from the southwestern islands to the village (in order to free the islands for nuclear testing), not realizing Aleuts are not Eskimos. Perhaps to compensate, BIA built a nice frame school. It's on insulated stilts so that heat from within the structure can't be transferred to the ground and melt the permafrost. If that occurred, the building would collapse.

At the time of one visit, I thought of Alaska as a "demographic 3." Total population was 300,000; roughly a third native, a third Caucasian, and a third military (thus, all colors). The native third could be roughly divided among the Indians in the Interior and southeastern panhandle, the Eskimos along the north and western coasts, and the Aleuts on the isles of the southwest that stretch perhaps a thousand miles from the mainland.

In lands of permafrost, underground water in sewage pipes freezes. So, in Nome, I'm introduced to water delivered in bottles and human refuse removed in "honey buckets." These chamber pots are emptied into the sea or, in winter, simply dumped onto the ice (pre-EPA!). (Nome, the last refueling stop for American airmen going west in WWII, showed the influence of Swedish missionaries. Covenant Church is the little town's focal point.) In Fairbanks, during my visit, streets were being dug up, perhaps to a depth of ten feet, the permafrost replaced with gravel. Workers laid sewer and water pipes deep in a layer of rock chips to provide water and sewage routes.

Folks at the University of Alaska in Fairbanks showed me genetics work with musk-ox, the strong bulky polar-region beast confined to Greenland and North America. (Some say it is also found in Russia's Siberia.) The beast's coat, the rarest and finest textile fiber, is pulled from the hide by hand.

The little Hotel Theodore Roosevelt in Fairbanks had no vacancies, necessitating my tossing the bedroll on the courthouse lawn and sharing the stars in the short-night sky with a host of hippies. The long-hair lads had come north to fight fires in the tundra and taiga, but the Bureau of Land Management rejected their services. And without $300 in Canadian cash they couldn't get into the Maple Leaf country on the Al-Can Highway to return to the Lower 48. Stranded, they camped on the government's lawn, resulting in a gender war. Men of the town, many of them pipeline workers, would prefer to line the hippies up to whip; their spouses, in contrast more sympathetic, encouraged the menfolk to build before winter a pavilion outside town to provide cover. The calendar reminded them the long-night cold season was on the way.

The small lodge at Mount McKinley (now Denali) National Park was also full, so I hiked to the foot of Mount Healey's 5716-foot peak and unrolled my bed. To the east lay the Susitna Valley. Awakened in the gloom of the midnight sky, I photographed the caribou intruding on my bivouac.

As for carrying capacity for wildlife, Alaska rates low. My limited observations suggest that much of the large game animals must be pastured in Denali, at the foot of Mt. McKinley. Indeed, a single moose consumes 50 pounds of willow a day. Moose population has declined due to wildfire exclusion, good habitat requiring a burn every 25 years to assure adequate food within browsing height. Winters are too severe for sustaining deer in the Wrangell Mountains. Mule deer, however, have been introduced into the islands of the southeastern Panhandle.

(Why, pray tell, did the Department of the Interior's Place Names Committee change the Park (a man-made division) name and not the mountain's name? Native-Americans, no one knows how long, had called the highest peak in North America "Denali," meaning the tall one. The natural phenomenon, not the political area, should bear the aboriginal name.)

I left the Park, as I arrived, on the Toonerville Trolley, built by the Army in WWII and now operated by the U.S. Department of Transportation. Engineers stop the train on the conductor's signal to allow fishermen to cast hooks in the cold crystalline waters of their favorite creeks. Fishermen then later flag down the locomotive to be taken on board.

When the Secretary of State during the Lincoln administration negotiated with the Russians to pay about $7 million for the territory, opponents called the deal Seward's Folly. What a deal—in minerals and timber. (Russian influence in church architecture and racial features remains evident.)

Veins of milky quartz, the source of placer gold (panning can be exciting) are common; Copper River's color tells of a vast deposit of the ore under thick covers of ice; tallus outwashes expose iron ore; and of course the oil from the North Slope, the Kenai Peninsula, and elsewhere encourage continued exploration.

To this inventory, geologists add coal, the deposits made famous to foresters by the Ballinger-Pinchot controversy in the early months of President Taft's administration. Of thirty-three coal claims, each a 640-acre section in size, 32 were considered fraudulent "dummies."

The 800 miles of oil pipeline were under construction during one of my visits. The 40-foot sections of 4-foot diameter steel were fabricated in Japan, welded on-site into 80-foot lengths, and hoisted onto cross-arms supported by piers that were sunk into the permafrost. Thus, about one-half of the line is above ground. To avoid a catastrophe, the pipes in the ground are equipped with refrigerants in case heat should escape from the oil through the steel pipe, its insulation, the jacket that holds the insulation, and the steel crossarm, to the piling. The Russians have been laying pipe over permafrost for a long time. They simply rest the pipes on bundles of wooden crossties. What better insulation than wood?

Contractors promptly revegetated disturbed sites, using a hydroseeder invented for the task. Seeds and fertilizer were included in a mulch distributed from tanks on trucks or aircraft.

Sometime in the 1960s, Washington bureaucrats figured we might need a safe haven for a military command post. They chose a site at the face of a gold mine, and with giant augers cut horizontal tunnels a thousand feet or more through the permafrost. Even with refrigerating equipment, the ice, at less than 0° C, began to melt and the aged organic matter, frozen in a buried lake of ice until then, commenced to decay. What a stench. CREEL, for Cold Regions Environmental and Engineering Laboratory, was abandoned.

Whale skins provide the fabric for fishing boats at Kotzebue, and Eskimo kids learn to be entrepreneurs at air strips. They await arrival of the planes. Every "toy" was $1.50. From one boy I bought an "Eskimo yoho," two hide-covered balls in which are sewn grain or seeds, the balls connected by a string. With a little practice they twirl in opposite directions. From another, I purchased a wolf-scarer carved from a 6-inch strip of the thin balleen bone of a whale and attached to a string. When flung, it makes an erie sound. Eskimos use them while fishing on the ice.

Deans' Tours

Annually the South's regional forester conducted a tour for forestry school deans through various forests under his care. Sometimes we'd escape Dixie to observe management in other regions. One such trip, in 1974, escorted us through the Southwest's Kaibab, Coconino, and Tonto national forests in Arizona. We reviewed silviculture in high pine country, deep canyons of oak, and colorful red rocks of the deserts.

Later I went back to the Grand Canyon, hiked to its bottom in the warm winter sun while snow fell to a depth of five inches on the rim. I began then a one-man lobby to exclude the messy burros from the hiking trails; if you can't walk the trail, stay at the top (unless disabled).

On another occasion I trudged to Supai, the Indian village a long day's hike below the rim in the Havasupai Indian Reservation within the Grand Canyon National Monument. I reported the social and political observations of the visit to Supai in the *Journal of Forestry*. At the time the Indians wanted more land from the federal monument holdings and the Public Domain, and Senator Barry Goldwater of Arizona was anxious to oblige. I was curious as to how the turf would be utilized, as the irrigation system for home gardens, installed by the Corps of Engineers at the request of the Bureau of Indian Affairs, didn't appear to be much used. Finally a native patriarch, only grunting to my questions and observations as we sat on a log bench on the village street, sent me on my way: "Are you some kind of sociologist or psychologist sent down here to study us?" I picked up my pack and, over a long trek, dropped 2000 feet in elevation to the Colorado's River to set up camp.

The 1973 Dean's Tour took us through eastern Tennessee. At a visit to the University of the South in Sewanee, we were entertained by the organist, playing a fabulous "king of instruments," in the chapel wherein is a stain-glass window dedicated to Gifford Pinchot, the Father of American Forestry, and where Dean George Garratt of Yale began his forestry teaching career. Sewanee sits on the Cumberland Plateau, a vast area of almost pure Virginia pine (*Pinus virginiana*) then being harvested by a British pulp manufacturer. Not far from there to the Hiwassee River where most of us took our first rubber boat float trip, a sport soon to become popular. Our buttocks, bruised from bouncing on the boulders, recovered. Serious study of TVA problems (including an old wooden flume on the Ocoee River), watershed management, and fire simulator experience accompanied fun and games.

On another occasion, the regional forester piped us aboard the good ship Cotton Blossom, out of New Orleans, for the annual opportunity to see Forest Service work, and thus pass this information on to students. On these occasions we'd also debate and pass resolutions, sometimes involving FS management or federal research budgets.

Silviculture Instructors

Silviculture instructors met in connection with the national annual meeting of the Society of American Foresters. As SAF conventions rotate around the country, faculty members of the host region regularly put on a several-day show-me tour. In time we considered the technical aspects of silviculture in every clime. A trip to the Sierra-Nevada on one of these jaunts reminded me of an observation about the bigtree (*Sequoia gigantea* to us older folks and *Sequoidendron giganteum* to the younger crowd) that I omitted in *Forests*. Giant sequoias, by which bigtrees are also known, seem to appear in three age classes: the Methuselahan-age stems that arose from minute seeds about 4000 years ago, mature-appearing stands 500 to 800 years of age, and some stands bearing boles numbering 80 or so years. Why the absence of a continuum? Why not some trees somewhere of every age? To my knowledge, no one has attributed the long time gaps to fire, disease, insects, storm, or harvest.

Death Valley and Beyond

Paul Burns and I were bored at an SAF meeting in Las Vegas. Imagine that? So the two staid Presbyterians rented a small car and, before daybreak in the desert, headed for Death Valley National Monument just over the state line in California. With all the significant sites in the park visited and photographed, and much of the day still remaining, we checked the map. The White Mountains rose to the northwest; we could travel the gravel road across the sands of Death Valley, leave the gravel for a shortcut over some sandy ruts in the Inyo Mountains, and come out on highway 168. Not far then to the Ancient Bristlecone Pine (*Pinus aristata*) Forest. Agreed, and the gas tank still reading nearly full, although no place here to top off the tank, we left the shore of *Badwater*. At -282 feet, it is the lowest dry land in the Western Hemisphere. Enough of the morbidity of *Funeral* Peak (6384 feet), *Last Chance* Range, *Devil's* Paintpot, *Furnace* Creek, and *Ghosttown*. We imagined the haunting mystery of these "points of interest" to the 20-mule-team wagon drivers—as they steered loads of borax, the white gold of the desert—and the joy of relief when they reached the saddle of *Jubilee* Pass.

We made it to the site of the "oldest known living things," high on the east-facing slope of the mountain in the Inyo National Forest, counted over 4000 rings in downed trees, and left for Vegas. From Badwater to the summit of the White Mountains, at 14,242 feet, our journey had taken us the greatest vertical distance per horizontal distance of any place in the world. If we saw *the* oldest living thing, we could not know; for the map with the X that marks the stem is stashed in a bank vault. I'm told only its discoverers have keys, vandalism being a great concern if the specific tree became public information.

The gas tank now reading E, we came down the mountain without benefit of power brakes, power steering,

Stems in the Ancient Bristlecone Forest in the Inyo National Forest of California's White Mountains. Edmund Schulman dated this "oldie" with the thin lifeline of green tissue at over 4,000 years. John Wicker took the picture not long before Paul Burns and I checked out the site (USFS photo).

or low gear. There was no choice. On California 168 we were relieved to see a highway road repair barn, but on its fence a big sign shouted, "No Gas." Soon, a kid on a bicycle told us of Deep Springs College, a little-known school. The folks there sold us five gallons, barely enough fuel to power us to a town with pumps. Relieved is an understatement.

Paul, director of the School of Forestry at Louisiana State University, never transpired a drop. Famed among foresters as both pianist and tennis player, his tie was still neatly in place when we arrived back at the hotel late that night.

Mt. St. Helens

The eruption of May 18, 1980, blasted a cubic mile of rock and ice from the top of Mt. St. Helens and raised Spirit Lake a couple hundred feet in elevation. Had the wind been out of the east, Portland, 50 miles away, might have become another Pompeii, like the buried city in Italy below Mt. Vesuvius. Instead the west wind carried the black and gray ash plume over fields to the east, providing natural fertilizer for farmers near Spokane.

Within seconds, an explosion equal to 500 Hiroshima atomic bombs devastated 200 square miles. Timber, estimated at 2 billion board feet, enough to build 150,000 homes, was shorn over tens of thousands of acres.

We could only come within a few miles of the new summit, as the belly of the mountain continued to belch. Loggers salvaging timbers on the Gifford Pinchot National Forest and Weyerhaeuser Co. holdings pointed their trucks downward and listened over portable radios for signals to escape in the likely event of further volcanic activity.

Harry Truman (that *was* his name), a crusty old proprietor of a lodge, served notice that nothing, not even volcanic eruption, would budge him from Spirit Lake. He's buried there, under giant Douglas-fir logs that cover the lake like great rafts, one upon another two and three deep.

Astonishingly, a couple years later rapid recovery of animal and plant life was evident. Fireweed, a splotch of color, became a welcome sight on the ashen terrain. Evergreen seedlings soon emerged from seeds, the polycotyledons unfolding in the heat of the warm sun reflected from the dark-colored soil.

Shortly, the Forest Service designated 35,000 acres around the mountain a Geological Area. Eventually, Congress created the Mt. St. Helens National Monument, the only such unit under FS jurisdiction. Interior's Park Service administers all other national monuments.

Consulting Jobs

Earlier chapters reported on consulting jobs abroad and the long retainer for the National Plant Food Institute. Now a few paragraphs about consultancies of a more local nature, principally timberland management, research assignments, and forensic forestry. I liked to practice forestry in the woods. Doing so made my research and instruction practical and experiential.

For forestland owners fearful of investing in the services of a practicing forester, two tales need to be told. The first took place near Washington, Georgia, in the 1950s. A sawmill had mailed a widow, an absentee owner in the North, a nice check for a sawtimber thinning. Several months later she got another draft, this time from a pulpwood dealer for the tops and inferior thinnings. Then, as the pulpwood foreman was scrapping up, checking on his own work, he met "a little old lady walking the logging road." Of course he stopped to inquire, learned she was looking for her land, spread his maps on his pickup's tailgate, and asked her name. "Why lady, you're in the middle of your land." "Oh no, this can't be my land; all my trees are gone." But it was her land, and she still had a well-wooded land, though she never dreamed thinnings would bring such a price. Previously, without the benefit of a forester's aid, she had been cheated.

Tale two: A school teacher in South Texas who owned land in the state's Big Thicket was sent a cashier's check and a contract to sign for all of his trees 10 inches in diameter at a 10-inch stump height. He returned both papers, saying, "No, but I'll take twice this amount." Promptly the absentee owner received a new check and contract. At that point he called for advice. I had to say, "You've made a counter offer and it was accepted. You've sold your trees." Wanting to know what his stupidity cost him, two students and I spent a Saturday cruising his land

and reluctantly told him he was still getting only half of the appraised value. Folks will hire lawyers, tax preparers, and accountants but shy away from "consulting" foresters. The high-sounding title, suggesting high fees, scares some.

A Southern Appalachian bald, the cause for the ecological oddity yet unknown, studied while on assigment for the Highlands Foundation.

As for forensic forestry, South Carolinians can blame or thank me for the periodic check of every mile of state road for "leaners," trees bent toward the road more than one foot from the ground for the lower six feet of bole. I showed that a tree that fell during a storm across the hood of a car, decapitating its driver—who happened to be a prominent attorney in a liability-specializing law firm—was high risk. The compression wood that formed on the lower side of the bole exhibited distinctive anatomical and physical characteristics, confirming the claim.

Other cases related to whether wooden playground equipment impregnated with CCA preservatives is safe (it is), whether flooding from an effluent with unknown ingredients from an army ordinance plant killed trees downstream (it did), whether overflow from an oil well saltwater settling pond killed trees (it did), and whether longleaf pine given more thorough drying would be less likely to quickly erupt in flame once ignited (it wouldn't).

22

THE ORGANIZATION MAN

Reclaimed surface-mined land will add to the inventory of wetlands (TU Services photo).

The Society of American Foresters

As in William Whyte's book by the title given this chapter, I believe commitment, and thus loyalty, to organizations with which one is affiliated are of no little moment. It is true, I expected every faculty member to belong to the Society of American Foresters. Without the SAF, the school would not have existed; without the SAF, we would have no standards with which to guide us in educating our students; and without the SAF, we'd have no professional continuity to remind us of our past and to lead us into the future. My response to folks who threaten to quit the SAF over some event is this: Don't hold against an organization what someone within the organization has done to you.

As a graduate student and a member of SAF, I visited its office in downtown Washington in Prof. Paul Graves' class. Henry Clepper, who became a cherished friend, was its sole professional employee. (He became the "official" historian for all things forestry for the American Forestry Association).

At Athens, I was the Secretary-Treasurer of the three-state Southeastern Section. There the national Natural Areas Committee challenged me to locate and describe every known virgin tract in Alabama, Florida, and Georgia. The account was published in the *Journal of Forestry*. In a cordial meeting with Frederick Vosburgh, editor of *National Geographic* magazine, I learned that if I had interviewed people, the *Geographic* would have been interested in publishing such an account. Now, when I read the magazine, people, not natural resources, jump from the page. It is a journal about people.

Texas days found me involved in SAF activities: vice-chairman of the section, annual meeting chairman, and a multitude of menial chores. Nationally, I was the first chairman of the Forest Ecology Working Group. The SAF

Council appointed me its representative to the Second North American Forest Soils Conference and to the Council of the American Association for the Advancement of Science. While Glenn Seaborg, the famous nuclear physicist, signed as president in 1972 the certificate showing my election to Fellow of the AAAS, world-renowned anthropologist Margaret Mead served as its prexy during my stint on the organization's Council. The little-bitty woman always carried her long, crude hiking stick, even in the lobbies of the finest hotels. The crooked staff was her identification. I felt she let her left-leaning attitudes encumber her leadership of the AAAS.

Serving on the AAAS Council when the centennial of Franklin Hough's address to the assembled AAAS members rolled around, I introduced a resolution to commemorate the event that gave rise to the US Forest Service. Following is the revised resolution on forest management as adopted on March 1, 1974, by the Council of the American Association for the Advancement of Science:

> *Whereas* the American Association for the Advancement of Science in 1874 addressed a memorial to the Congress of the UnitedStates on the seriousness of the potential timber shortage in the United States, and
>
> *Whereas* that communication alerted the nation, and its President, as well as Congress, to the critical situation, resulting in the creation of the Division of Forestry in the Department of Agriculture, to which Franklin Hough, who brought the matter to the attention of the AAAS, was appointed the first head, and
>
> *Whereas* that memorial initiated a program of forest protection, management, and research that 100 years later produces high yields of wood fiber while providing for abundant wildlife, water for man and industry, healthy range, and aesthetically pleasing landscapes, therefore
>
> *Be It Resolved* that the American Association for the Advancement of Science takes note of this occasion by urging (1) that the policies of wise management and use of the national forests for long-term national needs that have

> developed over the past 100 years be continued; (2) that these policies not be set aside for short-term political or economic considerations; (3) that the US Forest Service and other concerned governmental agencies fully comply with the spirit and the letter of their legal obligation to prepare and publish environmental impact statements whenever significant changes in forest management policy are planned; (4) that biological and related research on forests be greatly intensified by the US Forest Service and other governmental agencies; and (5) that in establishing and carrying out forest management policies, all governmental agencies give full consideration to the multiple use of the nation's forests for watershed protection, erosion control, recreation, aesthetic enjoyment, and wildlife habitation, as well as production of timber and other forest products.

As the Professional Forestry Employment Committee chairman for the national organization, I spent two weeks in Washington, principally in tow by Donald Theoe of the SAF staff, trying to understand employment problems for foresters in the Forest Service, Soil Conservation Service, Bureau of Land Management, Department of Defense, Civil Service Commission, and other agencies. Don and I published a series of articles on the subject in the *Journal of Forestry*. I felt that if we could get foresters employed, we could solve SAF membership (financial) problems and confront the nation's natural resource issues more effectively. I recommended that the Council put all its eggs in this one basket. With only one-half of America's foresters in the SAF, we couldn't speak with a united voice; and without an upfront pursuit of employment for forestry graduates, there was little interest in enrollment by potential members. On the other hand, never have we been able to successfully convince many foresters that the motivation for membership is not "what's in it for me," but what's in it for others. In fact, the SAF constitution mentions only fellowship as a possible self-serving benefit for its members. To provide anything more would encourage the wrath of the IRS and subject its nonprofit classification to scrutiny.

In 1974, Bill Mann, a research forester in Alexandria, La., who had served on the SAF Council, wanted to be a candidate for the organization's presidency. Bill was exasperated when he learned that his boss, Keith Arnold, the FS deputy chief for research, was going to run for the post. Bill called me to see if I would run. I agreed, knowing the chance for winning against a highly esteemed forester was slim. I suppose that it was Bill who got the necessary petitions signed to place my name on the ballot. Before the election, Keith retired from the FS and joined Steve Spurr at the University of Texas as the institution's director of research. Steve called to tell me that Keith was coming to Texas, that he was not there to begin a forestry program, and now we would have two Texans vying for the SAF post.

But West Coast lumbermen, unhappy about the FS monopolizing the president's office, got a petition up for James Bethel, dean of the College of Forestry at the University of Washington, to run. Jim told me later, when we chanced upon each other in Indonesia, that he didn't know of my nomination when he agreed to pursue the office. With collegians and non-feds split, Keith was a shoo-in. He was a good leader for SAF. Harry Murphy, from Birmingham, long my supporter and a fellow Penn Stater, was, I judge, among my chief sponsors. State foresters invited me to speak to their national convention in Portland as part of the campaign, presenting a cut, polished, and mounted thunder egg—formed 60 million years ago in the lava flow of central Oregon—as an expression of appreciation. I'm confident none of us spent a penny—even a postcard—to campaign for the office. That wasn't done in those days.

I should have known better than to be drafted, for already I twice failed to be elected to SAF's Council. Texas had never had a representative, and some friends thought it was time to change that. The Texas electorate, teamed with Louisiana's and Mississippi's superior numbers, would wait another dozen years to elect an SFA alumnus, James Hull, associate state forester, as its councilman.

I served with Charles Connaugton—once director of the Southern Forest Experiment Station, regional forester for the South, and regional forester for the Pacific Northwest—on the Gifford Pinchot Building fund-raising project. The committee worked diligently to procure the Grosvenor Estate called Wildacres, just outside Washington, as a headquarters facility for SAF. Thus, it was discouraging to see much of the property wrested from the SAF by the Renewable Natural Resources Foundation in a costly litigation finally decided in the RNRF's favor by arbitration. I read the SAF's side of the argument. Hardy Glascock, formerly the SAF executive vice president, and by this time an executive of the RNRF, sent me, upon request, a long personal letter stating his position. I remain confused as to what really happened that caused the dispute and why it was settled as it was. I am unable to summarize in a few paragraphs anything that would make sense from the documents sent me by both organizations. Could the whole dispute have been over power, personality, and prestige? "Setting aside the legal ramifications," I wrote Hardy, "It looks to me like a case where I'd been tricked by the kid down the street into giving him 'my little red wagon' that I worked so long and so hard to buy." My respect for Hardy's integrity remains undoubted. For Charlie Connaughton, one of the most respected people in American forestry, it was a low blow.

The Gulf States Section honored me in 1968 for "Distinguished Service to Forestry." In 1971, I was elected an SAF Fellow, among the last to be elected by living Fellows. Fellows Don Canterbury, retired chief forester for Champion Paper, and E.O. Siecke, retired Texas Forest Service director, lobbied on my behalf. One other Texan may have been an SAF Fellow at that time.

SAF Fellows were referred to as a retirees club, about three-fourths of them so classified. When Ralph Hall noted in the 1979 Journal that Fellows made up less than two percent of the membership and that only 316 of the 700 nominated since the first six in 1918 were elected, the membership changed the rules. Section members now play a large role in electing Fellows.

Sadly, while Steve Spurr was presiding at a SAF national meeting, as president, he suffered, as he wrote later, a "vascular spasm in my brain, but there was no damage done." By then he was a professor in the University of Texas' LBJ School of Public Affairs. Although troubled with speech, his brilliance continued to be reflected in answers to questions. Chatting in his office, he'd promptly pound out on his electric typewriter the response to a comment, rip off the page, and pass it to me. I joined in nominating Steve for the Sir William Schlich Memorial Award in 1979. Meanwhile I plugged for Walter Myers of the Forest Farmers' Association, at the urging of IP's executive Fred Gragg, for the SAF prexy. Walt, too, served well.

The *Journal of Forestry* published in 1978 my research on the value of licensing and/or registration for the profession. I spelled out the pros and cons. In a volley of responses, some likened governmental control to licensing plumbers; and many failed to understand that licensing is to protect the public, not to enhance the image of the practitioner. At that time about 12 states had such laws on the books. California's was the toughest, requiring a thorough examination following two years of apprenticeship. Privy to the questions on one exam, I felt SFA grads could do nicely in qualifying for a shingle in that state. Forestry consultants (they should use a less elitist name, like forest managers) endorse licensing if it will shut out public-service foresters doing work for free on small ownerships. To get a feel for the subject, I stood the examinations for a Texas real estate license; any forester should easily pass.

The Texas chapter asked me to draw up Growth and Drain and National Forest Management Practices policy statements. The Policy Statement of 1980 on Growth and Drain expressed concern over the diminished land base devoted to forest production; lack of planned regeneration on harvested, non-industrial, private land; lack of effective cultural practices; and the ecological succession of undesirable hardwoods on sites more suitable for producing pine. The Chapter's policy on national forest management urged consideration of natural silvical characteristics of the species of the forest type indigenous to the site when deciding upon a silvicultural system. "The even-aged system...is silviculturally sound and in keeping with scientific forest management principles...."

Eight years later, the forest supervisor was in federal court over implementation of even-aged management systems. My deposition as a defendant's witness, answering the charges of the Sierra Club, the Wilderness Society, and the Texas Committee on Natural Resources, put it this way:

Clearcutting is good ecological practice for the southern pines. For natural regeneration, all species of these hard, yellow pines require that (a) the mineral seedbed be exposed to receive the seed, (b) full sunlight be available to the crowns in order for adequate growth to occur, and (c) a source of seed be readily available. With clearcutting in blocks of 20 to 40 acres, the seeds come from the trees in the walls of timber surrounding the opening. As clearcutting logging operations scarify the soil, the mineral seedbed is exposed to receive the seed, enabling germination. Because clearcutting removes the overtopping canopy, seedlings have adequate sunlight to grow to sapling size, and on to maturity. One observes young southern pine seedlings under dense canopies because of an inherited silvicial characteristic which allows photosynthesis to take place under low light intensities at an early age, but this innate capacity deteriorates as the trees pass from seedling size and seedling age. Clearcutting is a legitimate natural regeneration procedure, developed by German foresters well over a hundred years ago.

Unevenaged management utilizing the selection harvest system is silvically inappropriate for the southern pines. Poor present-day stands throughout the South, including those on national forest lands, attest to the undesirability of this system, widely employed into the 1950s on both private and public forest lands. With this procedure, seed crops are not abundant, due to inadequate sunlight to the crowns of trees, and therefore inadequate carbohydrate production by the foliage. The few seeds that do reach exposed mineral soil, there to germinate, produce seedlings that grow poorly; soon they die. Selection harvests are often confused with thinning, a procedure solely to enhance the vigor of residual stems.

Unevenaged management is appropriate for some species, such as the beech—birch—maple type, many spruce and fir species, and some oak—hickory types. These species are climax forest covers, their seeds able to germinate on leaf and needle litter, and their radicals growing downward through the litter to mineral soil. As climax species, they grow adequately in the shade of overtopping trees of the

> same or other species. As climax in ecological succession, these forests are perpetuated *ad infinitum*, provided man's manipulation, fire, and devastating winds do not occur.

Apparently Judge Parker didn't believe me.

Jimmy Hull, always the enthusiast, asked in 1979 for a statement on "Why I feel the SAF has provided me with 'a golden opportunity.'" My answer had to be a disappointment. Here's its gist:

> Apart from great fellowship with professional peers who, I am convinced, are special folks among society's breeds, SAF contributions to my life have been indirect. Specifically, these contributions have been continuation of my education through gatherings and publications (which benefit directly my employer and society) and the psychological self-esteem associated with membership in a profession with high ideals for caring for the world's people by caring for their resources. (Without SAF there would be no forestry profession.) Not what we do, but the kind of people we are, has made forestry a prestigious profession.
>
> This, I think, is how it should be. SAF was founded for society's good, not to enhance the welfare of its members. The benefits that come our way in serving society through the Society are indirect but great. And they are rewarding.
>
> Others may have personally benefited through (a) SAF employment services and (b) in personal promotion through (1) exposure as an author in SAF publications, (2) incidental contacts at meetings, and (3) basking in the limelight of SAF leadership roles.

Even as I write, the professional organization is suffering turmoil. It's not new: Gifford Pinchot the utilitarian *vs* Horace Albright the sentimentalist; Pinchot the anti-industry forester *vs* his two successors as Forest Service chief, Henry Graves and William Greeley, who believed assistance of industrialists essential in developing American forestry; Chief Lyle Watts, the federal regulator *vs* Richard McArdle, Watts' successor as chief, who envisioned industry paying to rejuvenate America's forests; and Pinchot *vs* Richard Ballinger over the Alaskan coal lands. SAF endured these and many other hassles.

Boy Scouts of America

The patient reader has learned from an earlier chapter the debt I owe to the Boy Scouts of America. In East Texas I continued to serve, taking bus loads of youngsters to the Philmont Scout Ranch in New Mexico and to the Boundary Waters Wilderness Canoe Base in northern Minnesota. For these trips we rented an old school bus ("Philmont or Fizzle") from the superintendent of the black school district in the lumbermill town of Diboll. Willie Massey, the superintendent, drove the bus and, while we were on the Rocky Mountain trails, waited our return at the base camp. In those pre-integration times, no man ever earned, or received, more respect than did Mr. Massey, the African-American, from those scouts.

For a canoe trip, we rented only the rickety bus, piling in 32 kids. A senior student, Jeff Burch, and I took turns driving. With only two weeks available for the event, we left early in the morning, arriving at a state park in Iowa to drop our bedrolls that evening. The old bus made it to Ely in the Land of Lakes for the second night. We didn't speed; the bus had a dangerous speed-limiting "governor" that restricted us to 55 miles per hour.

When the scouts got too rowdy, we'd stop the bus. Everyone had to get off. Then I'd ask Mr. Massey or Jeff to wait for us one mile down the road. To the lads I'd say, "Anyone not on the bus when I get there will have to do a second mile." It worked. They ran, exhausted. They slept, and we had some peace—for a while.

After the second-day of speed paddling in northern Minnesota, the younger fellows were ready to go home. The rapid-paddling trip that followed the Voyageurs was now no fun for them. I called the guide assigned us aside and said, "You take two scouts with you and go as fast as you want. We'll follow at our pace. Take the evening's meal. And when we arrive at the (predetermined) camp site, have supper ready." The boys now enjoyed leisurely canoeing, taking their time to portage and stopping to study the forests of aspen (*Populus tremuloides*), white

New Mexico high-country Philmont Scout Ranch Expedition, 1965. Front row (l-r) Mike Haas, Steve Walker; Middle row (l-r) Mike Campbell, Steve Halls, John Stransky, author, Philmont staff ranger, Richard Kline, Charles Stockwell, Dan Williamson; Back row (l-r) Mark Garrison, David Teal, Don Angle, Tom Teal, Edsel West, Dyck Rich, Tom Taylor, Mark Wilson.

pine (*Pinus strobus*), red pine (*Pinus resinosa*), and birch (*Betula* spp.). The guide, the wise guy, got lost; he couldn't read the map. From that time forward we all got along well. Only upon returning to base did I learn that the guides competed for the title of who could cover the greatest distance over the summer.

We paddled down a stream, the center of which divided Canada from the United States, and entered into Basswood Lake. Soon we heard the call of a bullhorn from a small white building on the northern shore from which unfurled the flag displaying the sugar maple leaf. We were called to customs to declare the value of that which we would leave on Canuck soil. It wasn't much; only food.

Further along, a couple of shots whizzed over our heads. We stopped dead in the water. Checking the map, we found ourselves approaching too closely to a Canadian

Scout Canoe Trek on the Boundary Waters of Minnesota and Ontario, *ca* 1967. Front row (l-r) Wake Foster, Randy Hudgens, James Smith, John Sowell, Charles Olds; Middle row (l-r) Charles Mooney, Jeffrey Anderson, Barry Horton, Jeff Burch (forestry student and bus driver), William Rogers, Phillip Childs; Top row (l-r) Stephen Walker, James Olds, Bill Neathery, Louis Walker, author, Guide "Dave."

Indian Reservation. The natives didn't like our intrusion into the waters of Lac La Croix.

We learned how fast a storm can come up in those lakes. Leaving the shore of Loon Lake, we felt no wind. Midway to the far side, we "dropped anchor" to don life jackets. Arriving at the far side all wet, we were soon put to sleep in our lakside camp by the eerie sound of the loon and the splashing of the waves on the shoreline boulders.

As a longtime vice president of the East Texas Area Council, I saw various projects come my way, usually dealing with relationships between the movement and its local sponsors. When the forestry students who remained

Dr. David Hyink (l), Weyerhaeuser Company research executive, SFA graduate, and leader of the movement to preserve the BSA Boundary Canoe Waters. The author is at right.

active in the School's explorer post learned that a new award, the Distinguished Eagle Scout, was available, unknown to me they pursued it on my behalf. I had not heard of the honor that required the recipient to have earned an Eagle badge in his youth and to have made a "significant contribution" over the subsequent minimal 25 years. I was flattered. Senator Lloyd Bentsen, industrialist Ross Perot, Governor Bill Clements, and a few other Texans had preceded me in receiving the honor. The students also did the legwork for which I was the William T. Hornady Gold Medalist for 1986 "for distinguished service to forestry." Hornaday, a man of wealth and an early-day lay wildlife enthusiast, wrote *The Twenty Year War for Wild Life* (*sic*). (Latter-day wildlife biologists recognize wild life as that which goes on in some places on a Saturday night.) Hornaday willed his estate to the BSA to make possible conservation awards for troops, scouts, and the single annual presentation to an adult. Ben Love, chief scout executive and fellow member of the Association of Presbyterian Scouters, quoted Forest Service Chief Max Peterson in the presentation of the medallion (here greatly abbreviated): "His accomplishment, attitude,...zest for life epitomize Scouting ideals...role model for...young people." Thank You, Max.

George Fry, Penn State forester of the class of '33 and retired from a career in the National Park Service, was the fourth scout, in 1929, to receive the original Hornaday Gold Honor Badge for Wild Life Protection. The first recipient was an Indiana lad, in 1920. The BSA took over the program more recently.

The East Texas Area Council presented me the Silver Beaver in 1974, and the regional organization the Silver Antelope in 1984.

The Boy Scouts never fail to say "Thanks." That is a proper purpose for all the awards the organization passes out to its members and friends. For my induction into the Hall of Honor, someone even got Senator Phil Gramm, Congressman Charles Wilson, State Representative Jerry Johnson, and Mayor Judy McDonald to write nice letters. For the occasion in 1993, Charlie sent a flag that had flown over the nation's capitol. Judy presented a Resolution of the City Council and Jerry brought another from the State legislature. Jim Hull called me a patriarch. Yea, Jimmy, old bullfrogs are called patriarchs of the swamp.

The American Scientific Affiliation

ASA is an organization of scientists who are also Christians. During the period of my active participation, only one other "working" natural resource professional was a member. Oh, there were a lot of folks who knew all about natural resources, but they had never been required to meet a payroll from the utilization of natural resources or to regenerate a forest. I published two lengthy articles in the organization's prestigious journal, lectured three times at national meetings; but I think its members considered me an exploiter. Members did elect me a fellow.

For an annual meeting of ASA at Hope College (where Gordon Van Wylen, whom you've met earlier, was president) in Holland, Michigan, I organized and chaired a symposium on "Stewardship Dilemmas in Natural Resource Management." Following this experience, William Johnson, then dean of the School of Natural Resources at the University of Michigan, wrote, "It is rewarding to hear you speak on the frustrations of stewardship."

Development Groups

Leaders of eight counties that touch on the two big lakes dammed in the mid-1960s created the Deep East Texas

Arthur (Buddy) Temple, III, receiving the President's Award on behalf of the Deep East Texas Development Association in 1975. Temple was a state representative, railroad commissioner, and lumberman.

Development Association. Meetings rotated monthly among the counties, the host county required to feed the group. It was great fellowship. By the time of my entry into DETDA, its purpose seemed confused. Toledo Bend and Sam Rayburn (earlier called Magee and McGee Bend) reservoirs, now filled with water, needed no lobbying support. Was DETDA any longer useful? During my two terms (1974-75) as president, I endeavored to focus DETDA on all the resources surrounding us: trees, water, soil, recreation, fisheries, and wildlife. Our son, Stephen, designed a wood silhouette plaque which illustrated these interests. The folks at Center Plywood used computer-driven saws to cut the plaques which were presented to each member. We brought in people from Austin, and more locally, to speak on natural resource issues, and each county reported on its accomplishments. On one of these occasions I presented State Attorney General John Hill a plaque, upon which the Daily *Sentinel* reported that I "echoed General Hill's theme

The author presents to Congressman and Mrs. Wright Patman a DETDA award in 1974.

that we must have integrity, honesty, and truthfulness...." Turning from economic development to natural resources care, the next half-dozen presidents were foresters, the purpose of DETDA's existence becoming more focused. Charlotte Griffith, the faithful executive, kept things organized in the San Augustine office. D.N. Beasley, President Johnson's sidekick and manager of the Deep East Texas Electric Cooperative, and Ed Clark, LBJ's Ambassador to Australia and UT Board member, were strong supporters.

When Charles Wilson served in the State House as representative and senator, federal funds became available for establishing the Deep East Texas Economic Development District. My willingness to agree to Charlie's request to accept a slot on the Board related to my interest in furthering forestry in the region. More intensive forest management meant more forestry employment, which, in turn, meant more forestry students, more forestry faculty, and a more-favored forestry school. Senator Wilson was elected president of the new organization. I served on its executive committee when Neal Pickett, brother-in-law of Senator Ralph Yarborough and former mayor of Houston, was hired as executive director of what became the Deep East Texas Council of Governments. In time federal mandates required that only elected officials could serve on the Board. As the new name suggests, utilization of resources no longer was paramount: more socially oriented issues took over.

In this development task, First Chiefs Tic Ca Itche Bronson Cooper Sylestine and Kina Robert Fulton Battise of the Alabama-Coushatta Indian Tribes residing on reservation land east of Livingston, Texas, became good friends. One of them gave me a beaded bolo tie, fashioned in the design of the logo of the US bicentennial. Both men and their families were pillars of the Indian Village Presbyterian Church and worked with Boy Scout troops on the reservation. Chief Kina enjoyed talking to civic and school groups, "never understanding why those teachers tell the children that Columbus discovered America. Did you ever hear of such?" These leaders, as their kinsmen, worked in the woods as loggers, many for W.T. Carter and Bros., and in the nearby lumbermills until a government grant enabled them to attract tourists to the tribal lands in the Big Thicket. Western US Indians were hired to teach native dances, pottery-making, basket-weaving, beadwork, and cooking, the latter for a new restaurant. I always felt they weren't happy being on exhibit for the curious. The money would have been better spent erecting a modern sawmill. At that time, the Alabama-Coushatta (I think some Cherokees escaped from the Trail of Tears to join them) was the only Native-American reservation in Texas. Unlike those in other states, this tract (originally but 1280 acres) is state (not federal) land. Its acquisition for the natives is attributed to the influence of Sam Houston in appreciation for Indian help as guides in the war against Mexico for independence.

When federal funds became available for Indians to attend forestry schools, all expenses paid, I tried recruiting on the reservation. The effort bore no fruit. Indians, like African-Americans, think of forestry as the drudgery work in the woods their forebears had to do. Why would anyone volunteer?

I had worked with Charles Wilson, US Congressman since the late 1960s, on a variety of subjects of local forestry and social interest when the postman dropped off a surprise in 1992. In the box was a flag, a certificate signed by the Capitol architect authenticating its unfurling over the US Capitol, and a letter signed by "Good-time Charlie," as his opponents like to taunt him. Flags from this congressman are not uncommon; however, the letter accompanying the flag is too rich to pass over for this account:

> There are certain absolutes in nature, one being that if the river changes course there has to be a resulting change in the banks.
>
> Larry, in your life's ministry, you have inspired innumerable individuals to change the course of their river of life, causing their banks to house a richer, fuller, more productive stream.
>
> You have shared your love of these forests with whomsoever called the trees their friends. You have shared your unique communal with nature with your students, and others, causing all of our rivers to gravitate toward the sea of "Peace, Be Still."
>
> You have shared your finest gift with us, that of yourself.
>
> Larry, please accept the enclosed flag of our Country that has been flown over our Nation's Capitol in your honor. The caption on the certificate is certainly descriptive of the **MAN - LAURENCE C. WALKER - AN INSPIRING EDUCATOR AND NOBLE INDIVIDUAL.**
>
> As your river of life continues its journey toward the high water mark of fulfillment, may joy, peace and contentment be your constant companion.

While the congressman is a well-educated engineer

from the US Naval Academy, I never thought of this "ole East Texas boy" as a man of poetic expression. A little sleuthing unearthed the writer, long-time friend Judge Peyton Walters of Polk County, now the representative's administrative assistant. I wrote the congressman (in whose district I don't reside) that we could "now retire our boyhood family 48-star banner displayed on special occasions, and which is a bit frayed." My father acquired that flag at the conclusion of the First World War; it is now a part of the Fredonia State Bank's historic flag collection.

A call from A.L. Mangham told of being nominated for a place on the Nacogdoches County Chamber of Commerce in my first year in Nacogdoches. Then the Chamber executive mentioned a problem: "You aren't a member." The outfit got my dues. In those days the organization's thrust was community improvement, not so much on enhancing local business. Yet I saw the opportunity to utilize forest products. But first trees must be grown, and foresters would grow them.

TU's Environmental Steering Committee

The southern forestry deans had met and toured, as the guests of the regional forester, the hills of eastern Kentucky and Tennessee. Along the way, on this 1970 trip, we saw a lot of strip-mining for coal and reclamation of the sites from which seams of the black mineral substance had been extracted. Within a few days of returning home, I noted a column in the newspaper that Texas Power and Light Company was beginning to surface mine lignite on lands surrounding the Company's new Big Brown Steam Electric Station, near Fairfield, in the post oak region of Texas. The two events melded, observations in the Southern Appalachians and the clipping about Big Brown. So I addressed a letter to Louis Austin, president of TP&L, suggesting the willingness of the School of Forestry to assist the company in restoring the mine lands following coal removal.

That letter, I learned later, was passed to John Tilton and Richard White who apparently were obliged to ask me to join the company's newly created Environmental Steering Committee. Both White and Tilton were former zoologists with the Wildlife Division of the Texas Parks and

TU Services Environmental Steering Committee: standing (l-r) Jim Teer, Kirk Strawn, Moses Attrep, Thomas Hellier, Charles Groat, Lloyd Hossner, Richard White, Seated (l-r) Clark Hubbs, the author, John Tilton, Milton Weller. Scott Beasley joined the team later (TU Services photo).

Wildlife Department. How best to manage the fisheries of the warm-water power company lakes was their initial concern; they soon enlisted specialists in other fields to assist in the Committee's tasks. Meeting three times a year, the group's primary responsibility is to review proposals for funding graduate students and their research on environmental issues. Between 1971 and 1993, the Committee had allocated $3 million for 75 theses and dissertations relating to conserving natural resources. Subjects concerned the (a) restoration of surface-mined sites to pasture and forest, (b) fisheries and related ichthyology of the warm-water lakes at the outflow of electric plants, (c) ornithological and mammalian baseline studies, (d) mineralogy and soil genesis of surface-mined lands, and (e) the establishment of tree plantations and their management. These fellowships are perhaps the most lucrative in Texas institutions.

Members of the Committee, with their wives, have evolved into warm friendships over about 25 years of working together. In addition to Richard White, the coordinator, and John Tilton, now retired from the company, others and their areas of specialization are:

Moses Attrep (air and water chemistry) East Texas State

University and Los Alamos National Laboratory; Scott Beasley (forestry and watershed management) Stephen F. Austin State University; Charles Groat (mineralogy) University of Texas and Louisiana State University; Thomas Hellier (fisheries) University of Texas-Arlington; Lloyd Hosner (soils) Texas A&M University; Clark Hubbs (fisheries) University of Texas-Austin; Kirk Strawn (fisheries) Texas A&M University; James Teer (wildlife management) Welder Wildlife Foundation and Texas A&M University; and Milton Weller (wetlands) Texas A&M University. I am now *emeritus* status, Dean Scott Beasley succeeding me as the group's forester. Dick White, a TU Services vice president, continues as the Committee's coordinator, ably assisted by Paul Zweiacker.

The Committee, or TU Services because of the work of the Committee, has received numerous awards for its commitment to environmental care. Among the grantors of these honors are the Presbyterian Church, for corporate leadership in creating an environmental advisory board; the US Department of the Interior, for corporate environmental commitment; the Texas Mining and Reclamation Association; the Wildlife Society, for group achievement for national leadership in environmental protection; USDI's Office of Surface Mining, for sound environmental practices in land reclamation; the Texas Water Commission, for resource conservation; the Texas Water Conservation Association, for promoting multi-purpose use; and the Texas Railroad Commission, for land reclamation.

Research by the highest-quality graduate students has been exemplary. An example I came across while visiting a reclamation project near Vancouver, Washington, may suffice. Pat Angel, a forestry graduate student at SFA and among the first recipients of a TU fellowship, discovered that by mixing topsoil with lower strata material, a more productive media for revegetation results. Previously regulators required the horizons to be kept separate and replaced in the fashion excavated. To be able to mix the strata reduced the cost of a ton of lignite by 55 percent while improving growth of grasses and trees planted on the reclaimed lands. All over the country the results of this research reduced the cost of reclamation.

Perry Brittain, chief executive officer and chairman of the board of Texas Utilities, received SFA's Distinguished Alumnus Award in 1986. The recipient chooses the presenter, and Perry chose me. The privilege honored me, for Perry and Martha Nell, growing up in the woods of nearby Shelby County, had become good friends because of my service on TU's environmental steering committee. I noted that as a freshman in 1942 Perry had been kicked out of the boy's dorm on campus: the building became a barracks for the Women's Army Corps of the Second World War.

And Some Others

For about 25 years I served on the Corporation of Inter-Varsity Christian Fellowship, an organization previously mentioned. The 50-year chronicle of the organization, *For Christ and the University*, reports that some Texans formed a local committee to raise funds to underwrite the work in Texas. We tried to pick up the work so notably begun by Glen Zumwalt, by this time the Distinguished Professor of Aeronautical Engineering at Wichita State University. When in 1979 we invited Dr. John Alexander to visit SFA and to lecture to the faculty we had raised $180,000 from people interested in campus work. John, IVCF president, was internationally recognized as a geographer and professor at the University of Wisconsin-Madison when challenged to carry on this task.

I served as a member of the Research Advisory Committee for the Coordinating Board, Texas College and University System; and organizing chairman of the Piney Woods Conservation and Development District, a unit of the Soil Conservation Service. RC&D programs, established by a 1962 Act of Congress, provide resource and environmental protection in multi-county areas. Imported fire ant control, multifloral rose control, and lakeside erosion control were major matters of concern in the 1960s and 1970s.

Local tasks included the presidency of the County United Way. Drumming up enthusiasm for UW was difficult. Whether Little League should be included in the allocations and how much for each entity became major problems for board solution. I think it was this work that led Mayor A.L. Mangham and two former mayors to ask me to run for the City Council. No one on the faculty had ever served. They promised to keep any opponents at bay; but at the closing hour for filing, the brother-in-law of one of the ex-mayors' pitched his hat into the ring. He won by 89 votes. I say if 45 of the people who voted for Mike had voted for me, I'd have won; so I only lost by 45 votes out of 1700. That doesn't sound so bad. Then the sheriff's wife says to Anne, "I would have so liked to have voted for Laurence, but I've known Mike's mother all my life."

Membership in Rotary International enabled me to know local business and professional folks as well as other faculty members. Honored as a Paul Harris Fellow and Rotarian of the Year when historian Archie McDonald served as local club president, the presentations recalled

Erosion control on Sam Rayburn Lake, a major concern of the Piney Woods and Southeast Texas Resource Conservation and Development (RC&D) organizations, managed by the Soil Conservation Service.

service with RI's international president, Herbert Taylor (son of a lumberman), who also presided over IVCF.

Over the years I served on the Texas Forestry Association's Board of Directors and many committees, working closely with Ed Wagoner and with Ed's successor, Ron Hufford. Doris Goodrich Jones, daughter of "The father of Texas forestry," made me a Life Member of the Forestry Museum, sponsored by TFA, and overseen by Carol Riggs.

I wrote TFA, upon receiving the Association's research award in 1992 for the writing of *The Southern Forest: A Chronicle* (University of Texas Press):

> What a pleasant surprise.....! To Nancy and "Joe" Pase, whom I gather spirited the event, I am especially grateful. Others no doubt participated. May they know my gratitude. I appreciate, too, colleagues, past and present, at the laboratory bench and in the forest, on whose scientific endeavors I merely enlarged.
>
> Studying the southern forest for its aesthetic pleasure, its ecological fascination, and its beneficial utilization brings joy....

23

EMERITUS

Grand Ditch that carries water from the west to the east side of the Front Range in Colorado (© S.L. Walker).

I retired in February, 1988, continuing to teach part-time on the campus and at the state penitentiaries near Palestine. As I had been teaching five different courses each semester, totaling 15 semester-credit-hours, for several years, preparation didn't take much time. Retiring, I soon learned, was a mistake. I enjoyed the work, the students, and my faculty colleagues. And something was wrong with the system; it cost me 50 cents a day to continue to work. Retirement pension and social security together exceeded by that amount salary minus federal and state payments, respectively, for these purposes.

Retirement had been the second-most important decision of my life. When said that way, one fellow responded, "But you won't have to live with the second-most as long as with the first-most." I promised Anne that I married for better or for worse, but not for lunch. She sighed with relief.

To Higher Climbs

A call came from the University of Colorado to inquire about teaching forest ecology at the Mountain Research Station north of Nederland. At last I would get to work in the West. Several of our SFA students (Alma, the grandmother, and Kumar, the Indian, among them) enrolled in the summer course. Browsing through the station's files, I learned Dr. Harold Lutz, my thesis advisor at Yale, had preceded me in teaching this course.

We began our studies in oak (*Quercus* spp.) woodlands along the creeks, continued in the pinyon-juniper (*Pinus cembroides* and *Juniperus* spp.) stands at the lower elevations, then moved upward through the ponderosa pine (*Pinus ponderosa*) type, and further up the slope into the Douglas-fir (*Pseudotsuga menziesii*) forest. (When I was a student, this "false hemlock" was called *P. taxifolia*.) As we progressed toward the summit of the Front Range, students—from all across the country—dug holes to study soil profiles, and catalogued associated vegetation in stands of the spruce-fir (*Picea engelmannii* and *Abies lasiocarpa*) type. In the krummholz they found that trees 200 years old do not exceed 5 inches in diameter, banner trees occur, looking like flags carried by guidon-bearers at the head of the squad in the line of march. Other trees grow prostrate, the "crowns" of these wind-blown elfinwood of ancient spruce spreading out a hundred feet or more as they hug the ground. Seldom do heights exceed five feet. En route to the summit of Niwot Ridge, at about 11,000 feet, the students passed through groves of spruce as (a) upright spire trees in dense forests, (b) smaller, but still spire-like, trees as one nears the timberline, (c) banner trees standing out among stems in a grove, (d) smaller banner trees in

University of Colorado forest ecology class passing through krummholz to Niwot Ridge.

conifer islands, (e) ribbon and layering forests, (f) small islands of prostrate trees and of solifluction forms. Finally, the class arrived in the treeless tundra and alpine desert in the Indian Peaks Wilderness of the Roosevelt National Forest.

On these Rocky Mountain biome studies, I introduced students to bristlecone pine (*P. aristata*), blue spruce (*Picea pungens*), and, of course, the extensive stands of fire-climax serotinous lodgepole pine (*P. contorta*). Willows (*Salix* spp.) and cottonwoods (*Populus* spp.) of the streamsides also held our interest.

The University of Colorado forest ecology course considered the abrupt transition of the vegetation from a north- to south-facing slope in the Rocky Mountains National Park.

The Rocky Mountain National Park, once part of a national forest and just north of the research station, provided stands for wildfire successional studies, for observing damage done to aspen *(Populus tremuloides)* by elk, and for chasing the little pika, a non-hibernating herbivore, among the rocks at the highest elevations. My Penn State college classmate, Robert Krear, came out of retirement at Estes Park each year to guide us through the Park, for his doctoral dissertation at the research station had dealt with the little critter we watched with fascination.

I learned, even as I taught the students, something of the ecology of the east face of the Front Range of the Rocky Mountains. Joyce Gelhorn, who carried out doctoral research in the area, aided us. She had initiated successional studies following the 1978 Ouzel Lake fire, high along Bluebird Lake Trail, in the southern zone of the Park. Joyce tallied the initial flowers and fireweeds until the appearance of liverworts, then noted the arrival of aspen seedlings and suckers. Sexually produced seedlings displayed foliage with reddish tint; asexual suckers, as expected, grew faster and more bushier. Reduced stand density in the fire area encouraged the blasting of tree

The 1988 University of Colorado class at the site of the 1978 Ouzel Lake fire (Alma photo).

trunks by wind-blown sand and ice. (Park foresters had allowed the Ouzel to burn because the area had been designated "low risk.")

Students quickly noted the disappearance of aspen, especially in the Jack Gulch area, with fire suppression practices. To maintain these stands to provide food for elk requires harvests of encroaching conifers. We observed ouzels, thrush-like birds that build nests in rock "caves" and feed in water falls (and which someone called John Muir's "totem animal"), and Albert squirrels that change color with the seasons in their ponderosa pine habitat.

I insisted for these students of the Rockies, as I do for all forests, that the first question to ask when anticipating a silvicultural treatment is, "How will the practice affect the production of clear (free of sediment), clean (free of dangerous microbes) water?" So we went to Grand Ditch, the canal that carries water from the moist west side of the summit, eastward over the Continental Divide at 9500 feet, to provide water for the xeric municipalities and their industries in the rain shadow of the mountains along the lower slope of the Front Range. Grand Ditch leads water away from the long, winding Colorado River, the headwaters of which seep from the soil in the Never Summer Mountains of the Rocky Mountain Park. The stream's flow runs to La Poudre Pass for release to the east side. Use of this canal preceded construction of the Alva B. Adams and Moffatt tunnels, engineering feats dug for the same purpose. Water from the summit also cascades down the Cache la Poudre River which, the students quickly learned, encourages white-water rubber rafting.

The Station's environs also provided opportunities to see spruce budworm and mountain pine beetle damage. Entrepreneurs use the logs harvested from these infested forests for prefabricated log homes, even in Texas and the eastern states. Erection of nice homes and tolerating dilapidated shacks in tinder-dry woodlands, where only recently wildfires made national news, allowed for dynamic discussion. So, too, did dendrochronology, a science initiated by Leonardo daVinci to understand past climate. Living trees of the dry, cold Rockies lend themselves to charting dates by this technique. Woods from long dead trees also are dated, the earliest tally in the

US Southwest, about 300 AD, found in charcoal. The carbon mineral doesn't rot.

At the end of my second summer there, Anne noted I was huffing and puffing. I ignored the warning, for an internist with impeccable credentials (Harvard Medical School and participant in creating the US Forest Service step-test for fire-fighters) had, just before my leaving for Colorado, examined me. After the stress test, his parting comment was, "One thing you'll never have to worry about is heart trouble." But upon returning home and two days after turning 65 (when Medicare insurance kicks in), while mowing the grass, I felt the attack of the big grizzly squeezing the air out of my chest with its giant forelegs. Prompt open-heart surgery enabled me to return to my post at an elevation of 9500 feet the next summer.

Oh, I loved those mountains, so much so that on our last full day, a week before the heart attack, and the students having left for home, I strapped on my pack, camera, binoculars, and elk-call and told Anne, "I've got to have another last look at this grand view." Katherine Bates, a hundred years earlier in her national prayer hymn, saw the scene as "the purple mountains' majesty." I got to tree line by noon. Immediately upon walking into the krummholz of elfinwood, a great wind suddenly swept down from the summit of Niwot Ridge. No way could I proceed against its steady forces I judged to have exceeded 60 miles per hour. I retreated to the cover of the spruce and fir trees lower on the slope to wait it out, as one does a sudden rainstorm. It was quiet in the evergreens. Twice more I tried to press against the wind, but without success. Nothing else to do but to return to the cabin at the station. The next day, en route home, I was the visiting preacher at Loveland's First Presbyterian Church. Before the week was over, I had three bypasses. If I had had the heart attack on Niwot, the buzzards would have feasted on my tasty eyeballs and the black bears on my innards. The "wind" cheated them out of the meal. (Wasn't the Holy Spirit's appearance at Pentecost described as being like the blowing "sound...of a violent wind?")

The Faculty's Expression

I was unwilling to apply for *emeritus* status. As the

Latin word means "with honor," it seems inappropriate for one to request the respect that accompanies the title. But after three years of simply signing *ret.* on my letters, the faculty made the application and the president conferred the title. Bless their hearts! They put together a dinner for the occasion, did a bit of teasing, and gave Anne and me a double rocking chair for the porch. Anne wouldn't hear to putting it in the weather. Cushions were made on which visitors to my study sit to rock.

On that occasion in 1990, Nancy King Pase wrote my successor:

> As you probably know, Dr. Walker and the School of Forestry were an active and important part of my life for ten years. Although I left there in 1976 to get married, Dr. Walker and his family have continued to hold a special place in our hearts and lives.
>
> In a similar way, the School of Forestry will always be special to me. The lasting friendships I made among the students, the faculty, and the staff will be a constant reminder of the wonderful times God gave me during the decade I served as Dr. Walker's secretary. Being able to assist him in his success was a privilege for me.
>
> We appreciated the opportunity to participate in the "closing" of an important era in the history of the School of Forestry.

A new day dawned for me when Dr. Scott Beasley arrived as the dean. Scott had been in the senior regional silviculture class in my first year at SFA. He vividly recalled our field trips in 1964 to visit forest types not found in East Texas. Scott promptly provided typing privileges and moved me from a corner in a laboratory to a private office. Several professors had earlier insisted that I utilize their lab space in order to continue to be available to students. These were especially appreciated gestures.

Prison Walls

The men I taught at the three prisons lined up along the Trinity River some 80 miles from Nacogdoches, in turn, taught me much. Going through eight electric gates, locked in a cage between each two *en route* to the classroom, which the inmates called the schoolhouse, can be traumatic. On about my second day at that work I suggested, not facetiously, that if every high school freshman had to spend a half-day incarcerated in this prison, we could empty it in the next generation. "No," the warden replied, "You don't understand. The people we have here haven't got it made back home until they've been here. To spend time with us marks them as something special in their gangs." In time, I learned he was right.

I taught courses in conservation of natural resources and regional forestry for the SFA School of Applied Arts and Sciences, under Dean James Standley's direction

Once I was at the Beto II Unit to teach the afternoon course. The students were edgy and antsy. I couldn't hold their attention. The guard at the door had no reason to explain the situation, nor did other guards I queried on my way out through the cage doors.

When in sight of the Coffield Unit, I knew promptly what had occurred. Guards were milling around outside the main gate, as they can't go home if there is an escape during their shift. Nor could I get in; so no class that evening. The inmates at Beto II were better informed than their guards. Smoke Signals?

Escapees try to make it to the Trinity River. That's good. It is so infested with alligators and water moccasins that, when the guards row their boats through sloughs, they find the prisoners waiting their capture as they hang from branches in the trees.

Once I was taken hostage. By the Guards! The shifts had changed, the new sentries not knowing I was still in the classroom. A driver's license was of no value: "How do we know the real Mr. Walker isn't in a closet with a knot on his head. And everyone in here has a driver's license." A few telephone calls got me sprung. Sometimes guards opened my car trunk and even looked under the hood before letting me leave after the night class which concluded at about 9:30.

Yet, on my first day assigned to Beto II, I mistakenly entered Beto I, the pen for the worst offenders. I carried my box of teaching materials through all the gates, asking guards at each the direction to the classroom. Not until the guard at the schoolhouse door asked rather nonchalantly, "Who are you?" did it become apparent I was in the wrong pen. Directed to the proper prison a couple of miles down river, I walked out the same way. No one ever questioned me or looked into the box!

Wood for Energy

Agricultural engineers gathered in 1980 to discuss the use of biomass as an energy alternative for farmers. Following dialog on burning waste, like corn cobs, for fuel, the scientists moved to alcohol production. The press picked up the story as follows:

> Dr. Lawrence (sic) C. Walker, School of Forestry at Stephen F. Austin State University, said the United States would start using wood more as this country did in the past.
>
> He said until 1875 wood was the principal source of energy in the U.S., and it trailed only coal until about 1920 and was more important than water until 1945.
>
> •••
>
> He said until the 1930s all wood alcohol in the United States was produced from wood, but with the cheaper source from oil, this dropped. It is becoming more practical to start producing wood alcohol from wood again he added.
>
> •••

Trees loaded with gas? Strike a match to a freshly made gash in a post oak bole and watch the methane ignite.

The Elusive Franklinia

The Rotarian published an article entitled "The Great Green Hope" in 1979 that called attention to endangered and extinct species of flowering plants among the flora of the world. As the story seemed to blame timber extraction for the malady, I countered with a letter to the editor saying that

> ...there is no record of a case for which the blame can be attributed to forest management. This includes clearcutting as a silvicultural method essential for regenerating certain

tree species which require mineral seedbeds for seed germination and full sunlight for the growth of young trees. Land use and abuse for agriculture and for industrial and residential living space are another thing.

Hoping to entice the editor into further dialog on the subject (though unsuccessful), I continued:

> The illustrations accompanying (the) article show *Franklinia alatamaha* among the endangered species. This is a fascinating plant. It was first discovered by the father-and-son team of John and William Bartram, pharmaceutical explorers for the British Crown, before the American Revolution.
>
> Ten or twelve years later, William again found the small flowering tree, named for the Bartram family friend Benjamin Franklin. It is believed the geographic distribution of the tree, near the Altamaha (formerly Alatamaha) River in coastal Georgia, was an ellipsoid of only some twenty miles in its longest dimension.
>
> Moses Marshall next saw the tree in 1790 and John Lyons in 1803. Lost now until a geologist, Sir Charles Lyell, roaming the same ground and, seemingly intent on verifying the Bartram discovery, found the broadleaf evergreen with camellia-like flowers. The year was 1849.
>
> Lost again till the 1950's when the late G. Norman Bishop, dendrology professor at the University of Georgia, rediscovered *F. alatamaha* still growing vigorously. Never staked nor publicized lest it be "loved to death," it must now by still another person be rediscovered.
>
> In the meantime, the colorful summer-flowering tree, probably never more than 15 feet tall, is cultivated and, thus, far from extinct. Arboreta exhibit it, the seeds for their specimens first collected by the Bartrams and carried to Philadelphia for planting and distribution to horticulturists.
>
> There are only eight plants in the whole of the southern

forests listed by the Smithsonian Institute as extinct in the wild. No other tree falls into this category in spite of the logging and lumbering of the region for 150 years.

Elder Hostels

David Petty of the SFA Sociology Department lined me up to teach forest ecology at the ElderHostel programs he supervised on campus. Later others would enlist me to take these older people to the woods at the Piney Woods Conservation Center near Lake Sam Rayburn. Baylor University's gerontology program, headed by Ben Dickerson, asked me to help on ElderHostels in South Texas and in West Texas' Pecos Mountains. Some of these seniors were "Elderhostile" about cutting trees, so it was especially rewarding to receive notes like this: "I am a member of 'Save Our Forest' group, but some of my thoughts about proper care have changed by your excellent lectures." At these "lectures," we dug holes in the soil, measured humus layers, laid out seedling roots, and counted growth rings. I treasure a bolo tie displaying a hand-painted live oak on a ceramic base one ElderHosteller insisted I have. Letters still come as reminders of these events. Here are some lines from a Stephenville couple, Joe and Elain Frey:

> We will never forget our unusual week in East Texas driving down the fire lanes through the pines to sit on a stump and listen to you explain its past, present, and future to us.

These folks came from all over, even from abroad. Most were professional people and their spouses. Some were affluent. They were always a fun bunch to spend time with.

Still Writing

The *Texas Business Review* in 1979 requested an update since my earlier article in 1976, on the forestry situation in the Lone Star State. I canvassed industry leaders for their perception of the economic effect of the emergency ban on the herbicide 2,4,5-T, regulation of best management practices, clearcutting restrictions, productivity taxes, preservation of forests under the federal wilderness RARE II and Big Thicket legislation, and growth/drain (harvest) relationships.

Texas' largest live oak (*Quercus virginiana*) near Concan, measured during an ElderHostel class.

Consistent responses recommended that research, not emotion, should precede the ban on 2,4,5-T; that companies promote voluntary Best Management Practices for themselves and the small private landowner; that silvicultural systems other than clearcutting rate consideration; that *ad valorem* taxes should be based on site productivity, not highest market value; that a healthy growth/drain ratio depends on favorable federal capital-gains taxation, while a ratio of 1 must be sustained; and that the new Big Thicket National Preserve should satisfy preservation requirements, as further wilderness areas on the national forests in the state tie up enormous amounts of good timberland, taking them out of production. The lack of regeneration on the lands of individuals, they emphasized, calls for "a great deal of attention to reverse the shrinking of the southern forest."

An article I wrote for *Texas Parks and Wildlife* magazine evoked this response from O. C. Walker, which turned out to be a pen-name for Sara Dinyari, writer of a nature column for a Deep East Texas weekly newspaper: The Caddo Lake narrative was "a prose poem tribute to Harrison Bayou." Sara personified her claim that friends grow out of correspondence about an author's story. It is so.

Reviewing articles for editors of the *Journal of Forest History* and membership in the Forest History Society, its publisher, introduced me to interesting people. Lynn Day was one of these. Of the Weyerhaeuser clan, her parents' forebears were numbered with the branch of the family that stopped at Potlatch, Idaho, while others trekked on to Tacoma *via* James Hill's rails. Lynn visited Nacogdoches when Robert Maxwell of the History Department also served on our faculty as professor of forest history. By that time, Bob had been appointed to the Board of the Society. Lynn and I chatted about her cousin, C. Davis Weyerhaeuser, who served with me on the corporate board of Inter-Varsity Christain Fellowship. Of my authored *Axes, Oxen, and Men*, Lynn wrote, it "introduces us to a part of Texas foreign to our experience." And now the Potlatch Company is the second-largest landowner in Arkansas. The stories in the book that chronicles the early days of The Southern Pine Lumber Company would not be foreign to the razorbacks who "razed" those forests.

Anonymously reviewing manuscripts is a thankless task, and unappreciated by authors. I know the feeling when sarcasm reeks from the typed page like a putrid odor. Sometimes I'd try to help, as when an editor sent an article by an anonymous author. Here's how I answered:

> To be critical without suggestion is of no help to anyone. On the sloppy green comment copy are my notes...to suggest the need to "brighten" the account. A revision of the opening paragraph, thus, might read:
>
> Lone Star Pine: The Story of the Texas Thompsons
>
> The Thompsons were native Texans, not capitalists from beyond the bounds of the Lone Star State. Thus, they differed from most families participating in the early days of lumbering East Texas' piney woods. Among these families, few are as prominent and none is as venerable as the House of Thompson. Benjamin Franklin Thompson, the

> first of the clan in Texas, with his sons established a commercial sawmill in the mid-1800s. It was the second to produce boards in Texas, only 16 years after statehood. Nearly a century later—1942-1944—his grandson, Hoxie H. Thompson, assumed the presidency of the Texas Forestry Association. This story chronicles the events between those dates.

Just an idea! Regretfully, I never saw the article in print.

Ira Bray of the US Forest Service's Southern Region and I set up Writing Skills Workshops for Foresters in 1982 and later years in various southern locations.

SAF's Don Theoe lined me up for a boring and thankless task. The federal educationists wanted a "Taxonomy of Natural Resources for the National Center for Educational Statistics." Educational Management Services, Inc., brought the team together to put a number, title, and definition on everything involving natural resources. I've never seen any evidence the effort has been utilized.

Another Opinion

Glenn Chancellor corralled a committee on silvicultural practices in 1978. The Temple-Eastex Co. (as it was then called) executive wanted some opinions relating to the Texas scene. Committee members seemed reluctant to respond, so I chimed in:

> Thinning is not an easy solution to a complex economic problem; but it is an easy solution to the problem of how to increase Texas wood fiber production. Responsible thinning practices are essentially unknown for the present because "Who will do it for the available harvest dollar?" is a logical question.
>
> For it to be done, industry will have to defray part of the cost. Or, perhaps, industry could persuade Congress to include this as a practice of the federal Forestry Incentives Program.
>
> I'm surprised the forestry profession didn't get clobbered on this matter at the Tyler (clearcutting) trial (perhaps it did, but I haven't heard). For I believe that if we were

carrying out appropriate thinnings, there would be no need for clearcutting until the next century has begun. In the interim, "Third Forest" planting would be carried out on burned sites, denuded sites, severely understocked sites (after site preparation), and in the post oak belt.

An incentive payment to the logger (not the land-owner) would be in order. The landowner will benefit from the improved growth and vigor of the residual stands. And, too, southern pine beetle outbreaks (The bugs will, like General MacArthur, return!) should be greatly reduced by such silvicultural operations.

Then, too, to increase productivity, we need to convert much of the tension zone (the Post Oak Belt) to loblolly pine, using 24-inch tube-grown seedlings planted in holes made with post-hole drills mounted on tractors with a power take-off. Two hundred and fifty per acre could suffice, as survival would be high. In the meantime, managing the zone more intensively for post oak production would, I believe, be appropriate--if a wood shortage is really a threat. I like to think it is; because I like to think a lot of Texas wood will go into the manufacture of methane gas (for electric generation) and methanol alcohol (for autos) before another decade goes by.

24

"The Higher Good"

Flag-form subalpine fir near the summit of Colorado's Niwot Ridge (© S.L. Walker).

We come now to the meaning of the title of this book. The aged Dr. Carl Alwin Schenck was to lecture in 1952 on "Excelsior" at the New York State College of Forestry at Syracuse University. We graduate students were obliged to attend the oration, anticipating a speech on wood products, for the dictionary calls excelsior the wood shavings used for packing valuable china and glassware, the finely curled material forming a resilient mass.

However, the old forester got our attention with a more provocative theme than marketing wood waste or even forest economics. "Excelsior," he shouted at the end of each paragraph as he stretched his right arm to emphatically point heavenward. "Excelsior, the higher good!" The exclamation, in Saxon brogue, filled the air as the *forstmeister* translated the Latin word. ("Excelsior" also appears on the Great Seal of the State of New York.)

Who was this elder statesman of the forestry profession who compelled us to strive for "the higher good." Schenck, the Prussian friend of Gifford Pinchot since the latter's sojourn in Europe, organized the first forestry school in North America in 1898. The Biltmore Forest School, enrolling only men who possessed baccalaureate degrees, irrespective of subject, had the run of the North Carolina estate owned by the industrialist Cornelius Vanderbilt. Woodlands around Asheville and what is now the Pisgah National Forest served as laboratories for "practical instruction."

The German-trained Dr. Schenck operated the School, modeled after German *forstmeisters'* academies (rather like an internship), until a disagreement with the millionaire occurred in 1913. Then for a while the School was mobile, the men accompanying the forstmeister by train, visiting the nation's timberlands, both in the East and West. By ship they sailed to Europe to observe the work of silviculturists.

Never naturalized, the German returned to Europe to serve in the Kaiser's army during the First World War. Following the armistice, he found his way back to the USA. Again, he served his homeland's forces in the Second World War. And again he requested permission to reenter America in the early 1950s. I recall the circulating petition, but I recall not whether the paper bore my signature. Perhaps I was yet unforgiving, the Nazi nation served by the Biltmore School founder taking its toll upon my life. I'm glad now that the old man received sufficient support to enable him to come back across the Atlantic and, up and down the land, to tell the story of a forester's life.

That story spoke not just of teaching mensuration, surveying, silviculture, and timber utilization. It also told of a man of faith who, on the Sabbaths, supplied the pulpits of the little clapboard churches in the coves of the Southern Appalachian Highlands. He also recounted misunderstandings attributed, unknowingly to him, to the single greatest distinction between his Lutheran upbringing and the Calvinism John Knox sent with the Scots and Scot-Irish Presbyterians that peopled those hollows and "templed hills." Luther tossed out only that Roman Catholicism which conflicted with Scripture, thus maintaining many of Rome's traditions. In contrast, Calvin and his student John Knox retained only the Roman doctrines that could be supported by Holy Writ. Of Gifford Pinchot, the practicing Calvinist of French Huguenot ancestry, Schenck wrote, "Of all my friends, Gifford Pinchot was the most religious and the most abstinent." And this he said after they had publicly disagreed over forestry philosophy. The Biltmore School founder also recalled how Pinchot, while the consulting forester for the Vanderbilts, preached to mountain people "in the humble substitutes for churches in Pisgah Forest."

Excelsior was to be our motivation and, indeed, "the higher good" was the unwritten motto of the forester. From the beginning of the profession in America, foresters were glorified for their self-sacrificing efforts. Book after book praised the US Forest Service as the most efficient and dedicated agency in the federal bureaucracy.

In time "a different drummer" set the cadence, especially in the South. Now, foresters realized money could be made growing trees. For a while the same kinds of students who could be successful bankers joined the student ranks and graduated to take their places among the leaders of the wood-using industries and national forest administration. And it was true; with low interest rates, low labor rates, subsidies from the federal government, and higher prices from stumpage (the price of uncut trees in the forest), foresters could make money, for themselves or their employers, growing and marketing trees.

As the role of the forester proceeds to an end in American society, another kind of student, encouraged by the times, arrives on the scene. Now, one can be motivated by altruism and simultaneously recognize that wise use includes sensible management, and sensible management provides for a profit in a free enterprise system. Foresters and the landowners who employ them could reap financial rewards without raping the land. It was an exciting time to be a member of the professional fraternity and to provide instruction for the foresters of tomorrow.

Again, there occurred a sudden and dramatic change in events that would lead to another kind of student matriculating at the nation's forestry schools. These often would be motivated by selfish interests, led not by foresters but by self-serving and powerfully rich lobby groups. So-called conservation organizations flooded mailboxes with propaganda that rejected conservation in favor of biodiversity, ecosystem management, and other "buzzwords" with many possible definitions. Sustained yields of forest products no longer would satisfy; sustaining the ecosystem must prevail, even if folks of the future would live in cardboard boxes and read only from clay tablets.

Harry Wiant, whom you've met in these pages, expressed his concern apocalyptically in an editorial in the *Allegheny Forester*:

> In the latter days an anti-wise-use force will arise and will deceive many. It will reign for one generation. Mills will be closed, prices will rise, and once-productive forests will be filled with dead and dying trees. The sound of the saw and the axe will be heard no more. The woods will be the habitations of agitators, negotiators, and commentators. But this too shall pass. There will be wailing and gnashing of

teeth. The people will cry, "Why are we wasting this renewable resource? We cannot afford homes as our fathers before us, and lowly paper is beyond our means. Landscape ecology is a false god which none can comprehend. It is used by those destroying our means of production." And a new generation of foresters will come forth, once again guided by science-based reason and the knowledge that the stand has always been the basic silvicultural unit and timber the most important product of the forest. It will be understood that man cannot live by bread and shelter alone, but he surely can't live without them. Then the 150-year-old SAF will

(1) be greatly favored by all for having stood firm for wise-use conservation while other organizations were led astray by the sirens of feel-good forestry, or

(2) be assigned to the ash-heap of history for having led the people into a false forestry, full of words but devoid of meaning.

If the former prevails, the motivation to become a forester will have come around full circle. Through the pre-World War II years, young men were challenged by altruism. *American Forests* and the press generally were publishing accounts of timber famine then upon us. The Northeast, the Lake States, and the South had been cut out. Raging fires prevented successful natural regeneration and killed the trees in the few established plantations. And as long crosscut saws were beginning the harvest of the Pacific slopes, foresters expressed concern as to where our children and their offspring would find wood for homes and paper if the exploitation continued. From whence would come utility poles, railroad crossties, and posts for roadside barricades? Those who selected forestry as a profession were cut from the same cloth as those who chose to be men-of-the-cloth or medical doctors. Financial concerns were seldom considered a motivating force in those Great Depression years, nor in the preceding decades that spanned the Gay Nineties through the prosperous Roaring Twenties.

The Five Cs

I like to think of the characteristics describing American foresters as commitment, competence, consideration, continuity, and communications. The men and women I have known in this profession are *committed*, ofttimes fiercely so, to the tasks to provide wood, woodland recreation, water, and wildlife for future Americans. Only ill-informed extremists have questioned their *competence* to wisely manage the nation's timberlands. Yet, even in the face of a constant stream of abuse in the media and courts, these folks have been *considerate* of the desires of others. Perhaps, at times, too considerate, giving in to those with singular and selfish interests in the use of the forest. *Continuity* is important; indeed, some might say a great error in the management of the national forests has been the frequent transfer of personnel and, by that, breaking the continuity of oversight essential for crops that may take upwards of a century to mature. As for *communications*, since my earliest days in the profession, foresters have bemoaned their inability to get their messages to the public. Oh, they've tried, they've tried. But communication requires ears with a willingness to hear and learn as well as tongues with which to tell, and eyes willing to read as well as pens with which to write. The communication problem is not new. Pinchot complained about his fellow professionals in a popular journal for nonforesters, *The Saturday Evening Post*, in 1921:

> Foresters usually explain their profession, when they take the trouble to explain it at all, in such a way that the ordinary man is little the wiser. What they know so well they assume that he understands also, and so are apt to leave him as much in the dark at the end of the talk as he was at the beginning. Although the foresters of the United States know as much about their own work as the average civil engineer does about his, they have somehow failed to give the public a common-sense idea of what forestry is, of what it means to the rank and file, and of what is being done or left undone to put it into practice.

Trees: For Business or Pleasure

I addressed two uses of the woods in a 1967 article for *American Forests*, writing that:

> foresters get caught in the middle! They get stuck in the paint in the recreation picture because, in forestry, the recreation picture is a mosaic of policy rather than a diagram of procedure. To some it's frustrating; to others it's

> exciting to be involved in making, remaking, or recommending policies pertaining to outdoor recreation. The problem of parks is a problem of population. In 25 years, we'll have 65 percent more people. They'll *want* parks, but they'll *need* lumber, plywood, and paper. Parks are luxuries in any society, but wood for homes and paper for business and bibles are necessities. The wants and the needs are for the same people. It's not a case of preservationists versus forest-product manufacturers. The preservationist wants a paneled den and the industrialist likes to visit tall timber for fun and inspiration. Rather, it's a case of only so much land for both business and pleasure for the *same* people. And they, as a Society, must choose.

I concluded the article with a spike into a raw nerve: parks people and others need to recognize that much demand for parks is really for cheap motelling or a place to "slum." Nature education is of no concern, exercise is not desired, and spiritual values are beyond comprehension by those parked in air-conditioned cars, with radios blaring to rock-and-roll, and beer cans pitched into the very canyon the people set apart to preserve.

Foresters need the wisdom of Solomon in evaluating these matters. Lobby groups—in the appreciative sense of the word—like the National Parks Association, American Forests, the Sierra Club, and the Wilderness Society—all work hard at being convincing. It therefore behooves the decision-maker, politician, or parks practitioner to know the facts, and even here I have not given them all, because I simply don't have them all. As some wag said, "I've made up my mind—Don't confuse me with facts!"

Four Oldtimers Report

The reader realizes it is individual foresters and the profession of forestry to which I ascribe the phrase "*Excelsior*: the higher good." For bench marks I turn to several oral histories published in *Texas Forestry*. (Texts and transcriptions of these interviews and accounts of some wood-using industry personalities are filed in the Special Collections at SFA's Steen Library.) Paul Hursey was the first forester employed by industry in Texas, in 1937. Four others in the South worked for companies "looking

Gifford Pinchot, Calvinist and the "father of American forestry," attempted to steer the young profession to the altruistic "higher good" (USFS photo).

after trespass, fighting fires, marking timber, making management plans, and laying out railroad trams." In 1940, Hursey installed "a groundhog salvage mill to cut upland hardwoods in pine stands on the theory that they were taking up too much room for pines to grow. The pines came back in."

Hursey, a graduate of the University of Michigan, spoke of making a bad man good by paying him $10 a month to look after company acreage. That prevented fires, squatting, and timber theft. These "nesters" were too tough for sheriffs to handle.

When asked why the lumbermen were so eager to sell their lands to the federal government, the forester answered, "They didn't think that forests would come back." Hursey had to prove to his industry boss that you don't plant a pine cone, but a seed, to get a new tree. When convinced that 50 to 100 seeds could be obtained from a cone, the baron let his forester proceed to develop a tree nursery.

David (Andy) Anderson, a Penn State forester, joined the US Forest Service to supervise CCC crews, arriving in time to see the harvest of the remnant virgin pine forests near Ratcliff on the Davy Crockett National Forest. Soon he signed on with the Texas Forest Service. "It was a seven-day-a-week job in the fire season. Everybody was setting the woods ablaze." Andy located and set up the Alto nursery, begging an allocation of $50 for the grading table. He got old fabric belt from sawmills for moving the lifted seedlings along the table and bummed a motor from Texas A&M's engineering department to roll the belt. (The Alto nursery site had been a peach orchard.)

Andy, author of the highly respected *All The Trees and Woody Plants of the Bible*, set up 4-H and FFA forestry programs with help he got from the Sears, Roebuck Foundation. He also introduced the first (after Maryland) radio communications for fire control in the South.

During the Second World War, Andy looked for guayule in the West Texas Pecos Mountains: rubber tire manufacturers needed latex from the desert shrub. On another occasion he scoured West Texas for the Navy to estimate the availability of yucca as a substitute for imported hemp in rope manufacture. At that time, too, in 1942, cork oak (*Quercus suber*) seedlings grown from seeds brought from Spain were grown in the Alto nursery and widely distributed. They're seen around the state to this day.

Anderson described the TFS director, E.O. Siecke, who took the reins in 1948 or 1949, as "a strict disciplinarian, and demanded work from his people. He received it. He had, however, respect for all. He would do anything for anybody who worked for him. He had a heart. And, above all, he knew politics and how to get things done."

The state-employed forester in 1985 recalled how the early professionals had to teach and convince the industrialists what good forestry involved. They established planting demonstration plots with the four principle southern pines. These show-me plantations and management practices were laid out on state forests.

When asked for suggestions for present-day foresters, the elder statesman of the profession responded:

> Somehow I have the feeling that modern day foresters lack the crusading spirit of years past. When everyone was sort

> of "gung-ho," we had to sell people on planting trees, to manage timber, to curtail fires, and a lot more. During the fire season, foresters worked seven days a week. There was no let up. We weren't concerned about it. It was a job to do. If we had a society (Society of American Foresters) meeting, everybody went to it. We shared our resources. We may have had to sleep two in a bed or four on the floor; or we may have had to eat crackers and cheese, but everybody went. Throughout the Gulf States we knew every forester personally. I may be a little retiree-ish in such statements but, in that era when for many acres (it was appropriate to) "cut-out-and-get-out," we thought we were going to set the world on fire. We set our dreams pretty high. But didn't we do pretty good with our dreams? I hope the future will judge us on our accomplishments.

Don Young, a Yale Forester, was the first district ranger on the Davy Crockett National Forest, arriving in 1935. Times were testing for federal foresters, readily confused with revenuers in the minds of the citizenry. Don told of one squatter who promised to shoot him if he planted trees on "a certain 10 acres." "I sent word back that I'd be there the next morning to start planting, and reminded him that it was a federal offense to shoot a forest ranger. He did not show up the next day."

Timber trespass, especially tie hackers for hardwood crossties, was notorious. And open range laws meant anyone could graze hogs and cattle on the national forests. Eventually, farmers fenced leased pastures, the agency established carrying capacities, and charged a nickel a head for cattle.

Don also commented on the need, though it seemed ridiculous, to tell industrial landowners to practice forestry, "but that is what I did" after joining the Texas Forest Service. Yet, "they continued to cut timber to a diameter limit." Young believed there were about 10 foresters in Texas in 1930; by 1940, at least 75.

Young established a plantation, in 1936, at the entrances to the Ratcliff Recreation Area, the slash, loblolly, and shortleaf pines continuing to demonstrate species distinctions. He said, "The Regional Office chewed me out for planting slash pine in Texas." The species grew well until *Fomes annosus*, the fungus that causes a root rot, found its way on currents of wind to East Texas.

Joe McPherson, a University of Washington graduate forester, rounds out our visits with industrial, state, and federal foresters. Joe retired in 1968 as district ranger on the Sam Houston National Forest, coming to Texas in 1936 to head up the national forest planting program. The crews planted with dibbles, "each man carried his own trees, made the slit, planted the tree, and closed the hole." Twenty-four-man crews stretched across the fields of cutover, burned over, overgrazed "seas of waving grass, planting 700 to 800 trees per man in a 6-hour planting day." The rest of the day's time was in travel.

Joe spoke of the fall of 1942 when "folks (on the Sabine NF) were trying to burn us out; we had 20 to 25 fires a month.... Johnny Johnson and I fought fires (in February 1943) all night. We didn't stop fighting fires until October. We put out 290 forest fires in that year." "(N)o court would convict someone who had scratched a line around his field.... All we had was a flap, a rake, and a back-pack water can." A scratch was just *that*, not a plowed furrow.

Good timber was scarce on the Sam Houston. Joe "had to scour thousands of acres to find half a million board feet to cut (to meet the congressionally mandated cutting budget). But we had reproduction...from the clearcutting several years before. This timber grew fast." Joe never saw a red cockcaded woodpecker in his years on the Sam Houston. As I write, much of the national forest is "locked up" under court orders to protect the bird that makes its nest in living pines. Endangered, its comeback across the South is a remarkable story of the diligence of foresters.

These interviews, published in 1983, concluded with this recollection by Ranger Mac:

> It's a timber-growing country; you can say that much for it. All the national forests in Texas except the Crockett did not have much merchantable timber when the Forest Service took over. The Crockett Forest had some good timber. Our saw timber volume per acre on the Sam Houston was possibly (an average of) 2000 board feet when we bought it. When I left the Forest Service (in 1968), the average volume per acre was close to 7000 board feet. In the mean time, we

> probably cut 200,000 board feet of timber, maybe more (from the 161,000-acre forest).

What more need be said?

Jacks of all Trades

I was invited to a National Academy of Sciences program in 1971 to speak on "Undergraduate Education in the Sciences for Students in Agricultural and Natural Resources." I called attention to the relationship of *iconaea* of the Greek and stewardship of the Anglo-Saxon, this relationship requiring foresters to be "jacks of all trades."

> Only during the past half decade have I been much concerned about curricular matters in forestry education. But in this period, my position has changed—I like to think it has matured—from a feeling that every professional forester must be poured from the same mold to the conviction that diversity in curricula is essential to solving the problems ahead. Twenty years ago on the 180,000-acre Sabine National Forest there were but two of us professionals. It was fortunate for the Forest Service that we knew some physics, some economics, some report writing, some law, some geography, and a whole lot of silviculture. Neither of us knew much history, accounting, sociology, game management, recreation management, range science, or engineering. Perhaps we did not need to. Now, the staff of that same National Forest, but with 30,000 of its acres under water, includes nine professionals. Were they all cut to the same pattern as were we two earlier foresters, little would be done about recreation management on and around that fine new fishing water, the unmanaged cattle would still be in trespass, wildlife would yet be poached, and the ledger books would be a garbled mess. All nine of these people who now manage this forest are and should be foresters, yet among them must be the skills to manage all of the various resources of the multiple-use concept.

I expressed a concern for distinguishing silvibusiness and silviculture:

> Forestry educators recognized the wisdom of the agricultural educators' movement into business. A couple of

> (forestry) schools developed curricula that appear very much like branches of a collegiate business department. The danger, I think, is that too many students may be enticed into the business of forestry, arguing that forests can be managed from offices and computers, at the expense of "on-the-ground" management. Others believe the emphasis on the entrepreneurial aspects of forestry—or silvibusiness—may lead to exploitation that takes into account neither the responsibility for maintaining the resources nor for serving the public.
>
> Our students... must be prepared to work in the forests and to feel comfortable there. If there is any one thing I would want of a young forester, it is an ability to observe and then to diagnose the condition of the forests and, having done this, to prescribe for their management. The crucial issue may be managerial, or it may be nutritional, pathological, or entomological.... The program must be well-rounded, requiring basic and applied courses in these fields.
>
> We believe a forest-oriented program begins with the freshman in the forest, not in routine subprofessional laboratory exercises but in direct exposure to management problems, mensurational data collection, or studies of mycorrhizae on roots of seedlings.
>
> Silviculture is not an exercise in timber marking, but applied ecology.... The forest serves as the forestry student's laboratory, the timber type his practice cases, while the increment borer and the spade are his stethoscope and thermometer.

As I look back on the NAS conference, I find myself preaching:

> Man is unique in his ability to alter the environment. It is his privilege to reshape it to his needs. Hopefully, any such reshaping will be clean and orderly, because the land:man ratio is rapidly falling. If we labor well, we are masters of the ecosystem upon which we depend; if not, we are its slaves. Either way, we are part of a changing environment, with consequent alterations in the earth's resources on land and under the land, on water and in the water, and air. Some form of those resources is universal, in walled city and unfenced wilderness.

> Foresters are a breed of planner, the resource engineer who understands the biology of ecology, the human nature of ecology, the economics of ecology, and the manipulation of ecology.

And then I wrote of the future:

> Never has the future been more uncertain and, for the resource manager, brighter with opportunity. The Conservation Bill of Rights introduced into the House of Representatives calls for the "right of people to clean air, pure water...." It is the very uncertainty about our resources that makes the opportunities bright. Decisions must be made, policies established, and protection provided. To do this may call for the same kind of evangelism that motivated Gifford Pinchot and other early foresters.
>
> Today's faculties of forestry, mostly educated in the 1940's, were in college because of altruism. Timber famine was yet the cry, and many a city boy left his "ghetto dwelling"—though we did not call it that then and the front walk was scrubbed daily before school — to remedy the situation. Students in the fifties and sixties, in contrast, were the first generation of American foresters who could concentrate on managing forests for profit. For the seventies, foresters must again be motivated by an altruism that, in managing both commercial and noncommercial lands, recognizes paramount public interests.
>
> If the key phrase in the thirties was timber famine and in the fifties multiple use, perhaps in the seventies it will be *iconaea* (the root word for economics) — for foresters the bonding of man and land in resource management.

Forestry as a Calling

In a 1966 College bulletin designed to encourage enrollment of the best of the youth, I spelled out what I considered the requisites for success in forestry.

> Because of the technical requirements and diverse working conditions, young men must meet certain standards in order to enter the profession. Basic traits, such as intelligence, honesty, industry, initiative, ability to lead, and energy are

> necessary for success in most types of employment; but a number of others are essential in forestry. A man employed as a forester may expect to spend much of his time out-of-doors in all kinds of weather. Not only must he be physically sound in order to endure such a vocation but, for success, he must enjoy it.
>
> As forestry is intimately associated with people and communities, foresters should be able to work harmoniously with individuals and groups. Meeting people, developing friendships, and communicating ideas effectively are as important as technical skill.

I then described the work of the professional, trusting readers would find the tasks fascinating and rewarding, yet altruistic. These "Notes from a Forester's Fieldbook" follow:

> Forestry's fascination is its dynamic nature! It is a profession ever changing, led by men who will never see in the distance that the job is almost done. Gifford Pinchot broke "new ground" as the founder of the nation's oldest forestry school—at Yale, as the first chief of the U.S. Forest Service, and as the instigator of the practice of forestry on private lands in the United States. Foresters today, likewise, break new ground to achieve goals which are no more readily in their reach than were the aims of Pinchot in 1900. His objectives have been accomplished, but the stalwart old groundbreaker never dreamed of fertilizing forests, of controlling weed trees, insects, and disease with aerially applied chemicals of a complex nature; of computing growth and drain on mathematical robots; or of detecting fires by scanning vast areas with remote-controlled TV cameras. Nor did he ever conceive that the nation would one day have an abundance of wood which would jeopardize the economic health of communities.
>
> •••
>
> The young forester may mark timber, as did Pinchot himself, or he may plan the logistics and then supervise a chemical weeding operation. He may cruise timber for sale or purchase, as did Pinchot himself, or he may spend a few days in a data processing office, deriving from that cruise data the growth and drain of forest resources for

management decisions or tax assessing. He may instruct sub-professional men on the use of the alidade, as did Pinchot himself, or he may plan the assault by 500 men upon the flank of a raging chaparral fire, using radio-controlled ground-to-air communication, photogrammetry, fire-retardant chemicals, and recently derived equations involving the physics of fire behavior.

The forester is, by nature, frustrated. He wants to get the job done. He works and plans as though he can but, when the day is over, he realizes it won't be finished in his lifetime. Management techniques are always improving but markets must improve and products be developed to utilize the wood produced by better management. In turn, management is stimulated to grow wood better, faster, and at less expense. It is, in a sense, the cycle of competitive enterprise which made the nation great and which makes forestry one of the nation's great professions.

•••

A forester's role is not easily defined, and that—as indicated earlier—is what makes it fascinating. In a way, he is "all things to all men": counselor to lumberman, public relations man for his company or agency, personnel manager for subordinates, conservation specialist for local clubs and the press, public speaker in the schools for fire prevention and careers, disaster expert for the Red Cross and Civil Defense, and merit badge counselor for Boy Scouts.

On the job, the forester must be willing: willing to do the dirty job, willing to endure the unendurable. Society of American Foresters' Vice-President Ernie Allen, an industrial forestry leader, recently noted that sheer, unadulterated hard work is the only way one can outdistance the man who might have the edge in intellectual capacity. Ernie Allen has hired and fired enough men to speak from experience. The teacher attests to the wisdom of work overcoming wit. One must be alert, alert to the needs of the organization, alert to the possibilities of doing a job more efficiently, alert to the motivation and idiosyncrasies of others, and alert to the far-reaching implications of what's going on today. In a nut-shell, then, the reason for the practice of forestry is altruistic—whether in the regions of water economy as on the slopes of the Sierra Nevada; a recreation economy as in the Adirondacks of New York; a

sports game economy as in the high country of New Mexico; a range economy as in the slash pine forests of South Florida; or a timber economy as in East Texas. Thus, the betterment of mankind through environmental control, or improved economic opportunities, is the aim of the professional forester.
Students are frequently reminded that the forester keeps his eyes upon the crown of the tree. There he finds the indicators of its vigor. Always looking up, too, he finds the inspiration and the vision for being a forester.

"Older foresters took up the profession with a crusading spirit," expressed the sentiment of John Thompson of Natchez, Mississippi, in response to a 1969 article I wrote. Similarly, John Bradley, president of a southwide resource management firm, wrote in 1975 of his interest in establishing the evidence and the cause of his frequently observed extremely high correlation between one's "faith and the incidence of long-established, full-time practicing consulting foresters, particularly within the South." If I were pressed for conjecture," he penned, "I would have to think of the strong strain of the...work ethic and the principle of stewardship so clearly associated with Calvinism."

Oikos—The Household

In 1986, I wrote "Resource Managers and the Environmental Ethic" for the *Journal of the American Scientific Affiliation*. A few excerpts follow:

> The church generally has looked upon stewardship as the proper care of money and personal talents. But the well known narrative in St. Luke's Gospel (Chapter 16) in the *New International Version* of the Bible is entitled "The Shrewd Manager. "That shrewd manager is both the ecologist and the economist. *Oikos*, from the Greek, meaning house or household in the Lukian passage and elsewhere, is the basis for our word ecology. The earth is the house and its keeper is the manager or steward. Anglicizing the Greek has altered the spelling slightly, but foresters and other resouce professionals of forty years' service will recall ecology being spelled oecology; and the word defined as the study of

> the relationships of living things to each other and to their environments.
> The idea of stewardship introduces the Latin translation of *oikos*. The Fourth Century *Vulgate*, or *Commoner's*, Bible of that language gives *iconaea* as the equivalent of *oikos*. Change the initial *i* to *e*, and the root word for economics appears, suggesting the inseparability of ecology from economics and thus the managerial connotation of the steward of the earth's resources. Hence a forester, as an ecologist, is the managerial steward of resources that have been entrusted to him. So too is he the keeper of the watershed and the overseer of the range.

Ecology, of which foresters were the first to be knowledgeable, has been a troublesome word since it was popularized in the 1960s with the reincarnation of Henry David Thoreau. Of course, the English term, from the Greek *oikos*, postdates the Walden writer. When a Dallas *Morning News* scribe wrote in 1978,"...we wistfully wish for the ecology of the caveman," it was too much for me. I asked, Did the linquist, author of the column entitled "A Few Words," mean the environment of the caveman? The wordsmith didn't respond.

A week-long session at AuSable Trails Institute in a rustic setting in northern Michigan emphasized the lack of success in promoting wise use. This assembly, mostly theologians and philosopher types, was sponsored by a grant to Taylor University in Indiana. My paper, entitled "Selfishness and the Environmental Ethic," touched a raw nerve with this group. When one peer reviewer of the essay for publication in *The Environmental Crisis: The Ethical Dilemma* (edited by E.R. Squiers, in 1982) was aggrieved by my reporting of Robert Louis Stevenson's appraisal of Henry David Thoreau as a selfish person, I suggested this endnote, which the editor chose not to use:

> A peer reviewer objects to the inference that modern environmentalists are characterized as selfish and lacking in altruism. He writes, "if Christ wants us to do more than what's expected of us, how do we know that the 'more' isn't referring to (such things) as hand-weeding (the forest) as advocated by anti-chemical enthusiasts... Certainly we can't rely on simply economics.... (W)hen we posit an economic

basis for our actions and decisions, then economic externalities must also be incorporated in our analysis.

A medical doctor, Joseph Whelen, and his wife sat in on the deliberations as "peripheral observers." I was gladdened upon finding in the mail:

> ...really a breath of fresh air and a lone cry of reason in what has become an increasingly irrational political movement. Your appraisal and that of R.L. Stevenson of Thoreau is dead on target. I always considered him rather loutish, lacking in civility and basically hostile to those ideals and traditions that hold Western Civilization together. It is certainly not surprising that he became a cult hero of the counter-culture-woodstock-me generation.

At AuSable, on the cutover white pine and red pine forests restored to recreational beauty, I met Vernon Ehlers, physicist and now a US congressman and corresponding friend.

Too often political environmentalists have never faced down a water moccasin, endured trouser legs ripped by briars, picked a thousand seed ticks from their hiding places in the hide, been treed by wild dogs interbred with coyotes, cornered by razorback hogs or angry grizzlies or mad moose, spent a night lost in a swamp, driven nearly insane by fire ants, no-seeums, blackflies and mosquitoes, or encountered an alligator in its territorial bayou. The natural world then loses its serenity.

As a Christian and one who "does science," I also jotted essays for the *Presbyterian Journal* and the *Cumberland Presbyterian*. From the latter, an excerpt in 1963:

> Science means knowledge. God is omniscient, all knowing, as the catechisms say. To "do" science is to gain knowledge. To seek to gain knowledge is to seek to learn truth. So God as the all-knowing One is the God of science as well as the God of theology. He is truth; all truth is of God.

Developing a paper for the *Journal of Forestry* on "Foresters and the Environmental Ethic" brought about a long and fruitful correspondence with Arthur Greely, the former associate chief (and whose father William had been chief) of the Forest Service and in retirement a highly respected cleric in a Washington area church. Art wrote:

> Foresters are not, per se, religious people in the sense of a particular statement of faith. But foresters have been exposed to important teachings about stewardship of the land, foresters here and in other lands. Most members of this profession have a well-developed sense of stewardship. Perhaps we are not as aware as we should be of the full implications of that stewardship responsibility; and perhaps, also, we express it in terms that are meaningful to other foresters but not to non-foresters. Expressing that sense of stewardship in broader terms might be an exciting challenge for individuals in this profession to face.

Forests: For Enjoyment and for Timber

There is an old dichotomy that partitions man's attitudes about the forest and its wilderness character into two compartments, labeled "friend" and "foe." The primeval forest as a thing of beauty is, like the company of a friend, to be enjoyed. Valued friends are protected from all who would destroy them. To the tiller of the ground and the pioneer settler fearful of two- and four-footed adversaries, the forest is the foe. And the foe must be eliminated to protect those who fear it. Loggers and their bosses cutting the forests of early America have ofttimes been classed with the latter group. Conservation historians sometimes picture them as ruthless barons who exploited the wilderness as a conqueror does his enemy.

Recent care for the forest resources of the nation, however, suggests that these lands, whether viewed for their aesthetics or used for their wood-fiber, are to be appreciated as a friend of man. Thoreau doesn't seem to have noted it, but certainly the philosopher would have recognized the inconsistency of the friend-foe argument; for he wrote of sylvan nature while hewing logs, from the very forest of which he wrote, for the walls of the cabin in which he would live.

Eric Sloan, another New Englander and a master with pen at both the easel and the desk, entitled an historical book, *Reverence for Wood.* Drawings of products from trees are interlaced with his prose. Following a listing of things made from wood, he quotes Joseph Jenks, designer of the Massachusetts Bay Colony flag, "What better thing than a tree, to portray the *wealth* (italics mine) of our country?" Sloan, I believe, grasped the underlying attitude of our pioneering forebears who, while purchasing boards from the alleged barons, reverenced wood—and thus the woods—for its practical usefulness for survival.

The dichotomy in attitude is older than the chronicler's first claims of land abuse in North America. Roderick Nash attributes various views in the biblical narratives to man's desire to enjoy for its beauty the wilderness while fearing for his safety when wandering therein. Joel (2:3) described Eden as a garden, and later—though after The Fall—as a desolate wilderness. Whether The Fall resulted in the dichotomy of attitude is a brief for another time; but the wilderness, either forested jungle or parched desert, would now "do you in" if not subjected to man's manipulation for man's benefit.

But, we like to think, we are more advanced in our reasoning than those who roamed Joel's garden both before and after The Fall. The forest can be enjoyed—though at different seasons in its ecological cycle—for its beauty, its utility, and the beauty of its utility. No question of the beauty of trees or a tree. Joyce Kilmer ("Only God can make a tree") and Henry Longfellow ("Under the spreading chestnut tree") acknowledge for us this gift of pleasure. But the handle for a pick or the soft paper that replaced the rough cob may come from the trees in that same woodland cathedral. And are not a Stradivarius and a tennis racket from that same resource?

The forest is no longer the feared wilderness of Joel's time nor the pristine primeval woodland of Thoreau's lonely vigil and vision on the shores of a pond. Joel's jungle needed to be tamed; Thoreau's to be framed.

With wise management, the forest which is today harvested for the building of houses and the printing of books will, after a few tomorrows, be enjoyed for its beauty. And as population soars and the land base on which to grow necessary wood diminishes, the forest now loved for its scenic pleasure must take its turn in the rotation as a woodland to serve. Ugly for a while; but in the providence of a God who understands ecological succession far better than His creatures, beauty will return.

A Final Note

In the back of the book given me by Costa Rica's president José "Don Pepe" Figueres are lines which, when translated, read, "Voltaire told a visitor to his retirement farm that he was unable to do all he wished in one lifetime. 'But, it is true,' said the Master, 'I planted 400 trees.'" In a more figurative way, I like to think that, with my colleagues, we've "planted 4000 trees," and those 4,000 young seedlings have in their careers, each planted 4000 more.

In my first Christmas greeting to the forestry students at SFA, I scribbled a few lines of free verse. They somehow came to the attention of Lucy McBee, the college president's secretary. As Lucy's degree was in literature, and she liked the poem, I've retrieved it from the files and leave it for the patient reader to now ponder.

A PSALM OF THE FORESTER

The feed crib was of rough hewn wood.
    So was the cross!

We commemorate the birth with the
    living evergreen of the forest.
But the Place of the Skull was void
    of any living tree!

Wise men sought Him in the barn.
But He was sought by unwise men in a garden!

The star sang at the advent.
But thunder rattled the darkened day
    of despair!

The Light of the World appeared.
And the Shadows of the World could not dim it!

Thus the inspiration for "*Excelsior:* The higher good."

A genetically superior loblolly pine of the southern forest. Pollen from its male strobili will be introduced to female ovulate flowers to produce super seedlings that will rapidly grow to straight, tall, self-pruning stems with minimal taper and resistance to disease and insect damage. Foresters take on these tasks in order for future generations to have sufficient wood for homes and paper. Such efforts are for "the higher good."

APPENDIX A

Eulogy for Dr. M. Victor Bilan
Kiwanis Club District Conference
Nacogdoches, Texas
September 24, 1992

How a fellow with 900 wives and 300 concubines could know anything about masculine friendship is beyond me; but it was wise old Solomon who said, "If a man would have friends, he must show himself friendly; and there is a friend who sticketh closer than a brother." While the king's pronouncement was prophetic of a future redeemer—the friend who is closer than a brother to the Christian—the proverb provides a starting point for me to tell you of my friend, Victor Bilan.

I met up with Victor's reputation long before I met my friend. Dean Clarence Korstian, in the 1950s dean of Duke University's highly regarded School of Forestry, and I were collaborating on a silviculture text. I, a very junior member of the University of Georgia's forestry faculty, journeyed to Durham to work on the book with America's most respected elder statesman among the forestry profession. Dr. Korstian and I spread the pages on a bed in the guest room, and there among comments on ecology and tree physiology, I learned of the gardener from the Ukraine. Victor's green thumb was especially appreciated by Mrs. Korstian as the student cared for her flower gardens while working on master and doctor degrees at Duke. The Korstians were like parents to Victor, continuing to exchange greetings as long as the old dean and his wife survived.

Next, I met up with Victor *via* an SFASU bulletin which so impressed me that I addressed a letter to President Steen to commend his researcher, never imagining that someday I'd join Victor at SFASU. To this day, his paper remains an important contribution to the study of tree root physiology, a subject I was addressing in Georgia.

Then I came to Nacogdoches. As chairman, I replaced a wood technologist, a subject in which I was ill prepared. To complicate matters, the silviculturist of the 6-

person faculty had just left SFA. So we hired a wood technologist and I tried to teach all the silviculture. Meanwhile, Victor was salaried for full-time research. My request that Victor pick up some of the silvicultural instruction (and share the limited research budget) was the occasion for our first shouting match—which made us truly like brothers. Within a year, maybe two, Victor acknowledged it was one of the best things that ever happened to him. He thanked me profusely for "making" him teach, and, as you know, he became an outstanding classroom professor.

So close were we as friends that when attending meetings we shared a room. Once I mistakenly used his toothbrush. Victor didn't toss it in the trash, as I would have; he simply rinsed it in hot water and cleaned his teeth.

Victor's handling of his cancer seemed to confound the physicians' five-step procedure: denial, anger, bargaining, depression, and finally acceptance. When talking to him from his Baylor Hospital bed, he appeared to accept the prognosis, telling me that he had had a happy, productive life, and was ready to go on. Then anger entered and a doctor called for my help. I tried to get my friend to look at this as "the Last Great Adventure;" that the local oncologist's name is Ventura, Italian for adventure, is not necessarily a coincidence.

I think Victor never bargained—negotiating with God for healing, but he did deny his tumor's presence to almost the end. He dressed to teach his class only a day or two before he went into a coma. That day, to console, Hershel Reeves and I drove him in Hershel's pickup to Hershel's tree farm. My brother-friend seemed satisfied, commenting on the beauty of the East Texas countryside.

Why would I know this about denial, anger, bargaining, depression, and acceptance? I was expecting to sail beyond the sunset ahead of Victor. In a few months he'll be waiting at the Pearly Gates for ole Larry. If there's a delay, I can hear Victor now: "Those Scots shouldn't have sent John Knox to Geneva to learn Reformation theology from John Calvin, who sent the Scot back to Edinburgh to make Presbyterians of the Celts. That's why Larry isn't here."

"Oh, Victor (I'll respond), I was at your funeral. The opening hymn was written by Isaac Watts, son of the church of England cleric during the days of Elizabeth's reign." (Watts angered his dad by saying everything in rhyme. Once when sent to the belfry to learn why the bell frequently chimed faintly, the 6-year-old returned with the answer: "A mouse, for want of better stairs/Ran up the rope to say its prayers.") Watt's immortal hymn includes a verse often ignored.

Time, like an ever-rolling stream
Bears all its sons away.
They fly, forgotten, as a dream
Dies at the opening day.

We'll not soon forget Victor.

Victor had no problem with my simplistic faith: *soli fidelis, soli gratia*--faith alone, grace alone. And I can appreciate the adornment of the gospel as expressed in Father Young's conduct of a Christ-centered funeral. It was as it should be, a time of joy.

In conclusion, may I be personal: When I last visited Victor, he was unconscious. *I* was distraught. Elizabeth and his daughters comforted *me*. I missed out, or so it seems, the stages called denial, anger, and bargaining. But not depression! The night following Victor's death my pillow was as though it had been dipped in John Bunyan's Slough of Despond. I cried and prayed, not for self, nor for Victor. As for Elizabeth, she needed not my prayers. She said, "I'm tough." And what a sense of humor in adversity. When the kids complained on Wednesday that they needed to be back at work on Monday, Elizabeth said, "Well, go tell your father to die." My prayer was for courage like that of Elizabeth for my wife. The answer was prompt. The next day Anne asked, "Where can I find a Scot bagpiper to play *Amazing Grace* at your funeral?"

So Watts concluded his hymn, as we did at Victor's burial,

"O, God, our help in ages past,
Our hope for years to come,
Be thou our Guide while life shall last
And our eternal home."

"Victor, you and I met at the foot of the cross. My friend, we'll meet again at the Throne of Grace.'

Appendix B
BIBLIOGRAPHY OF THE AUTHOR

(Chronologically arranged by classification)
(I = Invitational)

Books

Ecology and Our Forests. A.S. Barnes & Co. 1972.

Reporting Technical Information. Kendall/Hunt Company. 1973, 2nd ed., 1975.

Axes, Oxen and Men. Diboll Free Press. 1974. (I)

Trees. Prentice-Hall, Inc. 1984.

Farming the Small Forest. Miller Freeman Co. 1987.

Forests. John Wiley & Sons. 1990; reprinted the University of Texas Press, 1995.

The Southern Forest: A Chronicle. The University of Texas Press, 1991.

The Southern Forest: Geography, Ecology, and Silviculture. Submitted.

***Technical Research Papers* (*refereed journal articles)**

*Influence of eastern redcedar on soil in Connecticut pine plantations (with R.A. Reed). *Jour. For.* 48:337-339. 1950.

Forest walls retard young longleaf (with V.B. Davis). *So. For. Notes* No. 93. Sept. 1954.

*Early scrub-oak control helps longleaf pine seedlings. *Jour. For.* 52:939-940. 1954.

*Foliar analysis as a method of indicating potassium-deficient soils for reforestation. *Soil Sci. Soc. Amer. Proc.* 19:233-236. 1955. (I--translated into German, French, and Italian, and reprinted in English by International Potash Institute 22:2. Dec. 1955).

*Influence of white birch in restoring potassium to deficient soils. *Jour. For.* 53:451-452. 1955. (I--translated into German, French, and Italian and reprinted in English by International Potash Institute 22:4. Oct. 1956).

*Seed trees retard longleaf pine seedlings. *Jour. For.* 54:269. 1956.

Studies of chlorobenzoic acid. *Proc. So. Weed Conf.* 9:134-141. 1956.

*Complete frilling essential for hardwood control (with

E.T. Carnes). *Jour. For.* 54:340. 1956.

*Foliage symptoms as indicators of potassium-deficient soils. *For. Sci.* 2:113-120. 1956. (I--translated into German, French, and Italian, and reprinted in English by International Potash Institute 22:7. Feb. 1957).

A new woody plant killer. *Proc. So. Agr. Workers* 53:154. 1956.

Extraction of carbon 14-tagged 2,4-D and 2,4,5-T from woody plants (with E. Beck). *Proc. So. Weed Conf.* 11:68-71. 1958.

Isotope tracer methods in tree nutrition. *Proc. N. Amer. For. Soils Symposium* pp. 25-30. Mich. State Univ. 1948. (I)

Seasonal influences on effectiveness of frill applications of growth regulators in hardwood control (with H.V. Wiant, Jr.). *Proc. So. Weed Conf.* 12. 1959.

*Planting following prescribed fire (with E.V. Brender). *Jour. For.* 57:123-124. 1959.

*Brush control in the Georgia Piedmont. *Jour. Range Mgt.* 12:16-18. 1959. (I)

*Absorption rates of 2,4,5-T tagged with carbon-14. (with E. Beck and E. Dumbroff) *For. Sci.* 5:128-137. 1959.

Silvicide screening (with H.V. Wiant, Jr.) *Ga. For. Research Council Report.* 1959.

*Mineral spirits effective in hardwood control (with H.V. Wiant, Jr.). *Jour. For.* 58:324. 1960.

Effectiveness of a mechanical tree girdler (with H.V. Wiant, Jr.). *Ga. Agr. Res.* 1(4):7. 1960.

*Flooding and drainage of slash and loblolly pine forests (with L. Green and J. Daniels). *For. Sci.* 7:2-15. 1961.

Effects of chlorobenzoic acids on *Carya* spp. *Ga. Agr. Res.* 2(1):10 *et fol.* Summer 1960.

Black plastic "mulch" for pine planting. *Tree Planter's Notes* 45:1. 1961.

Movement of 2,4,5-T in Hardwood Seedlings. *Proc. 14th So. Weed* Conf. 218-220. 1961.

Movement of 2,4,5-T in Hardwood Seedlings (with D. Purdom and W. Spivey). *Ga. Agr. Res.* 2(4) Spring 1961. 13 *et fol.*

*Variable Response of Diffuse and Ring-Porous Species to Girdling (with H.V. Wiant, Jr.). *Jour. For.* 59(9):676-677. 1961.

Phosphorus 32 uptake in pine tree roots (with J.C. Arnold). *Ga. Agr. Res.* 3(2) 11. 1961.

*Water and Fertilizer Effects on Loblolly Pine and Slash Pine Seedlings. *Soil Sci. Soc. Am. Proc.* 26:197-200. 1962.

*Response to slash pine of nitrogen and phosphorus fertilization (with C.T. Youngberg). *Soil Sci. Soc. Am. Proc.* 26:399-401. 1962.

Detecting tree feeding roots with radioactive phosphorus. *Advancing Frontiers of Plant Sciences* 4:179-188. 1963. (I)

*Movement of nutrients in coastal bermuda grass (with J. Giddens and H. Perkins). *Agron. Jour.* 54:379-382. 1962.

Fertilization. Chapter in *Slash and Loblolly Pine Plantation Management*. Ga. For. Research Council. 1963. (I)

*Natural Areas of the Southeast. *Jour. For.* 61:670-673. 1963.

Fertilization of Slash Pine (with C.T. Youngberg and J. Hamilton) *Ga. For. Research Council Paper* 17. 1963.

Fertilizer Response with Forest Trees in North America. (with R.L. Beacher). National Plant Food Inst. 24 pp. mimeog. 1963. (I)

*External applications of silvicide paste kill oaks (with H.V. Wiant, Jr.) *Jour. For.* 62(7):445. 1964.

*The nation's oldest plantation? *Jour. For.* 62:269. April 1964.

*Humus Types of the Highlands Area of North Carolina. *The Jour. Elisha Mitchell Sci. Soc.* 80(1):24-29. 1964.

Comparison of two silvicide injectors for killing oaks. *The Consultant* 9(14)16, Oct. 1964.

Frill and silvicide treatments for killing oaks during summer and winter (with H.V. Wiant, Jr.). 18th Proc. So. Weed Conf. 329-332. 1965.

Fertilizing a 21-year old slash pine stand. (with J.C. Morcock, Jr.). *Castanea* 30:81-84. 1965.

*Variation in the ability of slash pine progeny groups to absorb nutrients (with R.D. Hatcher). *Soil Sci. Soc. Am. Proc.* 29(5):616-621. 1965.

Fertilizers applied to natural regeneration (with A. Leiser). *Castanea* 30:231-237. 1965.

Silvicide comparisons. *The Consultant*. Oct. 1966.

Delaying applications of silvicides (with H. Wiant & R. Rinehart). *The Consultant* 14(2):17. 1969.

Cacodylic acid silvicides for thinning loblolly pine and

controlling hardwoods (with H. Wiant). *Proc. So. Weed Conf.* 260-262. 1969.

The Blue Elbow Swamp: A General Analysis. Univ. Tex. at Austin Natural Area Survey, Part V. 25-45. 1975. (I)

America's Forests: An Energy Enterprise (w/R. Hicks) In *Agriculture and Energy* p. 523-534. Acad. Press. 1977. (I)

The Forest as an Originator of Non-Point Source Pollution. *Proc. Water Quality Planning Seminar*, Univ. Tex. at Dallas. 1977. (I)

*Licensing of Foresters -- pros and cons. *Jour. For.* 76(11) 728-731. 1978 (I)

*Natural Resource Mangers and the Environmental Ethic. *Jour. Am. Sci. Affiliation.* 38(2)96-102. June 1986.

*Indonesia: Forestry by Decree (with J. Hoesada). *Jour. For.* 84(11)38-43. 1986.

***Popular Research Articles* (Principally reporting author's studies)**

White birch helps restore potash-deficient forest soils. *Better Crops With Plant Food* 37:13 *et fol.* Dec. 1953. (I)

Young longleaf growth retarded. *So. Lumberman* 189:126. Dec. 15, 1954.

Release longleaf seedlings early. *So. Lumberan* 189:169-170. Dec. 15, 1954.

Low potash soils in Adirondack forests. *Better Crops With Plant Food* 40:24 *et fol.* Feb. 1956. (I)

Weed tree problems in Piedmont pine stands. *Proc. So. Weed Conf.* 8:273-278. 1955.

Weed tree problems in Piedmont pine stands. *So. Lumberman* 191:61-62. Sept. 15, 1955.

Solving weed tree problems in Piedmont pine stands. *So. Lumberman* 191:76-78. Oct. 1, 1955.

"The Quirks of Blazening Pens." *Ga. Agriculturist* 47:12. Oct. 1955.

Leaf analysis: a potential forestry tool. *Better Crops With Plant Food* 40:25 *et fol.* 1956. (I)

Our experience with growth regulators (with E.T. Carnes). *For. Farmer* 16:8. 1957. (I)

Radioisotopes in forest research. *For. Farmer* 17:7 *et fol.* 1958.

Draining pine forests. (with R.L. Green). *For. Farmer* 20(6):8 *et fol.* 1961.

Will Fertilizing Forests be Good Business? (with W.T. Moss).

Pl. Food Rev. 7(4):9-11. 1961. (I) Reprinted in *The Consultant* 8(3):8-11. July 1963. (I)

Slash Pine responds to nitrogen (with C.T. Youngberg). *Crops and Soils.* 94(3-4):18. 1962. (I)

Fertilizing Flooded Forests. *Better Crops With Plant Food* 46(1):40-43. 1962. Reprinted in *Forest Farmer* 21(13):10-11. 1962. (I)

Breeding trees for fertilizer efficiency. *Farm Chemicals* 126(1):45. Jan. 1963. (I)

Trees for City Deserts. *American Forests* 69(3):28-30. March 1963.

Nitrogen benefits slash pine for five years. *Tree Planters' Notes* 18(1):21-22. 1967.

Effects of water and fertilizer combinations on loblolly and slash pine seedlings. *Tree Planters' Notes* 18(1). March 1967.

Trees--for business or pleasure? *Am. For.* 73(3):16 *et fol.* 1967. Also in 12th *An. Park and Recreation Training Inst. Proc.* 14-19. 1967. (I)

Blue Elbow Swamp. *Texas Magazine.* Houston Chronicle. March 30, 1980. p. 20 *et fol.*

Woodland Cathedral. *Texas Parks and Wildlife.* July 1981, p. 16 *et fol.*

A slice of Texas. *Texas Parks and Wildlife.* Feb. 1983. p. 6 *et fol.*

Popular Research Articles (Principally technology transfer type papers)

Fertilizing southern pines. *Commercial Fertilizer and Plant Food Industry* 95:27-29. Sept. 1957. (I)

From water to woods—the change at Watson Springs. Greensboro (Ga.) *Herald-Journal.* April 18, 1958.

Fertilizing forests—a practice for the future? *Better Farming Methods* 30:10-11. Dec. 1958. (I)

Fertilizing southern pines. *First Calif. Forest Soils Conf. Proc.* 15-23, 1948. (I)

Forest Research. In *Cypress Knee.* Univ. Ga. Sch. of For. Yearbook. 1959.

Forest fertilization—roundup of observations and concepts. *So. For. Soils Symposium Proc.* 86-95. La. State Univ. 1959. (I)

Observations on Forest fertilization. The use of Chemicals in Forest Management. *9th Am. For. Sym. Proc.* 54-62.

La. State Univ. 1960. (I)
Forest fertilization in eastern North America. *Proc. Fifth World Forestry Congress* Vol. 1:562-566. 1960. (abs. in French and Spanish).
Forest Fertilization—its effect on growth and wood quality. *Soil Sci. Soc. Amer.* 1960. Distr. by Nat'l. Plant Food Inst. Also in *Proc. 36th An. Meet. Council on Fertilizer Application Proc.* 82-83. 1960. (I)
What's New in southern forest fertilization. *Auburn Forest Forum Proc.* 51-62. 1962. (I)
Lime and legumes in European forests (with J. Stoeckeler). *Better Crops with Plant Food.* 47(2):30-38. March 1963.
Fertilizers for European forests (with J. Stoeckeler). *Better Crops with Plant Food.* 47(3):14-22. June 1963.
Pesticide use in forestry. *Farm Chemicals* 126(6):28 *et fol.* June 1963. (I). Reprinted in *For. Farmer* 23(1):20 *et fol.* Oct. 1963. (I)
The source of nitrogen. *Agr. Ammonia News* 48 *et fol.* Nov. - Dec. 1963.
Hardwood control not always essential. *For. Farmer* 22(13):9. Sept. 1963.
Is the South ready to fertilize forests? *Farm Chemicals* 127(1):16 *et fol.* Jan. 1964. (I)
Know your chemicals. *Forest Farmer Manual* 23(7):43-46. March 1964. (I)
Fertilizing shade and forest trees in parks. *Park Practice Guideline* 45-47. June 1964. (I)
Fertilization for pine plantations. *A guide to Loblolly and Slash Pine Plantation Management in Southeastern U.S.A.* Ga. For. Res. Council Rept. 14:292-297. 1965. (I)
Chemicals—their use in forest farming. *Forest Farmer Manual* 92-95. 1966.
Fertilizing the farm forest. *Prog. Farmer.* 81(7):17. 1966. (I)
Forest fertilization in North America. *Sixth World Forestry Congress Proc.* 2660-2663. Madrid, Spain. 1966.
The National Plant Food Institute Clearing House for Forest Fertilization Research. *For. Fertilization* 294-295. Tenn. Valley Authority. 1968. (I)
Troubles of Evergreens. *Proc. Midwestern Chap. Int. Shade Tree Conf.* 24:7-11. 1969. Reprinted in *Aborist*, May 1969. (I)
Forest fertilization in North America. *Proc. Seventh World*

Forestry Congress. (paper 202). Madrid. 1972.

Use of Chemicals in Forest Farming. *For. Farmer Manual* 68-72. 1975. (I)

Texas Forest Economy. *Tex. Bus. Review.* 50(3)64-68. Mar. 76. (I)

Wood, The Wonder Material. *Proc. Southwest Parks and Recreation Inst.* 62-74. 1978 (with V. Euteneuer). (I)

Use of Chemicals in Forestry. *Forest Farmer Manual* 46-48b. March 80. (I)

The Texas Timber Industry in the 1980's. *Texas Business Review* 54(4)226-229. July-August. 1980. (I)

Wood from the Five Forests in Texas. *Texas Natural History* 2(1) 12-16 Summer 1985. (I)

Thinning the Southern Pine Plantation: An Overview. *Proc. Workshop, Forestry and Harvesting Training Center* pp 1-10. 1982. (I) Rev. 1984. (I)

Chemicals Used in Forestry. *Forest Farmer Manual* 42(5)44-47. March 1983. (I)

Chemicals Used in Forestry Operations. *Forest Farmer Manual.* 25ed. 40-43. March 1985. (I)

Renewing the Forest. *Texas Forestry.* 30(4)6 *et fol.*, April 1989.

Articles in *Farmer-Stockman*

Are foresters really destroying wildlife? Dec. 1971. (I)

Forest fertilization—get your wood-worth. Feb. 1972.

Prescribed forest burns present modern dilemma. Apr. 1972. (I)

Bugs take heavier toll than man and fire together. Oct. 1972. (I)

Foresters, too, care about the environment. Jan. 1973. (I)

Root rot fungus takes toll in East Texas timberlands. Apr. 1973. (I)

Popular use of the woods—recreation. Aug. 1973. (I)

Will timber prices last? Oct. 1973. (I)

Timber productivity. Apr. 1974. (I)

Texas' timber resource. July 1974. (I)

FIP—forestry incentives program. Oct. 1974. (I)

Taxes on timber. Nov. 1974. (I)

Water in the woods. Jan. 1975. (I)

Measuring Forest Land Productivity. Apr. 1975. (I)

Preparing the Land for Tomorrow's Forest. Aug. 1975. (I)

Planting Forest Trees. Feb. 1976. (I)

Improving the Forest. Mar. 76. (I)
National Forests, Their Meaning. Nov. 76. (I)
Trees—A Source of Energy. Dec. 76. (I)
Texas' Forest Economy—An update. Jan. 77. (I)
Texas's Unique Hiking Trails. Feb. 77. (I)
Forestry Practices—Planting. July 77. (I)
Forestry Practices—Weed-Tree Control. Oct. 77. (I)
Forestry Practices—Thinning. Oct. 77. (I)
Forestry Practices—Pruning. Dec. 77. (I)
Forestry Practices—Harvesting. Jan. 78. (I)
Dust Storms and Tree Planting. Feb. 78. (I)
Here's How the Forestry Incentive System Works. November 78. (I)
Trees for Shelter in the Wind. March 78.
Sheltering Soil from the Wind. Apr. 78. (I)
New Appraisal of Texas Forests. May 78. (I)
Timber, Location, Soil, Roads Figure Into Forestland Appraisal. Apr. 79. (I)
The Value of Timber and of timberland. Jan. 79. (I)
Wilderness in Texas Woods. March 78. (I)
Timber Sale Contracts. Nov. 1980. (I)
Bugs in The Pineries. July 80.
Small Forest Landowner: Costs and Income. Apr. 1981. (I)
Wildlife and Woods Management. June 1984 (I)
Trees for Ranch and Farm. April 1983. (I)
Tree Failures Cost Money. Dec. 1983. (I)
Trees for Tomorrow. June 1982.
Trees and Taxes. Dec. 1981.
Dwarf and Giant: Longleaf Pine. March 1985.

Technical Bulletins

Controlling Undesirable Hardwoods. Ga. For. Research Council Report 3. 24 pp. 1956. (translated into Turkish by I. Atay, Univ. Istanbul.)

Forest soils and Silviculture in Georgia. (with H.F. Perkins). Ga. For. Research Council Report 4. 36 pp. 1958.

Silviculture of Shortleaf Pine (with H.V. Wiant, Jr.). Stephen F. Austin State Univ. For. Bul. 9. 60 pp. 1966.

Silviculture of Longleaf Pine (with H.V. Wiant, Jr.). Stephen F. Austin State Univ. For. Bul. 11. 100 pp. 1966.

Silviculture of Minor Southern Conifers. Stephen F. Austin State Univ. Sch. For Bul. 15. 106 pp. 1968.

Silviculture of Slash Pine. Stephen F. Austin State Univ.

Sch. For. Bul. 16. 113 pp. 1968.

Geography of the Southern Forest Region. (with L. Collier). Stephen F. Austin State Univ. Sch. For. Bul. 18. 68 pp. 1969.

Silviculture of Southern Upland Hardwoods. Stephen F. Austin State Univ. Sch. For. Bul. 22. 137 pp. 1972.

Silviculture of Southern Bottomland Hardwoods. (with K. Watterston). Stephen F. Austin State Univ. Sch. For. Bul. 25. 137 pp. 1972.

Published Experiment Station Reports (I)

TSI in south Alabama and West Florida. Ann. Rept. So. For. Expt. Sta. pg. 33. 1953.

Seed trees, forest walls, and longleaf reproduction in south Alabama. Ann. Rept. So. For. Expt. Sta. pp. 41-42. 1953.

Yields of longleaf pine best where thinning was light. Ann. Rept. So. For. Expt. Sta. pp. 30-32. 1954.

From Ag. Research Briefs. U. Ga. Col. Agr. Expt. Sta.

Controlling undesirable hardwoods.

2,3,6-trichlorobenzoic acid.....vol. 1; no. 3; July 1955.
survival of loblolly pines........vol. 1; no. 4; Sept. 1955.
water hardness..........................vol. 1; no. 5; Nov. 1955.
complete frilling......................vol. 1; no. 6; Jan. 1956.
Diesel oil ....vol. 2; no. 1; Mar. 1956.
prescribed fire..........................vol. 2; no. 3; July 1956.
prescribed fire..........................vol. 3; no. 3; Aprl 1957.
Fertilizing pines.......................vol. 4; no. 2; Mar. 1958.

Controlling scrub hardwoods. Ann. Rept. Ga. For. Research Council. pp. 13-14. 1955.

Controlling scrub hardwoods. Ann. Rept. Ga. For. Research Council. pp. 13-14. 1956.

Forest physiology. Ann. Rept. Ga. For. Research Council. 1957.

Forest physiology. Ann. Rept. Ga. For. Research Council. 1958.

Inferior species are major management problem. Ann. Rept. Ga. For. Research Council. 1959.

Forest physiology. Ann. Rept. Ga. For. Research Council. 1960.

Forest physiology. Ann. Rept. Ga. For. Research Council. 1961.

Forest physiology. Ann. Rept. Ga. For. Research Council. 1962.

Book Reviews

The Physiology of Forest Trees. *Amer. Sci.* 47(2):147A-148A. June 1959. (I)

Mineral Nutrition of Trees. *For. Sci.* 6(2):109. 1960. (I)

Natural Resources. *Amer. Sci.* 48(4):389A-390A. 1960. (I)

The Forest Needs Lime. (*Der Wald Braucht Kalk*). *For. Sci.* 7175. 1961. (I)

Man, Mind and Land, A Theory of Resource Use. *Amer. Sci.* 49(3):298A. 1961. (I)

The Cloud Forest. *Amer. Sci.* 50(3):304A. 1962. (I)

An Ecological Approach to Conservation. *Jour. For.* 63(8):626. 1966. (I)

Breaking New Ground. *Jour. For.* May 1973. (I)

Forestry and its Career Opportunities. *Jour. For.* 71(12):782-783. December 1973. (I)

Book Reviews. (regular feature) *Tex. For.* (formerly Texas Forests and Texans). 1967 to 1977.

All the Trees and Wood Plants of the Bible. *Presbyterian Jour.* p.20. July 2, 1980.

Oral Histories

Hursey, First Industry Forester in Texas. *Tex For.* 22(4) p. 8 *et fol.* Apr. 1981.

Young, Early Professional Forester in Texas. *Tex For.* 23(5)3 *et fol.* May 1982.

Joe McPherson. *Tex. For.* 24(4)4 *et fol.* April 1983.

W.E. Merrem, *Tex. For.* 24(9)5 *et fol.* Sept. 1983.

Prud'homme Tells of Early Days. *Tex. For.* 24(25)(3) 8 *et fol.* March 1984.

D. Anderson. Part I. *Tex. For.* 26(2) Feb. 1985.
II. 26(3) March 1986.
III. 26(4) April 1985.

Carter Caton. *Tex. For.* 27(6) June 1986.

Four Who Counted. East Texas Historical J. 29(2)61-70. 1991.

Essays

Facts from a forester's fieldbook. Chapter In *The Evidence of God in an Expanding Universe.* pp. 106-113. (International Geophysical Year publication.) G.P. Putnam's Sons. N.Y. 1952. (I).

On attitudes. *Sawdust* 1(4). April 1961. (I).

From a forester's fieldbook. In *Texas Tally.* 1965. (I)

Important to see both forest and trees. Chapter In *Beyond the Dim Unknown.* Putnam's Sons. 84-94. 1966. (I).

The Forests of Deep East Texas. SFASU Misc. Pamphlet, 11 pp. 1993. Abbreviated as "Introduction to *Deep East Texas*, by D. Gibson, Longstreet Press. 1993.

In The Presbyterian Journal

The nobility of work. 19(18):7-8. 1960.
Enjoying God. 1967.
It's just not fair. 26(23):9-11. 1967.
Your money or your life. 27(40):9-10. 1969.
Why the environment is polluted. 29(51):7-8. 1971.
Where's the center of reference. 30(23):13. 1971. (I).
The good that I would... 31(7):13 *et fol.* 1972. (I).
Christ, the motivator. 32:13. 1972. (I).
In the plan of God. 32(22):13. 1973. (I).
Dual role of the Christian. 32(43):13. 1974. (I).
InterVarsity Christian Fellowship. 33(35):13. 1974. (I).
Man and Nature. 33(52):13. (I). 1975.
The Adversary Society. 33(52)11. 1976.
Free Indeed! June 16, 1976. pg. 11 *et fol.* (I).
The Bible in English. 36(44):11 *et fol.* Feb. 26, 1978.
The Israeli Situation. 37(26):15. Oct. 25, 1978.
Inheritance of the Jews. 37(30):15 *et fol.* Nov. 22, 1978.
Beyond the Jordan. 37(32):15. Dec. 6, 1978.
The Coptic Evangelicals. 37(33):13. Dec. 13, 1978.
The Saints of Saloniki. 37(35):15 *et fol.* Dec. 27, 1978.
Forests: For Enjoyment and for Timber. PAVE of W. Va. Univ. 2(1). Aug. 1980. (I).
A Christian and the Environment. 39(13):11. July 30, 1980.

In The Cumberland Presbyterian

Environmental Stewardship. 154(9):6-7. May 1, 1982.
A Scientist and His Faith. 155(7):20. April 1, 1983. (I)
A Scientist and His Faith II. 155(10). May 15, 1983. (I).
One View of the Reformation. 156(1):13 *et fol.* Jan. 1, 1984.
Music in The Worship Service. 155(20):20. Oct. 15, 1983.
The Tie that Binds: The Church in Indonesia. July 1, 1985.
The Roots of Our Faith: A Scottish Experience. 158(3):4-7. Feb. 1, 1986.
The Crowning Glory, 1986. Oct. 15.
A Love That Would Not Let Go. Sept. 1, 1986.
Building for the Next Generation. March 15, 1989.
The Kirk in the Land of the Thistle. 162(10)30-34. Oct. 1990.
Christians and the Environment. 163(2)18. Feb. 1991.
Why the Earth is Polluted. 163(2)19-21. Feb. 1991.

The Church in South America's Land of Red Woods. 163(9)29-31. Sept. 1991.

The Church in China. 164 (5) 15-17. May 1992.

Non-Classified Publications

Red Cedar: good whittling wood. *Homecraft*. 296-296. June 1955.

Forest Fertilization research in the South (with S.L. Tisdale). Nat'l. Plant Food Inst. 50 pp. 1959. (I)

The coastal plain southern pine region. Chapter In *Regional Silviculture of the United States*. 246-295. Ronald Press. 1962.

Natural areas of the southeast (A report). *Forestry Newsletter* (Southeastern Section Soc. Am. For.) 19(2):5-7. June 1963.

The job ahead. Comments on *Proc. 5th Am. For. Congress* 180-182. Am. For. Assn. 1963.

Forestry department now flourishing at SFA. *Texas Forests and Texans* 5(3):9 *et fol.*, May-June 1964. (I)

What's new in forestry education. *The Consultantt* 7-9, July 1969.

Policies affecting today's foresters (with H. Wiant). *The Consultant* 15(3):64-66. 1970.

Technical/Professional Bugaboo. *Jour. For.* 69(12):865. 1971. (I)

Water in the woods. *Proc. Tex. Council Soil Cons. Soc. Am.* 1971. (I)

Costa Riça, forests and parks in the Rich Coast. *Explorer* 16(3):8-13. Fall 1974.

Southern Pine for Southern Playgrounds. *TRAPS* 22(3)21-21. July-August 78.

Preserving wooden playground furniture. *Park Practice GRIST* 25(1). Jan/Feb. 1981. (Dept. Interior)

Ecological concepts in forest management. *Jour. Am. Sci. Affil.* 32(4)207-214.

Selfishness and the environmental ethic. *In The Environmental Crisis: An Ethical Dilemma*. Au Sable Trails Institute of Environmental Studies. pp. 67-76. 1982. (I)

Trends in renewable natural resources curricula. In *Nat. Acad. Sci. Undergraduate Education in the Sciences for Studies in Agriculture and Natural Resources*. 36-41, 1971. (I)

Multiples use role of East Texas forest resources. *Proc. 23rd Gt. Plains Ag. Council.* 1971. (I)

How to promote research. *Aust. For. Tree Nutrition Conf.* 1971. 271-274-. For. ·and Tbr. Bur., Canberra. 1972. (I)

Forester employment in the Federal Government (with D. Theoe and S. Hopkins). *Jour. For.* 70(11):716-719. 1972. (I)

Forestry employment in academia (with D. Theoe). *Jour. For.* 71(1):56-57. 1973. (I)

Forestry employment in U.S. industry (with D. Theoe). *Jour. For.* 71(8):537-538. 1973. (I)

Forestry employment in consulting work (with D. Theoe). *Jour. For.* 71(10):672-673. 1973. (I)

Forestry employment by the states (with D. Theoe). *Jour. For.* 70(12):771-772. 1973. (I)

Career Corner. McNaught Syndicate (In 250 U.S. daily papers) March 4,`73. (I)

SFASU educates foresters. Tex. For 14(4):17 *et fol.* Reprinted in *Sawdust* 1(9):6-7. July 1974. (I)

Foresters in association employment. (with D. Theoe). *Jour. For.* 72(6):378. 1974. (I)

Eco-Echo. *Proc. An. Meet. Soc. Am. For.* 272-275. 1974. (I)

Canyon Scablands. *Am. For.* 81(5)2 *et fol.* 1975.

The Professional Forester. SFASU Pamphlet. 1975, repr. 1976.

Foresters: How many and what kind. *Gulf States Section (SAF) Newsletter* 18(2)5 *et fol.* Second Qtr. 1976. (I)

The southern pine region. In *Regional Silviculture in the United States*, 2d ed. Wiley. 1980.

Technical Forestry Training-*Encyclopedia of Am. For. and Cons. Hist.* (I). p. 237 Macmillan Publishing Co. 1983.

Technical Forestry Training-*Encyclopedia of Am. Forestry and Conservation History* (I). p. 239-240. Macmillan Publishing Co. 1963.

Forest Influences-*Encyclopedia of Am. Forestry and Conservation History* (I). 38(2)96-102, p. 215-217. Macmillan Publishing Co. 1983.

The Professional Forester. Texas Soc. Am. Foresters 1986 (I).

South America's Red Wood Country. *Forest World* 6(4)42-47. Summer 1991.

The Southern Pine Region. In *Regional Silviculture in the United States*, 3rd ed. John Wiley & Son. 1994.

27th Evac. Hosp. 72
52-20 Club 79,82
65th Signal Battalion 60
7th Army 70
84th Division 59

Abido, M. 298
acacia 362
Acheroid, J. 292
Adekiya, F. 298
AEC 195
AID 343,359-360
air boat 330
airplanes 199
Alaska 387-395
Alcock, J. 143
Alexander, J. 424
Allegheny Nat. For. 86-87
Allen, Ernie 457
Allozi, S. 361
Alsides, D. 297
Am. Sci. Affil. 415-416,458
Amazon 379
Amer. Assn. Adv. Sci. 404
American Forests 5,17
American University 79
Amish 385
Anderson, C. 199
Anderson, D. 450-451
Anderson, J. 413
Angel, P. 277-280,423
Angle, D. 412
Appalachian Trail 40-42
Araucaria 381
Aristotle 356
Arivantinos 355
Armstrong, F. 142
Army Military Institute 69
Arnold, K. 406
Arnold, R. 142
Arnold, S. 299
aspen 156,412
Aswan Dam 359
Atay, I. 188-189
Attrep, M. 421,422
Auffenberg, W. 373
AuSable Trails 459
Australia 334-445
Avido, M. 293
Baggett, D. 215
Baker, R. 207,211,234
bald, Appalachian 401
Balde, O. 298
baldcypress 99,123,125
Bali 344
Barber, B. 297
Barnes, M. 192
Barrens 83
Barrett, J. 152
Bartsch, P. 32, 33
Baun, D. 296
Bavaria 72
bear 131-132
Beasley, D. 417
Beasley, R.S. 303,421,422,433
Becton, W. 142
Bedouin 361-362,365
beech 66,156,321,366
Beede,V. 87-88
Bell, J. 109
Benny O. 53
Berst, J. 43-44
Bertesgaden 71
Bertrand, A. 254
Bethel, J. 11,406
Better Crops 323
Bible 458,459,462
bibliography, author 468-481
Big Bend Park 298-299
Big Thicket 124
Bilan, V. 207,211,236-237,465-467
Biltmore Forest School 443-444
Biltmore stick 13
birch 156,157,413
Bird, A. 38
birds 131-132
Birmingham, Suzanne 299
Bishop, G.N. 184
Bitche 70,71
Black Forest 66,323
Black, C. 299
Blanco, J. 370
Blanding, Captain 61
Bloom, R. 240
Bloomsburg Normal School 24
Blue Elbow 263
Blue Jay 86-89
Blysma, D. 101
boat-builders 127
Boggs, W. 390
Boggus, T. 260
Boles (Bowles) Field 114-117
Bolton, Dr. Mary 138
Bond, R. 251
Bonneau, M. 323
bonus marchers 23
bootlegger's car 103-105
Borneo 343
Bos, W. 272,308-310
Bosworth, Supervisor 119-120
Boundary Canoe Base 412
boxcar, 40 & 8 73

Boy Scouts 30-45,276,348,411-415
Bradley, J. 458
Bramble, W. 86
Bray, I. 440
Brazil 377-382
Brender, E. 175-176
Brewton, Ala. 166-171
Briegleb, P. 171
Brittain, P. 423
Britton, A. 240
broom, scotch 156
Brown, J. 15-16
Brown, R. 254
Browne, B. 200
Bryan, W. 54
Bryant, B. 142
Burch, J. 413
Bureau of Indian Affairs 390
Burgess, Rep. S. 109
Burkart, L. 207
Burkholder, B. 8
Burns, P. 142,397,399
Burns, R. 212
Burrow, D. 259-260
Burrow, T. 296
Bussa, Mary 257
Butts, Coach 200
Byrd, H.L. 223
C.C.C. 51,98
cadam *see* Kadambi
Caddo Lake 125
Caddo State Park 97
Cadet Band 47
Cahuita Coral 372-377
Calahan, J. 254
Callison, C. 31
Calvin, J. 309
Camp Bliss 44
Camp Bowie 61
Camp Brittain 99
Camp Claiborne 59
Camp Fannin 57,82
Camp Roosevelt 38,40
Camp Wilson 34-35
Campbell, M. 412
Campbell, W. 174,205
Cann, J. 101
Canterbury, D. 234,247,407
Cape Cod 52
Carlson, W. 165
Carvell, K. 142
Casale Monferrato 325-327
Castle, Young 105
Castro, J. 374
cedar 363
cerado 381
Chamber of Commerce 424
Champion Paper Co. 327
Chancellor, G. 440
Chang, M. 240
Chapel 15
Chapman, H. 135,136
Chen, P. 339
cherry 89,156
chestnut, Am . 266,339-340,356
Childers, Mrs. R. 228,229
Childs, D. 413
China, lectures in 337-341
Chisman, H. 11,83
Chovet, J. 85
Christy 23
church 3,15,21,22,58,69,107-108, 145,168,238,308-317,322,340, 357,363,378,392,422,432,444, 458,466,467
Church of the Pilgrims 35,38
church, Rock Springs 309
church, Scottish 313-316
church, Tennessee 312-313
church, Trawick 310
Civil Service Commission 95
Clark, D. 86
Clark, E. 417
Clark, R. 7,17,142,143
Clark, W. 151-152
clearcutting 408-410
Clepper, H. 17,403
Cliff, E. 231,249
Coalburn 314
Coast Artillery 43-44
Cole, K. 296
Cole, R. 296
Coll. of For. N.Y. State 146-165
Colman 64-65
Communism, Mexican 383
Companhia Vale do Rio Doce 377-382
Condit, G. 143
Conklin Hall 5,12
Connaughton, C. 101,248,249,406-407
consulting 400-401
Continental Divide 426,431
Cooper, John 185
Coordinating Board, Texas 216
Cope, H.N. 4,7,9,13,14,88
Copper Canyon 384
coral 337
Costa Rica 215,368-375,385
Coster, J. 240
cottonwood 276,318-327,429
Coweta 196

Craig, D. 272
Craighead, J. & F. 47
CREEL 395
Crews, N. 211, 241
Croker, T. 167-171
Crowley, C. 299
Cub Scout 24
Cunningham, B. 297
cut-out-and-get-out 126-130
Cutter, B. 240
cypress 363,370
D'Andelet, J. 51
Dachau 68-69
Daniel, T. 361
Danube River 71
Darling, O. 143
Davenport, M. 223
Davies, G. 298-299
Davis, Jerry Ann 211,235
Davis, V. 168
Dawson, J. 289-290
Day, Lynn 439
DDT 68
Dead Sea Scrolls 366
Death Valley 395,399
death, student 300
deBin, J. 295
Decker, J. 159-160
Deep East Texas 122-132
DeLoach, G. 175
Dembitz, Mrs. 51
Denali Park 393
dendrochronology 431
Dendy, F. 101
Denkhaus, Gail 297
Denman, J. 228
Denmark 320-321
der Matterhorn 328
DETCOG 416-417
DETDA 229,416-417
Devall, W. 267-268
Devers, General 64
Dickerman, M. 221
Dickerson, B. 240,437
Dickey, W. 211
Dickinson, F. 138
Dierks 391
Dillingen 71
Dimond, A. 137
Dinyari, S. 438-439
dipterocarps 332-346
Dixon, M. 101
dogwood 389
Douglas, C. 297
Douglas, Justice 261
Douglas-fir 191,306,366,427
Drake, L. 32
Dreka Tower 99
Dwight, T. 145
Eagle Scout 33,40,414
Early, B. 8,240
Eaton, J. 101
ecology 458-459
Ecology and our Forests 38
Edwards, Prof. 84
Effenberger, L. 207
Egypt 357-361
Ehlers, Cong. V 460.
Eisokovitch, D. 364
ElderHostel 437
Ellenberg, E. 120
Elliot School 33
Ellison, L. 199
Elson, E. 69
energy, wood 435
environmentalists 356
Eriksson, H.C. 101
Erminger, O. 292
Escambia Exp. Forest 166-171
eucalyptus 334,363
Experimental Forest 224-226
Fain, V. 214,275
FAO 324-327
Farnsworth, G . 152
Faulkner, H. 296
feminists 278-280
Ferguson, J. 85
Fernow, B. 159
Ferrell, Elin 299
Ferrer-Figueres, Pres. 368, 370-373,463
fertility 192
fertilization 318-330
Fiji 336
Findeisen, T. 289-290
Finland 323
fir 91,428,442
fire 129-130,430
fire, Borneo 344
fire, prescribed 113
fire, wild 334-335
First Amendment 268
Five C 446-447
Florestas Rio Doce 377
Foggy Bottom 54
Ford V-8 31-22
Ford, J. 215
forensic forestry 401
Forest History Soc. 439
Forest Service, US 220-222,224
Forrester, I. 313
Foster, C. 152

Foster, W. 413
Fountain, Sheriff 168
Fox, T. 304
France 323
Franklinia 139,435
Fraser Forest 190
Frazier, F. 245
Fredenburg, F. 296
Fredonia Bank 420
French Corps 64
Frey, J. 437
Friend, T. 83
Frodin, R. 165
Fry, G. 415
Ft. Meade 57
Ft.Yukon 390
G.I. Bill of Rights 79-80
Gaines, J. 263-264
Gara, R. 240
Garrett, G. 135, 138
Garrett, Mrs. 99
Garrison, M. 412
Garasic, W. 277
Garstka, W. 234
Gauthey, L. 83
Gelhorn, J. 430
Georgia, tour guide of 177
Georgia, Univ. of 172-202
Gerber, J. 205,207,237
Germany 322
GI Bill 148
giant sequoia 386,396
Gibraltar 330
Gibson, D. 132
Giddens, J. 195
Ginkgo 20,27,29,139
Glascock, H. 407
Goddard, M. 15
Goldberg, M. 294,365
Golden, E. 240
Goodrum, P. 223
Gragg, F. 233,408
Grand Ditch 426,431
Grant, B. 184
Graves Registration 63
Graves, H. 60,135
Graves, P. 152,403
Gray, J. 142,143
Great Salt Lake 80
Greece 354-357
Greeley, A. 249
Greeley, W. 60
Griffith, C. 229,417
Grimes, J. 294
Groat, C. 421,422
Grosenbaugh, L. 142,143,171
grouse 90-91
Grove, A. 11
Guion, Sergeant 62
gum, tupelo 125
Guttenberg, S. 162
H, E, & WT 98
Haas, M. 412
hacienda 370-372
Hacker, D. 296
Hagenstein, B. 213
Haggard, R. 296
Haig, B. 8
Hall, R. 407
Halls, L. 222
Halls, S. 412
Hard Labor Creek 183
hardwoods, tropical 381
Harper, B. 277
Harper, S. 299
Harris, J. 295
Harris, Sheriff 168
Harrison Bayou 263
Harvey, G. 11,14,17
Hashwey, Rev. A. 363
Hastings, E. 230
Hatcher, R. 228
Hawaii 333
Hawkins, G. 47
Hawley, R. 135,136
Hayman, K. 294
Hearnberger, S. 235
heart attack 324
Heczko, C. 101
Heefner, S. 11
Heiberg, H.H.H. 162
Heiberg, S.O. 144,152,158,162
Hellier, T. 421,422
hemlock 89,389
Henry, C. 90
herbarium 295
herbicides 102,192-193
Herrick, A. 184,1856
hickory 141
Hicks, R. 240
Highlands Station 182-183,401
Hindmarsh, G. 142
HMT Esperance Bay 60,61
Ho, K. 298
Hoesada, J. 342
Hofman, J. 16-17
Holers, L. 296
Holland, A. 294
Holstener-Jorgensen 321
Honduras 375-377
honor society 201,302
Hoover Dam 81

Hopkins, W. 167,168,170
Hopson, M. 101
Hornaday award 299,415
Horsfall, Mrs. 4
Horton, B. 413
Hosner, C. 421,422
Hotel Capacabana 378
Hough, F. 404
Howard, J. 240
Howarth, S. 297
Howes, Linda 296
Hubbs, C. 421,422
Hudgens, R. 413
Hudson, J. 101
Hufford, R. 425
Hull, J. 219,296,410
Hunt fund 270
Hunt, E. 207-211
Hunt, K. 276,423
Hunt, L. 207,232,248,280
Hursey, D. 448,450
Hussein, King 363
Hyink, D. 276-277,414
Ickes, H. 151
iconaea 453,455
Iguacu Falls 380
Indians 129-130,418-419
Indonesia 323,342-350
Int-Hout, P. 101
Inter-Varsity Fellowship 89-91,145,164,192,276,423
iron ore 379
Irwin, J. 223
Israel 363-367
Itaipu Dam 380-381
Italy 318-327
Jackson, Ala. 168
Jackson, W. 299
Jenkins, J. 187-188
Jerico 364
Johnson, E. 142,143
Johnson, F. 190
Johnson, Harlow 110
Johnson, J. 98,100
Johnson, L. 101
Johnson, Rep. J. 415
Johnson, W. 269,270
Johnson, W. (U. Mich.) 416
Jones, Doris 425
Jones, W. 223
Jordan 361-362,363,364
Jordan, Mr. 35
Jorgensen, D. 143
Juncker, Fl. 319
Juneidi, M. 361
junior foresters 101
Juniperus 321
junk shop 25
Justice, W. 268
Kadambi, K. 240,329,335,371,382
Kalimantin, East 344
Kallsen, T. 237
Karnack 359
KDET 109
Keer, D. 80
Keijean, R. 323
Keller, W. 83
Kelley, L. 101
Kelly, J. 84
Kemp, A. 207
Kent, J. 8,17
Kerr, D. 74
Ketchikan 388
Kew Gardens 20
King, G. 199
King, Nancy 238,241-245
Kingman School 24
Kirby Corporation 201
Kirby, J. 252-253
Kit Carson Forest 283
Kitchens, J. 289
Kline, R. 412
Knotts, D. 296
Korstian, C. 465
Kotzebue 391-392
Kozlowski, T. 237
Krear, R. 430
Kribs, D. 11
Kroll, J. 240
krummholz 428
Kulhavy, D. 303
Kwiatkowski, Diane 299
Laman, C. 299
lamplighter 25
land-line controversy 117-118
Landback, Fannie 43
Lanier, S. 173,202
larch 351,389
Lawrence, Captain 267
Lay, D. 222
Legg, M. 240
Leipzig 72
Lemons, B. 294
Lenhart, J.D. 240
Lewis, J. 269
Lindsey, S. 223
Linthicum, Mary Jo 235,236
Lipps, L. 181
litigation 266-272
Little, S. 142
Liverpool 61
Lock, O. 221

locust trees 363
Loftus, Rev. 35
Longwell, J. 8
Loughmiller, C. 262
Louisiana-Pacific Co. 256
Lowry, G. 240
LST 62
Luhat, E. 298
Lumpkin, J. 223
Lutz, H. 139,234
Lynch, D. 262
Lyon, France 72
MacAdam 316
MacPeak, M. 240
Magadan 350-353
Magruder, Miss 51
mahogany 347
Malin, P. 298
Malone, S. 214-215
Maltby Lake 140-141
Mangham, A. 420
Mann, B. 406
map-maker 27,28
maple 156
Mar:Müller, K. 320
Marian 49
Marion, Maid 52
marriage, student 299-300
Marshall Forest 181
Marshall, G. 50
Martin, J. 240
Mason, M. 259
Massie, W. 411,412
Maxwell, R. 240
McAdoo, D. 47
McBee, L. 467
McCaney, F. 142
McCarty, H. 240
McColloch, Lacy P. 35-39
McDonald, A. 415
McDonald, Mayor J 415.
McElroy, J. 101
McGee, R. 237
McGrath, T. 240
McIntire-Stennis fund 219,235
McIntyre, S. 240
McKibben, R. 237
McKnight, S. 142,143
McMullen, J. 256
McNary Act 99
McNutt, P. 53
McPherson, J. 452-453
McSherry, M. 53
McSherry, Mrs. F. 38,52
Mead, M. 404
mentors 44-45
Mergen, F. 237
Mexico 382-384
Meyer, H. 83
Meyer, W. 137,141
Milburn, Gen. F. 70
Miles, B. 219
Miller, E. 237
Miller, E. 240
Miller, H. 297
Mim 52
Mixed-hardwood stand 2
moccasin 99
Moir, Captain 239
Moller, M. 11,14
Mont Alto 2-19
Mooney, C. 413
Moore Plantation 114
Moore, A. 231,277
Morocco 330
Moss, B. 186,187
mothers 238-239
Mount Wheeler 283
Mountainair 284
Mt. Gilboa 367
Mt. McKinley 393
Mt. St. Helens 399-400
Mullikin, H. 296
Munich 68
Murphy, H. 406
museum 228
Mustain, A. 101
Myers, W. 220,270,408
Nacogdoches, Tex. 204-205
Nash, R.
Nat. Acad. of Sci. 453-456
Nat. Plant Food Ins. 319
National Geographic 14
National Guard 21,43-44
National Park Service 40-42
natural areas 179-183,403
Nature Conservancy 181
nazareno 369
Neal, J. 151
Neathery, B 413.
Neff, A. 142
Neid, T. 287 -289
Nelson, A. 142
Nelson, T. 142
New Zealand 333-334
Newton, E. 47
Nicholas, I. 94,98,102,110
Nickless, Rev. L. 143
Nile River 359-360
nitrogen 321
Niwot Ridge 428,442
Nixon, R. 283,285

Norman, L. 235
Norris, J. 296
nuclear 194-195
oak 102,141,167,301-302,427
Oak Ridge Institute 194-195
oikos 458-459
Okefenokee Swamp 182
Olds, C. 413
Olds, J. 413
organizations 402-425
Ossabaw Island 182
Ott, C. 101
Outhouse 6-7
Ouzel Lake 430
Overgaard 320
Owens, R. 205,207
P.L. 16 79
Pack Forest 150,153,163-164
palm 381
Palmer, H. 294
Papalexious, N. 355
parents 287
Paris 73
Parker, Judge 410
Parker, R. 186,187
Pase, J. 245,425
Pase, N. 425,433
Pate, O. 100
Patman, Rep. W. 418
Patterson, A. 184,185
Payne, Mrs. 106-107
Peabody, D. 292-293
Peace Corps 368-375
Pearl Harbor 51
pecan 289
penitentiary, Texas 433-434
Penn State College 78-92
Perkins, J. 183,195
petrified palm 184-185
Petty, D. 437
Pfeiffer, W. 11,13
Pfleuger, O. 11,12
Philmont 411,412
Pickett, N. 417
pika 430
Pikes Peak 80
Pinchot, G. 135,316,444,447,455
pine 46,59,94,99,114,123132,144, 157,167,172,190,192,193,301-302,321,333,339,351,355,375, 395,399,412-413,427,429
Pine Log 271
pinyon-juniper 427
pipeline, Alaska 394
pirogue 124
plantation, oldest 172
plowing 105
Po Valley 318-327
poison ivy 144
policy statements, SAF 408-410
poplar 354
Poplar Institute 324-327
Popular Homecraft 42
post oak sale 105
potassium 153-159
POWs 116-17
Prater, P. 101
Prendergast, F. 240,297
Presbyterian Journal 216
Prevosto, Dr. 318
printer's devil 27
prison instruction 300-301
programs, SFA 216-219
Psalm of the Forester 463
Public Law 16,72,145,146
Puerto Rica 382
Pweiacker, P. 422
Querengaesser, F. 323
Rabun Gap School
Raft, M. 144
Ralston, J. 143
Ramanos, L. 356
Rampart Dam 391
Randolph, P. 290-291
Ray, H. 144
RC&D, SCS 424
Read, R. 139,144,151
reclamation 402
recruiting 213-216
redcedar-red pine 139-140
Redskins 50
Reeves, H. 240
Regulators/Mod. 108-109
Reines, M. 138
Reinhart, B. 282
Renewable Nat. Res. Found. 407
Res. Mgt. Int. 347
research, herbicide 253
retirement 426-435
Reunion 15
Reynolds, R 162
Rhine River 64
Ribeauville 70
Rich, D. 412
Richer, A. 84
Riggs, C. 425
Riley, B. 292
Rinehart, D. 101
Rink, G. 240
Rio de Janeiro 377
Rockwell Foundation 247
Rocky Mountain Park 429

Rogers, W. 413
Roosevelt, F. 72,79
Roosevelt, T. 230,303
Rosier, L. 261
Ross, Ann 297
Rossoll, H. 250-252
Rotary 358,424-425
ROTC 10
Rothrock, J.
Ruark, E. 175
Rudder, E. 208
Russell, B.
Russia 73,330-353
Sabine National Forest 94-121,301
Samford, V. 275
Samson, N. 211-213
San Augustine 97,106-108,150
Sandeen, B. 296
Sanders, K. 260,299
Santa Rosa Park 372
Saturday Evening Post 447,449
scarab 360
Schenck, C. 323,443-444
Schmidt, E. 294
Schneider, G. 207
Schulman, E. 398
Schutzer, J. 293
Schwartzwald 66
Scott, D. 141
Second World War 10,56-76
SEGABA machine 64
Seine River 61-62
Sentinel 214,375
Shay locomotive 128,254-257
Sheely , F. 59
Shenandoah National Park 95
Shipman, R. 78
Shirley, E. 281-285
Shirley, H. 149,152,165
Short, H. 240
Siberia 330-353
Siecke, E. 247,407,451
Siegfried Line 65,70
Sierra Madre Oriental 383
Sinclair, Anne 58,82,97
Singer, J.R. 211,240
Skinner, L. 11
skunk-hunting 8
Sloan, E. 462
Slocum Heights 149-150
Smith, D.M. 137
Smith, D.V. 240
Smith, Gov. P. 235
Smith, J. 413
Smith, K. 240
Smithsonian Institution 33
Soc. of Am. For. 212-213,230,231, 403,457
Solomon 92
Solution to Pollution 307
Somberg, S. 240,267-270
Sonderegger, V. 136
south trip 161-162
Southampton 61
Southern Pine Co. 228
Southland Paper Co. 254
Sowell, J. 413
Spain 328-331
Sparks, Mutt 7 377-382
special-use permits 117
Spedale, K. 302
Spencer, Rev. J. 108
spruce 90,156,321,351,389,428
Spurr, S. 91,141,248,271,407-408
Squiers, E. 459
squirrel, Albert 432
St. Avold 70
Staebler, G. 335
Staebler, R. 335
Standley, J. 434
Stanley, G. 201
Starks, H. 101
Starostka, R. 319
Steen, R. 205,207,208,210,213,221, 237,269
Stein, C. 290
Stephen F. Austin U. 204-265
Stevenson, D. 335
Stockwell, C. 412
Stoeckler, J. 323
Stone Valley 83
Stone, E. 159
Stone, J. 101
Stransky, J. 222,412
Strawn, K. 421,422
Stripling, R. 107
students 274-305
Sukwong, S. 350
Sul Ross Univ. 217-218
summer camp 294
Supai 395
Sweden 322
Switzerland 327-328
sycamore 359
Sykes, C. 101
Sylvans 257,302
Syracuse University 146-165,179
T.R. Miller Co. 169
Talmadge, G. 175
Talmadge, H. 174-176
tamarac 389
Tamm, C. 319

Tau Phi Delta 85
Tauberbischofsheim 70
tax collector, forester as 345
Taylor, H. 425
Taylor, T. 412
Teal, D. 412
Teal, T. 412
Teate, L. 295
Tel Aviv Univ. 364
Temple, A., III 417
Tenaha, Timpson, Bobo, Blair 98
Terry, B. 297
Testa, Christina 48,49
Testa, L. 47,52
Tex. utilities 402
Texas A&M Univ. 208,216,219
Texas National Forests 94-121
Texas Tally 224
Thailand 350
Theoe, D. 440
Thessalonica 356
thinning 102
thinning, incentive for 440
Thomas, A. 168
Thomas, E. 18
Thompson Co. 439-440
Thompson, J. 458
Thompson, M. 277
Thompson, W. 101
Thoreau, H. 460
ticks 118-119
Tilton, J. 421
timber trespass 102
Todd, W. 269
Tolbert, F. 215
Tortello, S. 293
Toumey, J. 135,139
Townsend, D. 219
transmigration 349
trolley cars 26
Troop 20 31-35
Troop 72 35-40
Truman, Margaret 42
TSI 102
Tsoumis, G. 356
TU Services Com. 402,421-423
Tucker, K. 296
Tucker, R. 241
Tufts, D. 142,143
tuliptree 141
Turner, E. 217,218
Turner, W. 237
Tyler, TX 57
ulcers 161
United Nations 338
United Way 424
Univ. of Alaska 392
Univ. of Colorado 429-435
Univ. of South 396
Universite de Paris 73
Urbanovsky, E. 282,289
Vacinium 389
Valdosta, State College 199
Van Cleve, P. 284
Van Wylen, G. 416
VanGoor, C. 322
VanWylen, G. 91
Verne, J. 33
Verrall, A. 211,240
Viele, D. 161
Viro, P. 319
Volcan Poaz Park 372
Volvo 320
W.R. Grace Co. 319
W.T. Carter Co. 254-257
W3GXI 60
Wade Plantation 180-181
Wadsworth, F. 382
Wagoner, E. 425
Walker family 244
Walker, Anne
142,144,149,161,432,467
Walker, Hobart, Jr. 67
Walker, Hobart, Sr. 21,59
Walker, Hope 9
Walker, Jean 20
Walker, Larry, MD 245,311
Walker, Laura 22,23,24,25
Walker, Louis 413
Walker, M. 4,41-42
Walker, Michael 299
Walker, S. 412,413
Walters, P. 420
Warp and Woof 184-185
Wash. & Jeff. Coll. 58
Washington Airport 28
Washington, D.C. 20-29
water 426,431
Watergate 54
Watermellons 21
Watterston, K. 211
Watts, I. 467
Weaver, G. 240
Webb, B. 142
Webb, C. 240
Webb,R.143,240
Weddell, D. 174
Weeks Law 99
Weitz, R. 294
Weller, M. 421,422
Wendling, R. 240
West, D. 142

West, E. 412
western forests 280-290
Western High School 47-55
wetlands 196,197
Weyerhaeuser Co. 345-346,399
Wheaton College 97,244
Whelen, J. 460
White Mountains 397-398
White, D. 144,152,158
White, E. 47
White, J. 85
White, R. 421,422
White, Z. 142
Whiting, R.M. 240
Wiant, H. 211,218,445-446
Wiestling Hall 4
wildfire 102,112
Wilford, W. 211,240
Williams, D. 142,143
Williams, J.K. 219
Williamson, D. 412
Williston, H. 142,143
willow 389,429
Wilson College 6
Wilson, M. 412
Wilson, Mrs. Woodrow 127
Wilson, P. 258
Wilson, Rep. C. 417,419-420
Wimmer, B. 297
Winer, H. 142
Wishart, J. 142,143
Wittich, W. 322
women foresters 278-280
wood imports, China 341
Woodland Cathedral 263
Woodward, P. 300
World Forestry Congress 328
Worrell, A. 200
Wrangell Mountain 390
Wright, Peggy 269
Xi Sigma Pi 78,92,148,302
XXI Corps 66-70
Yale University 134-135
Yarborough, Sen. R. 220-223,418
yaupon 124
Yee, C. 240
Yellowstone Park 80
Young, D. 451-452
Youngberg, C. 191-192
Yunk, S. 300
Zealand 320
Zermatt 328
Zumwalt, G. 138,434
Zweiacker, P. 442